Salmonella

METHODS IN MOLECULAR BIOLOGY™

John M. Walker, SERIES EDITOR

419. Post-Transcriptional Gene Regulation, edited by *Jeffrey Wilusz, 2008*
418. Avidin–Biotin Interactions: *Methods and Applications,* edited by *Robert J. McMahon, 2008*
417. Tissue Engineering, Second Edition, edited by *Hannsjörg Hauser and Martin Fussenegger, 2007*
416. Gene Essentiality: *Protocols and Bioinformatics,* edited by *Andrei L. Osterman, 2008*
415. Innate Immunity, edited by *Jonathan Ewbank and Eric Vivier, 2007*
414. Apoptosis and Cancer: *Methods and Protocols,* edited by *Gil Mor and Ayesha B. Alvero, 2008*
413. Protein Structure Prediction, Second Edition, edited by *Mohammed Zaki and Chris Bystroff, 2008*
412. Neutrophil Methods and Protocols, edited by *Mark T. Quinn, Frank R. DeLeo, and Gary M. Bokoch, 2007*
411. Reporter Genes for Mammalian Systems, edited by *Don Anson, 2007*
410. Environmental Genomics, edited by *Cristofre C. Martin, 2007*
409. Immunoinformatics: *Predicting Immunogenicity In Silico,* edited by *Darren R. Flower, 2007*
408. Gene Function Analysis, edited by *Michael Ochs, 2007*
407. Stem Cell Assays, edited by *Vemuri C. Mohan, 2007*
406. Plant Bioinformatics: *Methods and Protocols,* edited by *David Edwards, 2007*
405. Telomerase Inhibition: *Strategies and Protocols,* edited by *Lucy Andrews and Trygve O. Tollefsbol, 2007*
404. Topics in Biostatistics, edited by *Walter T. Ambrosius, 2007*
403. Patch-Clamp Methods and Protocols, edited by *Peter Molnar and James J. Hickman, 2007*
402. PCR Primer Design, edited by *Anton Yuryev, 2007*
401. Neuroinformatics, edited by *Chiquito J. Crasto, 2007*
400. Methods in Lipid Membranes, edited by *Alex Dopico, 2007*
399. Neuroprotection Methods and Protocols, edited by *Tiziana Borsello, 2007*
398. Lipid Rafts, edited by *Thomas J. McIntosh, 2007*
397. Hedgehog Signaling Protocols, edited by *Jamila I. Horabin, 2007*
396. Comparative Genomics, *Volume 2,* edited by *Nicholas H. Bergman, 2007*
395. Comparative Genomics, *Volume 1,* edited by *Nicholas H. Bergman, 2007*
394. Salmonella: *Methods and Protocols,* edited by *Heide Schatten and Abraham Eisenstark, 2007*
393. Plant Secondary Metabolites, edited by *Harinder P. S. Makkar, P. Siddhuraju, and Klaus Becker, 2007*
392. Molecular Motors: *Methods and Protocols,* edited by *Ann O. Sperry, 2007*
391. MRSA Protocols, edited by *Yinduo Ji, 2007*
390. Protein Targeting Protocols, Second Edition, edited by *Mark van der Giezen, 2007*
389. *Pichia* Protocols, Second Edition, edited by *James M. Cregg, 2007*
388. Baculovirus and Insect Cell Expression Protocols, Second Edition, edited by *David W. Murhammer, 2007*
387. Serial Analysis of Gene Expression (SAGE): *Digital Gene Expression Profiling,* edited by *Kare Lehmann Nielsen, 2007*
386. Peptide Characterization and Application Protocols, edited by *Gregg B. Fields, 2007*
385. Microchip-Based Assay Systems: *Methods and Applications,* edited by *Pierre N. Floriano, 2007*
384. Capillary Electrophoresis: *Methods and Protocols,* edited by *Philippe Schmitt-Kopplin, 2007*
383. Cancer Genomics and Proteomics: *Methods and Protocols,* edited by *Paul B. Fisher, 2007*
382. Microarrays, Second Edition: *Volume 2, Applications and Data Analysis,* edited by *Jang B. Rampal, 2007*
381. Microarrays, Second Edition: *Volume 1, Synthesis Methods,* edited by *Jang B. Rampal, 2007*
380. Immunological Tolerance: *Methods and Protocols,* edited by *Paul J. Fairchild, 2007*
379. Glycovirology Protocols, edited by *Richard J. Sugrue, 2007*
378. Monoclonal Antibodies: *Methods and Protocols,* edited by *Maher Albitar, 2007*
377. Microarray Data Analysis: *Methods and Applications,* edited by *Michael J. Korenberg, 2007*
376. Linkage Disequilibrium and Association Mapping: *Analysis and Application,* edited by *Andrew R. Collins, 2007*
375. In Vitro Transcription and Translation Protocols: *Second Edition,* edited by *Guido Grandi, 2007*
374. Quantum Dots: *Applications in Biology,* edited by *Marcel Bruchez and Charles Z. Hotz, 2007*
373. Pyrosequencing® Protocols, edited by *Sharon Marsh, 2007*
372. Mitochondria: Practical Protocols, edited by *Dario Leister and Johannes Herrmann, 2007*
371. Biological Aging: *Methods and Protocols,* edited by *Trygve O. Tollefsbol, 2007*
370. Adhesion Protein Protocols, *Second Edition,* edited by *Amanda S. Coutts, 2007*
369. Electron Microscopy: *Methods and Protocols, Second Edition,* edited by *John Kuo, 2007*
368. Cryopreservation and Freeze-Drying Protocols, *Second Edition,* edited by *John G. Day and Glyn Stacey, 2007*
367. Mass Spectrometry Data Analysis in Proteomics, edited by *Rune Matthiesen, 2007*
366. Cardiac Gene Expression: *Methods and Protocols,* edited by *Jun Zhang and Gregg Rokosh, 2007*
365. Protein Phosphatase Protocols: edited by *Greg Moorhead, 2007*
364. Macromolecular Crystallography Protocols: *Volume 2, Structure Determination,* edited by *Sylvie Doublié, 2007*
363. Macromolecular Crystallography Protocols: *Volume 1, Preparation and Crystallization of Macromolecules,* edited by *Sylvie Doublié, 2007*
362. Circadian Rhythms: *Methods and Protocols,* edited by *Ezio Rosato, 2007*
361. Target Discovery and Validation Reviews and Protocols: *Emerging Molecular Targets and Treatment Options, Volume 2,* edited by *Mouldy Sioud, 2007*

METHODS IN MOLECULAR BIOLOGY™

Salmonella

Methods and Protocols

Edited by

Heide Schatten

Department of Veterinary Pathobiology,
University of Missouri-Columbia,
Columbia, MO, USA

and

Abraham Eisenstark

Cancer Research Center, Columbia, MO, USA

HUMANA PRESS ✳ TOTOWA, NEW JERSEY

999 Riverview Drive, Suite 208
Totowa, New Jersey 07512

www.humanapress.com

This publication is printed on acid-free paper. ♾
ANSI Z39.48-1984 (American Standards Institute) Permanence of Paper for Printed Library Materials

Cover design by Karen Schulz

Cover illustration: Fluorescence/immunofluorescence micrograph of *S. typhimurium* after invasion into PC-3 prostate cancer cells (as described in Chapter 16). The microtubule cytoskeleton in the PC-3 cells is shown in green as detected with a fluorescein-labeled beta-tubulin antibody. The blue fluorescence staining represents nuclei in PC-3 cells (large blue spheres) and bacteria (small blue dots) as detected with DAPI. The insert shows *S. typhimurium* (blue) in Salmonella-containing vacuoles (SCVs) in another PC-3 prostate cancer cell. Green: microtubules; blue: DNA for cell nucleus and bacteria; red: phalloidin-TRITC. Photograph courtesy of Zhisheng Zhong and Heide Schatten.

For additional copies, pricing for bulk purchases, and/or information about other Humana titles, contact Humana at the above address or at any of the following numbers: Tel.: 973-256-1699; Fax: 973-256-8341; or visit our Website: www.humanapress.com

Printed in the United States of America. 10 9 8 7 6 5 4 3 2 1
eISBN 978-1-59745-512-1

Library of Congress Control Number: 2007931075

Preface

The study of *Salmonella* has generated enormous interest with the realization that many bacteria have developed resistance to the most common antibiotics. *Salmonella* infection ranks second among causes of food-borne illnesses, and every year millions of people worldwide become ill and thousands die as a result of infections caused by food-borne pathogens. *Salmonella* causes typhoid fever that normally can be treated with broad-spectrum antibiotics, including tetracycline, chlorotetracycline, oxytetracycline, demeclocycline, methacycline, doxycycline, minocycline, and a number of other semisynthetic derivatives. However, *Salmonella* resistance to antibiotics has become a problem in recent years, and new avenues are being explored to develop new drugs that interfere with bacterial proliferation while not harming their mammalian host cells. Basic research on *Salmonella* will undoubtedly advance the field and may uncover new approaches for the development of antibiotics or vaccines. Basic knowledge and techniques developed in *Salmonella* research will also be transferable to other enterobacteria.

Another avenue in *Salmonella* research is based on the ease of gene manipulations and molecular methods that have added to the possibilities for exploring *Salmonella* as a therapeutic vector to interfere with disease. Genetic modifications of *Salmonella* have allowed the development of auxotrophs that are attenuated for toxicity and do not cause septic shock. *Salmonella* is now being explored for its use in cancer research. The exciting finding that *Salmonella* selectively and preferentially replicates in cancer and destroys proliferating cancer tissue without harming normal tissue has resulted in a newly emerging field using *Salmonella* as a tumor-targeting vector and therapeutic anticancer agent, even against the most advanced stages of cancer. Further genetic manipulation and modifications will allow use of *Salmonella* as drug-delivering vectors to increase its therapeutic efficiency. Studies are underway to elucidate the molecular pathways through which *Salmonella* anticancer activity is achieved.

Salmonella: Methods and Protocols presents detailed methods on a variety of different aspects. We have selected those techniques that have become landmarks in advancing our knowledge of *Salmonella*. Topics include molecular genotyping, new rapid, accurate, and sensitive determination of antibiotic resistance profiles, basic research to characterize *Salmonella*-specific cell antigens, and many others. This volume also covers *Salmonella* host–cell interactions, another area of intensive research. Detailed methods are presented to study *Salmonella* motility, the molecular interactions during contact with host cells, invasion, and interactions with the mammalian cytoskeleton and host cell organelles that lead to destruction of host cells.

Each chapter provides a short overview of the topic followed by detailed techniques that are normally not described in regular research papers. Genetic manipulation, molecular methods, and molecular imaging are techniques that will be of interest to geneticists, cell and molecular biologists, microbiologists, environmentalists, toxicologists, public health scientists, clinicians in human and veterinary medicine, agriculture, and other researchers who want to become familiar and apply techniques that are commonly not available in research papers. This book will also be of interest to students for its unique perspective on the various directions taken by *Salmonella* researchers. Because no recent comprehensive literature in this format is available on *Salmonella*, this book will prove valuable to a wide spectrum of researchers.

It was a great pleasure and timely to edit this volume on *Salmonella* depicting specific methods that have impacted *Salmonella* research. The methods presented here are in demand and are expected to continue to be of value to incoming investigators on *Salmonella* in the future. We are indebted to Dr. John Walker for inviting this volume on *Salmonella* and to the publisher, with special thanks to Tom Lanigan. We thank the contributors for sharing their specific expertise and experiences with the wider scientific community and for revealing protocol details and practical insights that are not generally disseminated in regular research papers. Our sincere thanks to all for their excellent contributions.

Heide Schatten
Abraham Eisenstark

Contents

Contributors

HANS-W. ACKERMANN • *Department of Medical Biology, Laval University, Quebec, Canada*
BRIAN M. M. AHMER • *Department of Microbiology, Ohio State University, Columbus, OH*
BURTON W. BLAIS • *Research and Development Section, Ottawa Laboratory (Carling), Canadian Food Inspection Agency, Ottawa, Ontario, Canada*
ALISON DINO • *Cancer Research Center, Columbia, MO*
AUDREY DUMONT • *Centre d'Immunologie de Marseille-Luminy, CNRS-INSERM-Université de la Méditerranée, Parc Scientifique de Luminy, Marseille Cedex, France*
SHERRY A. DUNBAR • *Luminex Corporation, Austin, TX*
JESSICA L. DYSZEL • *Department of Microbiology, Ohio State University, Columbus, OH*
KELLY K. EDWARDS • *Global Biostatistics and Medical Writing, Kansas City, MO*
ABRAHAM EISENSTARK • *Cancer Research Center and Division of Biological Sciences, University of Missouri, Columbia, MO*
MARTINE GAUTHIER • *Research and Development Section, Ottawa Laboratory (Carling), Canadian Food Inspection Agency, Ottawa, Ontario, Canada*
JEAN-PIERRE GORVEL • *Centre d'Immunologie de Marseille-Luminy, CNRS-INSERM-Université de la Méditerranée, Parc Scientifique de Luminy, Marseille Cedex, France*
MICHAEL HENSEL • *Institut für Klinische Mikrobiologie, Immunologie and Hygiene, Friedrich-Alexander Universität Erlangen-Nürnberg, Erlangen, Germany*
JAMES W. JACOBSON • *Luminex Corporation, Austin, TX*
MARK A. JEPSON • *Department of Biochemistry, School of Medical Sciences, University of Bristol, Bristol, United Kingdom*
ROBERT A. KAZMIERCZAK • *Cancer Research Center, Columbia, MO*
RULA KHREIS • *Cancer Research Center, Columbia, MO*

PAULINA KONCZY • *Public Health Agency of Canada, Laboratory for Foodborne Zoonoses, Guelph, Ontario, Canada*
ANDREW M. KROPINSKI • *Host and Pathogen Determinants, Laboratory for Foodborne Zoonoses, Public Health Agency of Canada, Guelph, Ontario, Canada, and Department of Microbiology and Immunology, Queen's University, Kingston, Ontario, Canada*
JONATHAN C. KUHN • *Israel Institute of Technology, Kiryat HaTechnion, Haifa, Israel*
RUITING LAN • *Microbiology and Immunology, School of Biotechnology and Biomolecular Sciences, University of New South Wales, Sydney, New South Wales, Australia*
AMBER LINDSAY • *Department of Microbiology, Ohio State University, Columbus, OH*
SHU-LIN LIU • *Department of Microbiology and Infectious Diseases, University of Calgary, Calgary, Alberta, Canada*
MICHAEL E. MAGUIRE • *Department of Pharmacology, Case School of Medicine, Case Western Reserve University, Cleveland, OH*
MICHAEL MCCLELLAND • *Sidney Kimmel Cancer Center, San Diego, CA*
STÉPHANE MÉRESSE • *Centre d'Immunologie de Marseille-Luminy, CNRS-INSERM-Université de la Méditerranée, Parc Scientifique de Luminy, Marseille Cedex, France*
DUSTIN NEWMAN • *Cancer Research Center, Columbia, MO*
CHERYL A. NICKERSON • *Center for Infectious Diseases and Vaccinology, Biodesign Institute, Arizona State University, Tempe, AZ*
CHARLOTTE A. PERRETT • *Department of Biochemistry, School of Medical Sciences, University of Bristol, Bristol, United Kingdom*
CORNELIUS POPPE • *Laboratory for Foodborne Zoonoses, Public Health Agency of Canada, Guelph, Ontario, Canada*
STEFFEN PORWOLLIK • *Sidney Kimmel Cancer Center, San Diego, CA*
WOLFGANG RABSCH • *Robert-Koch Institut, Wernigerode Branch, National Reference Centre for Salmonellae and Other Enterics, Wernigerode, Germany*
PETER R. REEVES • *School of Molecular and Microbial Biosciences (G08), The University of Sydney, New South Wales, Australia*
HEIDE SCHATTEN • *Department of Veterinary Pathobiology, University of Missouri-Columbia, Columbia, MO*
NINA SCHROEDER • *Centre d'Immunologie de Marseille-Luminy, CNRS-INSERM-Université de la Méditerranée, Parc Scientifique de Luminy, Marseille Cedex, France*

JENEE N. SMITH • *Department of Microbiology, Ohio State University, Columbus, OH*

ALEXANDER SULAKVELIDZE • *Intralytix, Inc., The Warehouse at Camden Yards, Baltimore, MD*

JAMES W. WILSON • *Center for Infectious Diseases and Vaccinology, Biodesign Institute, Arizona State University, Tempe, AZ*

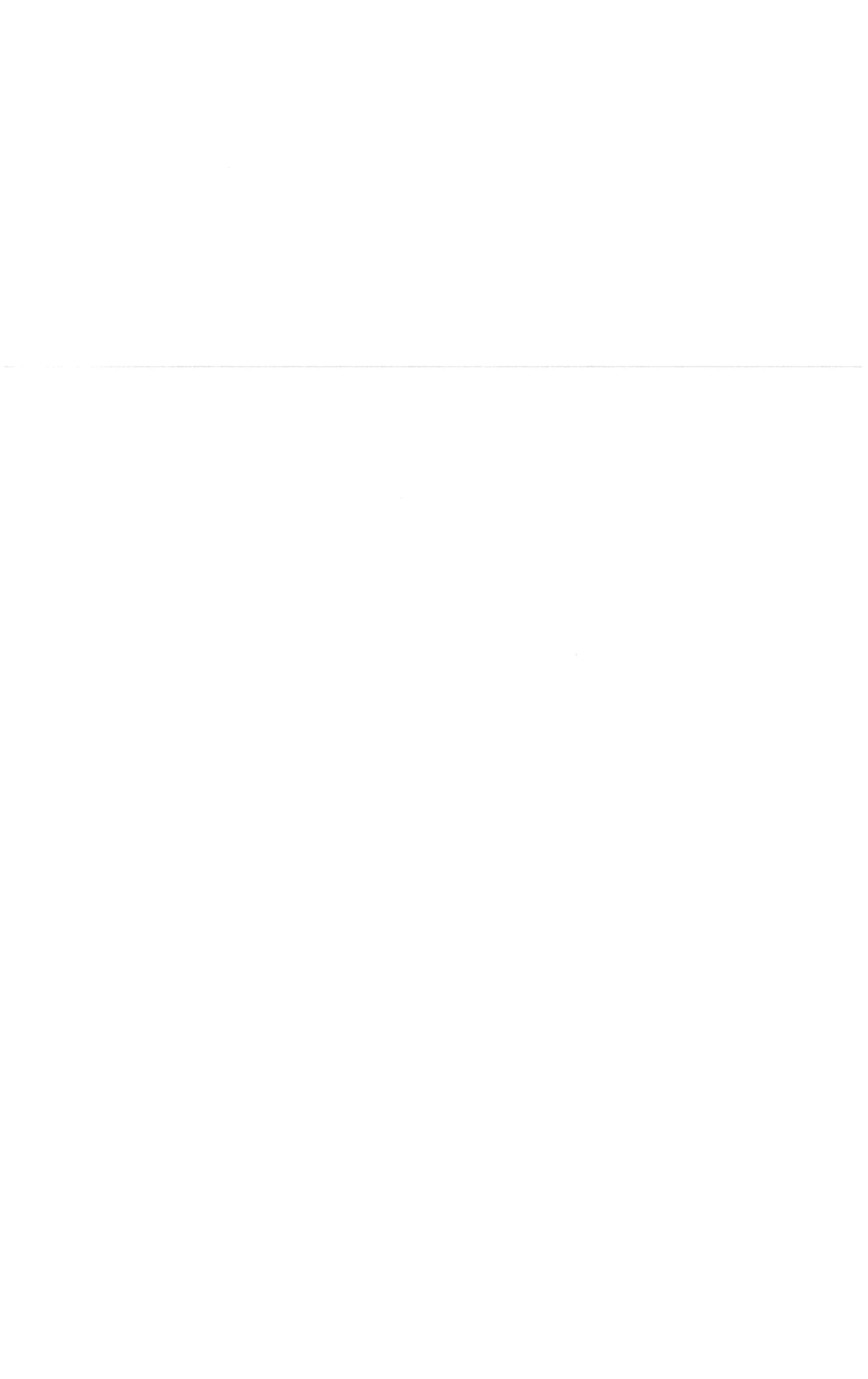

1

Quantitative, Multiplexed Detection of *Salmonella* and Other Pathogens by Luminex® xMAP™ Suspension Array

Sherry A. Dunbar and James W. Jacobson

Summary

We describe a suspension array hybridization assay for rapid detection and identification of Salmonella and other bacterial pathogens using Luminex® xMAP™ technology. The Luminex xMAP system allows simultaneous detection of up to 100 different targets in a single multiplexed reaction. Included in the method are the procedures for (1) design of species-specific oligonucleotide capture probes and PCR amplification primers, (2) coupling oligonucleotide capture probes to carboxylated microspheres, (3) hybridization of coupled microspheres to oligonucleotide targets, (4) production of targets from DNA samples by PCR amplification, and (5) detection of PCR-amplified targets by direct hybridization to probe-coupled microspheres. The Luminex xMAP suspension array hybridization assay is rapid, requires few sample manipulations, and provides adequate sensitivity and specificity to detect and differentiate Salmonella and nine other test organisms through direct detection of species-specific DNA sequences.

Key Words: *Salmonella*; bacterial detection; Luminex; xMAP technology; suspension array.

1. Introduction

Illnesses caused by foodborne and waterborne bacteria, viruses, and parasites affect as many as 80 million persons in the United States each year with an estimated annual cost of 7–17 billion dollars *(1–5)*. Whether present in the environment as free-living organisms or as the result of fecal or animal contamination, bacteria caused 80% of foodborne disease outbreaks of known

From: *Methods in Molecular Biology, vol. 394: Salmonella: Methods and Protocols*
Edited by: H. Schatten and A. Eisenstark © Humana Press Inc., Totowa, NJ

etiology reported from 1988 to 1992 ***(6)***. During the 1980s, *Salmonella enterica* serotype Enteritidis emerged as an important cause of human illnesses in the United States, primarily because of the consumption of undercooked eggs ***(7–9)***. Total *Salmonella* infection rates rose from 10.7 per 100,000 in 1976 to 24.3 in 1985 ***(10)***.

Rapid identification of *Salmonella* and other pathogens can help prevent foodborne diseases through better control of processed foods. Pathogenic bacteria that were previously isolated and identified by time-consuming, labor-intensive plating and biochemical testing procedures can now be detected quickly and reliably by rapid testing methodologies, including bioluminescence, cell counting, impedimetry, ELISA, and nucleic acid amplification ***(11,12)***. However, many of these rapid tests are costly and laborious. Molecular biological assays can be affected by interfering substances in the sample or may lack the sensitivity needed to detect very low levels of bacteria, precluding their direct application to food or environmental samples ***(13)***. In addition, multiple tests and test formats are often required to detect all of the different pathogenic species.

This study describes the development of a multiplexed direct hybridization assay using a microsphere-based suspension array technology, the Luminex® xMAP™ system, as a method for rapid, simultaneous detection of *Salmonella* and other bacterial pathogens. Microsphere-based suspension array technologies offer a novel platform for high-throughput molecular detection and are being utilized for microbial detection with increasing frequency ***(14–23)***. Some advantages of these technologies include rapid data acquisition, excellent sensitivity, and multiplexed analysis capability ***(24–26)***.

The Luminex xMAP system incorporates 5.6-μm polystyrene microspheres that are internally dyed with two spectrally distinct fluorochromes. Using precise amounts of each of these fluorochromes, an array is created consisting of 100 different microsphere sets, each with a specific spectral address. Each microsphere set can, in turn, possess a different reactant on its surface. Because microsphere sets can be distinguished by their spectral addresses, they can be combined, allowing up to 100 different analytes to be measured simultaneously in a single-reaction vessel. A third fluorochrome coupled to a reporter molecule quantifies the biomolecular interaction that has occurred at the microsphere surface. Microspheres are interrogated individually in a rapidly flowing fluid stream as they pass by two separate lasers in the Luminex® analyzer: a 635-nm, 10-mW red diode laser excites the two classification fluorochromes contained within the microspheres and a 532-nm, 13-mW yttrium aluminum garnet (YAG) laser excites the reporter fluorochrome (R-phycoerythrin or Alexa 532) bound to the microsphere surface. High-speed digital signal processing classifies the microsphere based on its spectral address and quantifies the reaction on the surface. Sample throughput is rapid, requiring only a few seconds per sample.

Direct hybridization is the simplest assay chemistry for single-nucleotide discrimination and takes advantage of the fact that, for oligonucleotides approximately 15–20 nucleotides in length, the melting temperature for hybridization of a perfectly matched template compared with one with a single-base mismatch can differ by several degrees ***(27,28)***. Nucleic acid detection assays using a direct hybridization format on the xMAP platform have been described ***(15,16,18–23)***. As in other assay chemistries utilizing a solid phase, the reaction kinetics can be adversely affected by immobilization of a reactant on a solid surface. The effects are less severe for a microsphere in suspension than for a flat array, but the diffusion rate of the immobilized capture probe can be slower and the effective concentration is reduced as compared with free DNA in solution ***(28,29)***. However, taking these factors into consideration during probe/primer design and assay optimization usually circumvents any potential drawbacks of the direct assay format. Here, we demonstrate the feasibility of this technology for bacterial detection and identification using *Salmonella, Escherichia coli, Listeria monocytogenes, Campylobacter jejuni, Shigella flexneri, Vibrio cholerae, Vibrio parahaemolyticus, Yersinia enterocolitica, Staphylococcus aureus*, and *Bacillus cereus* as the model organisms for the method. This assay was developed strictly for instructional purposes.

2. MATERIALS

1. Bacterial DNA samples (American Type Culture Collection, Manassas, VA) (*see* **Table 1**).
2. Oligonucleotide capture probes with 5′ Amino Modifier C12 modification, solubilize in dH_20 (*see* **Table 2**) (*see* **Note 1**).
3. Reverse complementary target oligonucleotides with 5′-Biotin modification (*see* **Table 3**).
4. PCR amplification primers with 5′-Biotin modification for target strand only (*see* **Subheading 3.2.**).
5. Luminex xMAP carboxylated microspheres (light sensitive, store at 4 °C).
6. Aluminum foil (for protecting microspheres from prolonged light exposure).
7. 1-Ethyl-3-(3-dimethylaminopropyl) carbodiimide hydrochloride (EDC; Pierce, Rockford, IL), store desiccated at −20 °C (*see* **Note 2**).
8. 1.5-mL Copolymer microcentrifuge tubes, #1415-2500 (USA Scientific, Ocala, FL) (*see* **Note 3**).
9. MES (100 mM), pH 4.5, filter-sterilize, store at 4 °C (*see* **Note 4**).
10. 0.02 % Tween-20, filter-sterilize, store at room temperature.

Table 1
Bacterial Strains

Organism	ATCC number	GenBank accession number	Genome size (bp)	*rrl* copy number
Bacillus cereus	10987D	NC_003909	5224283	12
Campylobacter jejuni	33560D	NC_002162	1641481	3
Escherichia coli	10798D	NC_000913	4639675	7
Listeria monocytogenes	19115D	NC_002973	2905310	6
Staphylococcus aureus	10832D	NC_003923	2820462	6
Salmonella enterica subsp. *enterica*	49284D	NC_006905	4755700	7
Shigella flexneri	29903D	NC_004741	4599354	7
Vibrio cholerae	39315D	NC_002505 NC_002506	4033464	8
Vibrio parahaemolyticus	17802D	NC_004603 NC_004605	5165770	10[a]
Yersinia enterocolitica	9610D	NC_003222	4615899	7

[a]The probe sequence used for *V. parahaemolyticus* is specific for *rrl* on chromosome I only.

11. 0.1 % SDS, filter-sterilize, store at room temperature.
12. TE buffer: 10 mM Tris–HCl, 1 mM EDTA, pH 8.0, filter-sterilize, store at room temperature.
13. 1.5× TMAC hybridization solution: 4.5 M tetramethylammonium chloride, 75 mM Tris–HCl, pH 8.0, 6 mM EDTA, pH 8.0, 0.15 % Sarkosyl (*N*-Lauroylsarcosine sodium salt); store at room temperature (*see* **Note 5**).
14. 1× TMAC hybridization solution: 3 M tetramethylammonium chloride, 50 mM Tris–HCl, pH 8.0, 4 mM EDTA, pH 8.0, 0.1 % Sarkosyl; store at room temperature (*see* **Note 5**).
15. Bath sonicator.
16. Vortex mixer.
17. Microcentrifuge.
18. Hemacytometer.
19. PCR amplification equipment, including thermal cycler and consumables.
20. HotStarTaq PCR Master Mix (Qiagen, Valencia, CA).
21. Costar® Thermowell™ Model P 96-well polycarbonate plates, #6509 (Corning, Corning, NY).
22. Microseal™ “A” sealing film (Bio-Rad Laboratories, Inc., Waltham, MA) or equivalent.

Table 2
Oligonucleotide Capture Probes

Probe	Organism	Modification[a]	Sequence 5′ → 3′	Microsphere set
BC	*Bacillus cereus*	5′-AmMC12	CGTAATGGTATGGTATCCTT	046
CJ	*Campylobacter jejuni*	5′-AmMC12	TATAGAGATATACATTACCT	031
EC	*Escherichia coli*	5′-AmMC12	TGTTTCGACACACTATCATT	010
LM	*Listeria monocytogenes*	5′-AmMC12	TCTTTAGTCGGATAGTATCC	019
SA	*Staphylococcus aureus*	5′-AmMC12	AGTTATGTCATGTTATCGAT	055
SE	*Salmonella enterica* subsp. *enterica*	5′-AmMC12	TGACTCGTCACACTATCATT	040
SF	*Shigella flexneri*	5′-AmMC12	TGATTCGTCACACTATCATT	021
VC	*Vibrio cholerae*	5′-AmMC12	TGCATAAGCAGTTACTGTTA	024
VP	*Vibrio parahaemolyticus*	5′-AmMC12	GTGCATAAGCACGTATCCTT	016
YE	*Yersinia enterocolitica*	5′-AmMC12	CAATTCGTTGCACTATTGCA	043

[a] 5′-Amino Modifier C12.

Table 3
Reverse Complementary Oligonucleotide Targets

Target	Organism	Modification	Sequence 5′ → 3′
CBC	*Bacillus cereus*	5′-Biotin	AAGGATACCATACCATTACG
CCJ	*Campylobacter jejuni*	5′-Biotin	AGGTAATGTATATCTCTATA
CEC	*Escherichia coli*	5′-Biotin	AATGATAGTGTGTCGAAACA
CLM	*Listeria monocytogenes*	5′-Biotin	GGATACTATCCGACTAAAGA
CSA	*Staphylococcus aureus*	5′-Biotin	ATCGATAACATGACATAACT
CSE	*Salmonella enterica* subsp. *enterica*	5′-Biotin	AATGATAGTGTGACGAGTCA
CSF	*Shigella flexneri*	5′-Biotin	AATGATAGTGTGACGAATCA
CVC	*Vibrio cholerae*	5′-Biotin	TAACAGTAACTGCTTATGCA
CVP	*Vibrio parahaemolyticus*	5′-Biotin	AAGGATACGTGCTTATGCAC
CYE	*Yersinia enterocolitica*	5′-Biotin	TGCAATAGTGCAACGAATTG

23. Centrifuge capable of centrifuging 96-well plates.
24. Streptavidin-R-phycoerythrin reporter (light sensitive, store at 4 °C).
25. Luminex analyzer.
26. 70 % Ethanol or 70 % isopropanol.
27. Distilled water.

3. Methods

The methods described include (1) design of oligonucleotide capture probes, (2) design of degenerate universal PCR amplification primers and PCR amplification conditions, (3) preparation of the probe-conjugated microsphere sets, (4) direct hybridization of biotinylated PCR amplification products, (5) direct hybridization of biotinylated synthetic oligonucleotide targets to the multiplexed probe-coupled microsphere sets, and (6) results and data analysis.

3.1. Design of Oligonucleotide Capture Probes

Species-specific oligonucleotide capture probes, corresponding to unique sequences in the 23S ribosomal RNA gene (*rrl*), were designed for each bacterial species for assay in 1× TMAC hybridization solution. TMAC stabilizes AT base pairs, minimizing the effect of base composition differences on hybridization ***(30,31)***. For oligonucleotides up to 200 bp in length, hybridization efficiency in TMAC is a function of the length of the perfect match and is less dependent on base composition. Hybridization buffers incorporating 3 or 4 M TMAC equalize the melting points of different probes and increase duplex yields, allowing probes with different characteristics to be used under identical hybridization conditions ***(32,33)***. The capture probes were designed to be matched in length at 20 nucleotides, complementary in sequence to the biotinylated strand of the PCR product, and with the polymorphisms located at or near the center of the probe sequence. Mismatches in the center are known to have a more profound effect on the equilibrium state than mismatches near the 5′ or 3′ end ***(34)***. The position of the polymorphism within the probe sequence was adjusted when necessary to avoid potential formation of secondary structures. Probes were modified with 5′ Amino Modifier C12 to provide a terminal amine and spacer for coupling to the carboxylated microspheres. Probe sequences used for this assay are summarized in **Table 2**.

3.2. Design of PCR Amplification Primers and PCR Amplification Conditions

For PCR amplification, degenerate primers corresponding to conserved regions of *rrl* (upstream 5′ → 3′, ADYCDDVGATDTCYGAATGG; downstream 5′ → 3′, RGGTACTWAGATGTTTCARTTC) were used to generate PCR products of 87–112 bp in length, depending on the bacterial species. The downstream primer was labeled at the 5′ terminus with biotin. Using a small target DNA (approximately 100–300 bp) minimizes the potential for steric hindrance to affect the hybridization efficiency at the microsphere surface. In some cases, we and others have used larger targets (400–1200 bp) successfully, suggesting that hybridization efficiency is also dependent on the sequence and overall secondary structure of target ***(21)***.

Targets were amplified by PCR in 50-μL reaction mixtures containing 1× PCR buffer (Qiagen), 200 μM of each dNTP, 0.5 μM of each primer, 2 mM $MgCl_2$, 2.5 units HotStarTaq DNA polymerase (Qiagen), and approximately 100 ng template DNA. PCRs were amplified in a DNA Engine™ thermocycler (Bio-Rad) by incubation at 95 °C for 15 min to activate the enzyme, followed by 5–25 cycles of denaturation at 94 °C for 30 s, annealing at 55 °C for 30 s, and extension at 72 °C for 30 s. Final extension was done at 72 °C for 3 min, and reactions were held at 4 °C. Production of the PCR products was verified by 4 % agarose gel electrophoresis, and the concentration was determined by densitometry of ethidium bromide-stained gels using the GelDoc-IT Imaging System and LabWorks 4.6 software (UVP, Upland, CA) (not shown). Calculations were based on comparison with standard curves using 1.25 μL 100-bp Ladder (Promega,

Madison, WI). These values were used to determine the copy number contained in 1–10 μL PCR reaction, which was used for hybridization (*see* **Note 6**).

3.3. Coupling of Amino-Modified Oligonucleotide Capture Probes to Carboxylated Microspheres

Microspheres should be protected from prolonged exposure to light throughout this procedure.

1. Bring a fresh aliquot of −20 °C, desiccated Pierce EDC powder to room temperature (*see* **Note 2**).
2. Resuspend the amine-substituted oligonucleotide capture probe to 1 mM in dH_2O (*see* **Note 1**).
3. Resuspend the stock microspheres by vortex and sonication for approximately 20 s.
4. Transfer 5.0×10^6 of the stock microspheres to a USA Scientific microcentrifuge tube (*see* **Note 3**).
5. Pellet the stock microspheres by microcentrifugation at $\geq 8000 \times g$ for 1–2 min.
6. Remove the supernatant and resuspend the pelleted microspheres in 50 μL 0.1 M MES, pH 4.5, by vortex and sonication for approximately 20 s (*see* **Note 4**).
7. Prepare a 1:10 dilution of the 1 mM oligonucleotide capture probe in dH_2O.
8. Add 2 μL of the 1:10 diluted capture probe to the resuspended microspheres and mix by vortex (*see* **Note 7**).
9. Prepare a fresh solution of 10 mg/mL EDC in dH_2O. Return the EDC powder to desiccant to reuse for the second EDC addition.
10. One by one for each coupling reaction, add 2.5 μL of fresh 10 mg/mL EDC to the microspheres and mix by vortex.
11. Incubate for 30 min at room temperature in the dark.
12. Prepare a second fresh solution of 10 mg/mL EDC in dH_2O.
13. One by one for each coupling reaction, add 2.5 μL of fresh 10 mg/mL EDC to the microspheres and mix by vortex.
14. Incubate for 30 min at room temperature in the dark.
15. Add 1.0 mL 0.02 % Tween-20 to the coupled microspheres.
16. Pellet the coupled microspheres by microcentrifugation at $\geq 8000 \times g$ for 1–2 min.
17. Remove the supernatant and resuspend the coupled microspheres in 1.0 mL 0.1 % SDS by vortex.
18. Pellet the coupled microspheres by microcentrifugation at $\geq 8000 \times g$ for 1–2 min.
19. Remove the supernatant and resuspend the coupled microspheres in 100 μL TE, pH 8.0, by vortex and sonication for approximately 20 s.
20. Enumerate the coupled microspheres by hemacytometer or other counting methods. If by hemacytometer:

    a. Dilute the resuspended, coupled microspheres 1:100 in dH_2O.
    b. Mix thoroughly by vortex.
    c. Transfer 10 μL to the hemacytometer.

   d. Count the microspheres within the four large corners of the hemacytometer grid.
   e. Microspheres/μL = (sum of microspheres in four large corners) × 2.5 × 100 (dilution factor). Maximum is 50,000 microspheres/μL.

21. Store the coupled microspheres refrigerated at 4 °C in the dark.

3.4. Verification of Microsphere Coupling by Hybridization to Biotinylated Reverse Complementary Oligonucleotide Targets

Microspheres should be protected from prolonged exposure to light throughout this procedure.

1. Select the appropriate oligonucleotide-coupled microsphere sets.
2. Resuspend the microspheres by vortex and sonication for approximately 20 s.
3. Prepare a working microsphere mixture by diluting coupled microsphere stocks to 150 microspheres of each set/μL in 1.5× TMAC hybridization solution (*see* **Note 5**). For each reaction, 33 μL working microsphere mixture is required.
4. Mix the working microsphere mixture by vortex and sonication for approximately 20 s.
5. To each sample or background well, add 33 μL working microsphere mixture.
6. To each background well, add 17 μL TE, pH 8.0.
7. To each sample well, add 5–200 fmol of biotinylated reverse complementary oligonucleotide and TE, pH 8.0, to a total volume of 17 μL.
8. Mix reaction wells gently by pipetting up and down several times with a multichannel pipettor.
9. Cover the reaction plate to prevent evaporation and incubate at 95 °C for 1–3 min to denature any secondary structure in the sample oligonucleotides (*see* **Note 8**).
10. Incubate the reaction plate at 45 °C (hybridization temperature for this assay) for 15 min (*see* **Note 8**).
11. Prepare fresh reporter mix by diluting streptavidin-R-phycoerythrin to 10μg/mL in 1× TMAC hybridization solution (*see* **Note 5**). For each reaction well, 25 μL reporter mix should be used.
12. Add 25 μL reporter mix to each well and mix gently by pipetting up and down several times with a multichannel pipettor.
13. Incubate the reaction plate at 45 °C for 5 min.
14. Analyze 50 μL of each reaction at 45 °C on the Luminex analyzer according to the system manual.

In a typical assay, the Luminex analyzer is set up to measure the reporter fluorescence for a minimum of 100 microspheres of each microsphere set present in the reaction (i.e., a minimum of 100 events are collected for each microsphere set). The results are reported as the median of the fluorescent intensity (MFI) measured for each microsphere set. The MFI values are used to generate standard curves for each oligonucleotide target to assess the relative coupling efficiency of each capture probe to its respective microsphere set. A typical standard curve is shown in **Fig. 1**.

3.5. Direct Hybridization of Biotinylated PCR Amplification Products to the Multiplexed Probe-Coupled Microsphere Sets

Microspheres should be protected from prolonged exposure to light throughout this procedure.

1. Select the appropriate oligonucleotide-coupled microsphere sets.
2. Resuspend the microspheres by vortex and sonication for approximately 20 s.
3. Prepare a working microsphere mixture by diluting coupled microsphere stocks to 150 microspheres of each set/μL in 1.5× TMAC hybridization solution (*see* **Note 5**). For each reaction, 33 μL working microsphere mixture is required.
4. Mix the working microsphere mixture by vortex and sonication for approximately 20 s.
5. To each sample or background well, add 33 μL working microsphere mixture.

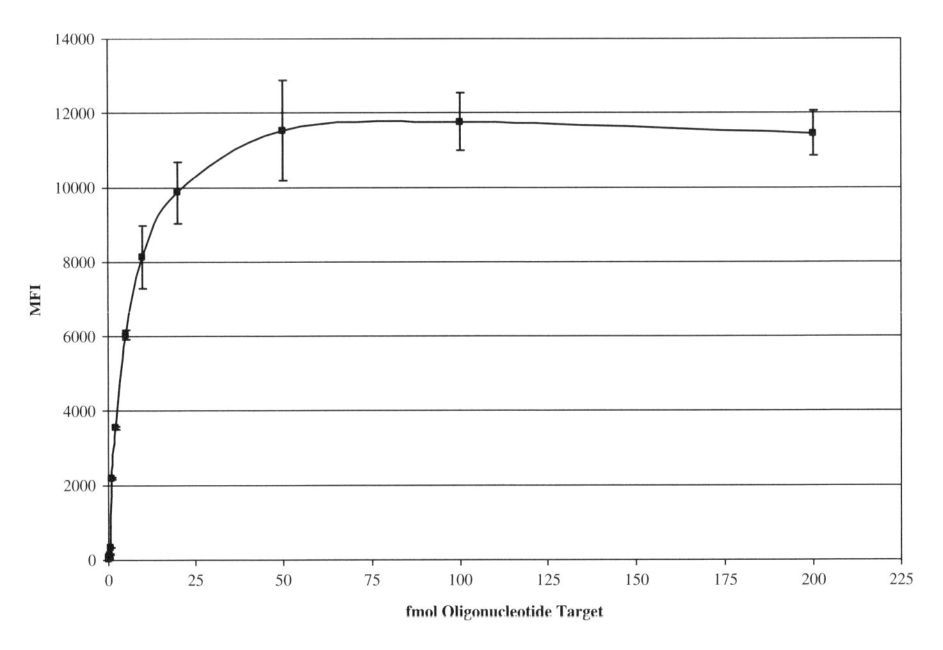

Fig. 1. Oligonucleotide target hybridization. A typical oligonucleotide hybridization curve is shown for the *Salmonella* probe-coupled microsphere set hybridized to the complementary biotinylated oligonucleotide target. Median fluorescent intensity (MFI) values are the average and standard deviation of duplicates from two independent experiments.

6. To each background well, add 17 μL TE, pH 8.0.
7. To each sample well, add 1–10 μL of the appropriate PCR reaction and TE, pH 8.0, to a total volume of 17 μL (*see* **Note 6**).
8. Mix reaction wells gently by pipetting up and down several times with a multichannel pipettor.
9. Cover the reaction plate to prevent evaporation and incubate at 95 °C for 5 min to denature the amplified biotinylated DNA (*see* **Note 8**).
10. Incubate the reaction plate at 45 °C (hybridization temperature for this assay) for 15 min (*see* **Note 8**).
11. Centrifuge the reaction plate at $\geq 2250 \times g$ for 3 min to pellet the microspheres (*see* **Note 9**).
12. During centrifugation, prepare fresh reporter mix by diluting streptavidin-R-phycoerythrin to 4 μg/mL in 1× TMAC hybridization solution (*see* **Note 5**). For each reaction well, 75 μL reporter mix is required.
13. After centrifugation, carefully remove the supernatant using an eight-channel pipettor to simultaneously extract the supernatant from each column of wells. Take care not to disturb the pelleted microspheres.
14. Return the reaction plate to 45 °C.
15. Add 75 μL reporter mix to each well and mix gently by pipetting up and down several times with a multichannel pipettor.
16. Incubate the reaction plate at 45 °C for 5 min.
17. Analyze 50 μL of each reaction at 45 °C on the Luminex analyzer according to the system manual.

3.6. Results and Data Analysis

For each well, a minimum of 100 events are collected for each microsphere set and the results are reported as the MFI of each microsphere set. The data are written to a .csv file, which can be imported into Excel or other analysis software for data reduction. A background well, consisting of all reaction components except a DNA sample, is used to determine the background reporter fluorescence associated with each microsphere set in the absence of DNA target. The background MFI values are subtracted from the sample MFI values to determine the net MFI. For this study, we used a cutoff value of net MFI greater than two times the background MFI to indicate a positive detection.

3.6.1. PCR Amplification Sensitivity

PCR amplification was performed for 5, 10, 15, 20, and 25 cycles on replicate reactions containing 124 ng of *E. coli* genomic DNA and 1–10 μL of the amplification reactions were hybridized to the probe-coupled microsphere sets to determine the sensitivity of the amplification reaction (*see* **Fig. 2**). The PCR product was detectable after 10 amplification cycles in hybridization reactions

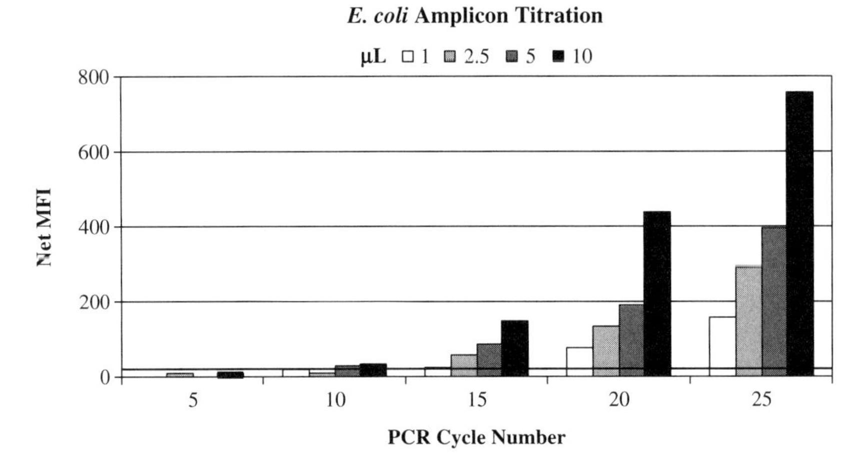

Fig. 2. Amplification sensitivity. Graphical presentation of net MFI for hybridization of various volumes of PCR-amplified *Escherichia coli* genomic DNA. PCR reactions were cycled for 5–25 amplification cycles. The cutoff value for a positive reaction (two times the background MFI) is indicated by the horizontal line (boldfaced).

containing 5 and 10 μL of the PCR reaction (3×10^6 copies/μL). At 15 cycles, 2.5 μL of PCR reaction (3.5×10^6 copies/μL) was detected. This suggests that greater sensitivity could be achieved if more PCR reaction volume (10–20 μL) was used for hybridization.

3.6.2. PCR-Amplified Target Titration

The results of the detection of the target sequences amplified from bacterial genomic DNA are summarized in **Table 4**. Results are representative of at least two (*V. cholerae*) or three (all others) independent experiments. The hybridization assay was capable of detecting the target sequence when at least 30 (*E. coli*) to 180 ng (*V. parahaemolyticus*) of PCR product was present. This corresponds to a detection sensitivity of 5.9×10^6 to 3.2×10^7 *rrl* copies and 6.7×10^5 to 9.1×10^6 genome copies. Further titration can be used to determine the limit of detection for each organism. Similar sensitivities were seen when hybridization was performed at 42 °C (not shown). The *V. cholerae* target was not detected at either 45 or 42 °C, but the data suggest that detection may have been possible if more PCR product had been hybridized. The assay was specific and could discriminate target sequences differing by only two nucleotides with a signal-to-noise ratio greater than or equal to 5.2 for each sequence. Cross-hybridization was observed for *Salmonella* to *S. flexneri* (SF) at 10 μL and for

Table 4
Multiplexed Detection of PCR-Amplified Targets From Bacterial Genomic DNA

Sample	Volume (μL)	*rrl* copies	Genome copies	Net MFI										S:N
				BC	CJ	EC	LM	SA	SE	SF	VC	VP	YE	
2 × Bkg				49	34	32	38	38	44	62	29	30	27	
BC	1	**8.03E+06**	**6.69E+05**	**70**	7	10	−1	0	4	−9	−5	7	7	7.0
	2.5	**2.01E+07**	**1.67E+06**	**147**	1	0	2	0	−2	−4	10	1	2	14.7
	5	**4.02E+07**	**3.35E+06**	**196**	2	4	8	4	2	2	8	5	2	24.5
	10	**8.03E+07**	**6.69E+07**	**477**	7	12	7	11	11	4	14	7	16	29.8
CJ	1	**2.72E+07**	**9.08E+06**	−3	**240**	7	6	−4	−4	−6	5	−1	7	34.3
	2.5	**6.80E+07**	**2.27E+07**	−12	**232**	−7	−12	−3	−7	−15	2	−4	2	116.0
	5	**1.36E+08**	**4.54E+07**	3	**446**	6	−1	5	1	−23	10	6	4	44.6
	10	**2.72E+08**	**9.08E+07**	9	**1651**	10	−9	7	7	−1	4	11	7	150.1
EC	1	**5.90E+06**	**8.43E+05**	6	−2	**272**	−2	4	−4	−5	1	11	9	24.7
	2.5	**1.48E+07**	**2.11E+06**	5	6	**535**	0	0	4	1	5	4	7	76.4
	5	**2.95E+07**	**4.22E+06**	3	10	**758**	0	−1	7	3	10	12	15	50.5
	10	**5.90E+07**	**8.43E+06**	10	5	**1208**	9	8	19	33	6	12	11	36.6
LM	1	**3.23E+07**	**5.39E+06**	3	5	11	**173**	0	3	−13	7	−2	9	15.7
	2.5	**8.08E+07**	**1.35E+07**	9	7	−5	**232**	8	10	−2	13	2	8	17.8
	5	**1.62E+08**	**2.70E+07**	−7	7	16	**334**	2	−5	−8	7	4	−3	20.9
	10	**3.23E+08**	**5.39E+07**	13	11	12	**952**	2	−5	5	11	6	8	73.2
SA	1	**1.08E+07**	**1.80E+06**	5	6	0	−4	**295**	−3	−14	8	−1	3	36.9
	2.5	**2.70E+07**	**4.50E+06**	10	1	−3	−1	**582**	−2	−10	6	3	2	58.2
	5	**5.40E+07**	**9.00E+06**	10	2	5	1	**748**	5	−8	2	2	1	74.8
	10	**1.08E+08**	**1.80E+07**	12	7	10	1	**1306**	0	−8	13	3	12	100.5
SE	1	**1.48E+07**	**2.11E+06**	4	4	4	0	3	**85**	−3	16	0	6	5.3
	2.5	**3.70E+07**	**5.28E+06**	3	7	−1	−9	2	**159**	9	19	9	6	8.4
	5	**7.40E+07**	**1.06E+07**	4	7	16	−4	8	**233**	21	8	6	6	11.1
	10	**1.48E+08**	**2.11E+07**	4	9	14	2	3	**541**	**94**	2	8	5	5.8

Table 4
(Continued)

Sample	Volume (μL)	*rrl* copies	Genome copies	Net MFI										S:N
				BC	CJ	EC	LM	SA	SE	SF	VC	VP	YE	
SF	1	**1.13E+07**	**1.61E+06**	−6	4	6	−4	−2	9	**154**	13	−2	1	11.8
	2.5	**2.83E+07**	**4.03E+06**	1	−12	−7	−3	−3	25	**239**	9	−3	−4	9.6
	5	**5.65E+07**	**8.05E+06**	12	1	14	6	3	**62**	**421**	5	8	5	6.8
	10	**1.13E+08**	**1.61E+07**	3	9	20	−6	2	**155**	**998**	13	9	8	6.4
VC	1	1.81E+06	2.26E+05	2	−2	8	0	1	5	−9	12	8	−3	
	2.5	4.53E+06	5.65E+05	−2	1	9	−2	8	6	−9	10	2	5	
	5	9.05E+06	1.13E+06	9	2	6	3	0	−4	−5	17	−5	11	
	10	1.81E+07	2.26E+06	9	7	18	3	6	11	−6	29	9	16	
VP	1	3.18E+06	3.18E+05	5	10	0	−3	8	10	−9	7	5	3	
	2.5	7.95E+06	7.95E+05	−1	−10	−3	2	2	−3	−6	1	16	8	
	5	1.59E+07	1.59E+06	10	3	6	0	7	9	−3	−1	25	9	
	10	**3.18E+07**	**3.18E+06**	13	7	13	0	2	3	−11	5	**68**	4	5.2
YE	1	3.36E+06	4.80E+05	1	−3	8	0	−4	−2	−3	14	11	21	
	2.5	**8.40E+06**	**1.20E+06**	−1	−2	−3	1	1	−3	−8	−2	−1	**46**	46.0
	5	**1.68E+07**	**2.40E+06**	11	2	6	−2	2	−10	−7	3	−2	**68**	6.2
	10	**3.36E+07**	**4.80E+06**	−4	−2	16	−5	5	−2	5	6	−1	**192**	12.0

Net median fluorescent intensity (MFI) data are shown for *Bacillus cereus* (BC), *Campylobacter jejuni* (CJ), *Escherichia coli* (EC), *Listeria monocytogenes* (LM), *Staphylococcus aureus* (SA), *Salmonella enterica* subsp. *enterica* (SE), *Shigella flexneri* (SF), *Vibrio cholerae* (VC), *Vibrio parahaemolyticus* (VP), and *Yersinia enterocolitica* (YE) where 1–10 μL PCR reaction was hybridized to the multiplexed probe-coupled microsphere sets. The target concentration in each PCR reaction (determined by densitometry, *see* **Subheading 3.2.**) was used to calculate the *rrl* copy number and corresponding genome copy number for each volume of each PCR reaction. The signal-to-noise ratio (S:N) for positive reactions was determined by dividing the net MFI on the positive microsphere set by the highest net MFI on the other microsphere sets. Twice the background MFI (Bkg × 2) was used as the cutoff to indicate a positive reaction. Positive reactions are indicated in boldface.

S. flexneri to *Salmonella* (SE) at 5 and 10 μL. Greater cross-hybridization was seen when the assay was performed at 42 °C, and increasing the hybridization temperature to 52 °C did not improve discrimination (not shown). This is not surprising as the probe sequences for these two organisms differ by only one nucleotide at nucleotide position 4. Previous studies have shown that maximum specificity is achieved when the single mismatched nucleotide is located at positions 8–14 of the 20-nucleotide probe ***(35)***. Redesigning the probes to position the mismatched nucleotide at the center was not possible in this case because of strong secondary structure in the region.

Although the signal-to-noise ratios between *Salmonella* and *S. flexneri* were 5.8 and 6.4, respectively, in practice these results could be misinterpreted as a mixed specimen. Ideally, each probe should hybridize to only one bacterial species and to all members of that species, but often two or more probes are required ***(36)***. Inclusion of additional probes to multiple target sequences can increase the confidence of the assay and facilitate the identification of isolates or strains within a species that show minor sequence variation in the target region. The assay will become more robust as more targets are added. The range of organisms that can be identified can also be extended by adding additional species-specific probes to the assay. An important feature of the multiplexed suspension array is that the panel can be easily expanded (up to 100 sequences per well) by adding additional probe-coupled microsphere sets to the reaction mixture.

4. Notes

1. Amine-substituted oligonucleotide probes should be resuspended and diluted in dH_2O. Tris, azide, or other amine-containing buffers must not be present during the coupling procedure. If oligonucleotides were previously solubilized in an amine-containing buffer, desalting by column or precipitation and resuspension into dH_2O is required.
2. We recommend using EDC from Pierce for best results. EDC is labile in the presence of water. The active species is hydrolyzed in aqueous solutions at a rate constant of just a few seconds, so care should be taken to minimize exposure to air and moisture ***(37)***. EDC should be stored desiccated at −20 °C in dry, single-use aliquots with secure closures. A fresh aliquot of EDC powder should be used for each coupling episode. Allow the dry aliquot to warm to room temperature before opening. Prepare a fresh 10 mg/mL EDC solution immediately before each of the two additions, and close the dry aliquot tightly and return to desiccant between preparations. The dry aliquot should be discarded after the second addition.

3. Uncoupled microspheres tend to be somewhat sticky and will adhere to the walls of most microcentrifuge tubes, resulting in poor postcoupling microsphere recovery. We have found that copolymer microcentrifuge tubes from USA Scientific (#1415-2500) perform best for coupling and yield the highest microsphere recoveries postcoupling.
4. MES (100 mM), pH 4.5, should be filter-sterilized and either prepared fresh or stored at 4 °C between uses. Do not store at room temperature. The pH must be in the range 4.5–4.7 for optimal coupling efficiency.
5. We use 5 M TMAC solution from Sigma (T-3411) for preparation of 1.5× and 1× TMAC hybridization solutions. We find that this TMAC formulation does not have a strong "ammonia" odor. TMAC hybridization solutions should be stored at room temperature to prevent precipitation of the Sarkosyl. TMAC hybridization solutions can be warmed to hybridization temperature to solubilize precipitated Sarkosyl.
6. The hybridization kinetics and thermodynamic affinities of matched and mismatched sequences can be driven in a concentration-dependent manner ***(38)***. At concentrations beyond the saturation level, the hybridization efficiency can decrease presumably because of competition of the complementary strand and renaturation of the PCR product ***(25)***. Therefore, it is important to determine the range of target concentrations that yield efficient hybridization without sacrificing discrimination.
7. The optimal amount of a particular oligonucleotide capture probe for coupling to carboxylated microspheres is determined by coupling various amounts in the range of 0.04–5 nmol per 5×10^6 microspheres. For this assay, we found 0.2 nmol per 5×10^6 microspheres in a 50-µL reaction to be optimal. The coupling procedure can be scaled up or down. Above 5×10^6 microspheres, use the minimum volume required to resuspend the microspheres. Below 5×10^6 microspheres, maintain the microsphere concentration and scale down the volume accordingly.
8. Denaturation and hybridization can be performed in a thermal cycler. Use a heated lid and a spacer (if necessary) to prevent evaporation. Maintain hybridization temperature throughout the labeling and analysis steps.
9. Whether it is necessary to remove the hybridization supernatant before the labeling step will depend on the quantity of biotinylated PCR primers and unhybridized biotinylated PCR products that are present and available to compete with the hybridized biotinylated PCR products for binding to the streptavidin-R-phycoerythrin reporter. This should be determined for each individual assay.

Acknowledgment

The authors would like to thank Lisa M. Sutton for expert technical assistance.

References

1. Archer, D. L. and Kvenberg, J. E. (1985) Incidence and cost of foodborne diarrheal disease in the United States. *J. Food Protect.* **48,** 887–894.
2. Bennett, J., Holmberg, S., Rogers, M., and Solomon, S. (1987) Infectious and parasitic diseases, in *Closing the Gap: The Burden of Unnecessary Illness* (Amler, R. and Dull, H., eds.), Oxford University Press, New York, pp. 102–114.
3. Todd, E. C. D. (1989) Preliminary estimates of costs of foodborne disease in the United States. *J. Food Protect.* **52,** 595–601.
4. Foegeding, P. M., Roberts, T., Bennett, J., et al. (1994) Foodborne pathogens: risks and consequences. Task Force Report 122. Council for Agricultural Science and Technology, Ames, IA.
5. Mead, P. S., Slutsker, L., Dietz, V., et al. (1999) Food-related illness and death in the United States. *Emerg. Infect. Dis.* **5,** 607–625.
6. Centers for Disease Control and Prevention. (1996) CDC surveillance summaries: surveillance for foodborne-disease outbreaks—United States, 1988–1992. *MMWR Morb. Mortal. Wkly Rep.* **45(SS-5),** 66.
7. Patrick, M. E., Adcock, P. M., Gomez, T. M., et al. 2004. *Salmonella* Enteritidis infections, United States, 1985–1999. *Emerg. Infect. Dis.* **10,** 1–7.
8. Rodrigue, D. C., Tauxe, R. V., and Rowe, B. (1990) International increase in *Salmonella* enteritidis: a new pandemic? *Epidemiol. Infect.* **105,** 21–27.
9. Mishu, B., Koehler, J., Lee, L. A., et al. (1994) Outbreaks of *Salmonella* Enteritidis infections in the United States, 1985–1991. *J. Infect. Dis.* **169,** 547–552.
10. Centers for Disease Control and Prevention. (2000) Public Health Laboratory Information System. CDC *Salmonella* surveillance summaries 1976–1999. U.S. Government Printing Office, Washington, D.C.
11. Swaminathan, B. and Feng, P. (1994) Rapid detection of food-borne pathogenic bacteria. *Annu. Rev. Microbiol.* **48,** 401–426.
12. De Boer, E. and Beumer, R. R. (1999) Methodology for detection and typing of foodborne microorganisms. *Int. J. Food Microbiol.* **50,** 119–130.
13. Feng, P. (1997) Impact of molecular biology on the detection of foodborne pathogens. *Mol. Biotechnol.* **7,** 267–278.
14. Fulton, R. J., McDade, R. L., Smith, P. L., Kienker, L. J., and Kettman, J. R. (1997) Advanced multiplexed analysis with the FlowMetrix™ system. *Clin. Chem.* **43,** 1749–1756.
15. Smith, P. L., WalkerPeach, C. R., Fulton, R. J., and DuBois, D. B. (1998) A rapid, sensitive, multiplexed assay for detection of viral nucleic acids using the FlowMetrix system. *Clin. Chem.* **44,** 2054–2060.
16. Spiro A., Lowe M., and Brown, D. (2000) A bead-based method for multiplexed identification and quantitation of DNA sequences using flow cytometry. *Appl. Environ. Microbiol.* **66,** 4258–4265.

17. Ye, F., Li, M.-S., Taylor, J. D., et al. (2001) Fluorescent microsphere-based readout technology for multiplexed human single nucleotide polymorphism analysis and bacterial identification. *Hum. Mutat.* **17,** 305–316.
18. Spiro, A. and Lowe, M. (2002) Quantitation of DNA sequences in environmental PCR products by a multiplexed, bead-based method. *Appl. Environ. Microbiol.* **68,** 1010–1013.
19. Dunbar, S. A., Vander Zee, C. A., Oliver, K. G., Karem, K. L., and Jacobson, J. W. (2003) Quantitative, multiplexed detection of bacterial pathogens: DNA and protein applications of the Luminex LabMAP™ system. *J. Microbiol. Methods* **53,** 245–252.
20. Cowan, L. S., Diem, L., Brake, M. C., and Crawford, J. T. (2004) Transfer of a *Mycobacterium tuberculosis* genotyping method, spoligotyping, from a reverse line-blot hybridization, membrane-based assay to the Luminex multianalyte profiling system. *J. Clin. Microbiol.* **42,** 474–477.
21. Diaz, M. R. and Fell, J. W. (2004) High-throughput detection of pathogenic yeasts of the genus *Trichosporon. J. Clin. Microbiol.* **42,** 3696–3706.
22. Wallace, J., Woda, B. A., and Pihan, G. (2005) Facile, comprehensive, high-throughput genotyping of human genital papillomaviruses using spectrally addressable liquid bead microarrays. *J. Mol. Diagn.* **7,** 72–80.
23. Diaz, M. R. and Fell, J. W. (2005) Use of a suspension array for rapid identification of the varieties and genotypes of the *Cryptococcus neoformans* species complex. *J. Clin. Microbiol.* **43,** 3662–3672.
24. Kellar, K. L. and Iannone, M. A. (2002) Multiplexed microsphere-based flow cytometric assays. *Exp. Hematol.* **30,** 1227–1237.
25. Nolan, J. P. and Mandy, F. F. (2001) Suspension array technology: new tools for gene and protein analysis. *Cell Mol. Biol.* **47,** 1241–1256.
26. Nolan, J. P. and Sklar, L. A. (2002) Suspension array technology: evolution of the flat-array paradigm. *Trends Biotechnol.* **20,** 9–12.
27. Ikuta, S., Takagi, K., Wallace, R. B., and Itakura, K. (1987) Dissociation kinetics of 19 base paired oligonucleotide-DNA duplexes containing different single mismatched base pairs. *Nucleic Acids Res.* **15,** 797–811.
28. Livshits, M. A. and Mirzabekov, A. D. (1996) Theoretical analysis of the kinetics of DNA hybridization with gel-immobilized oligonucleotides. *Biophys. J.* **71,** 2795–2801.
29. Peterson, A. W., Wolf, L. K., and Georgiadis, R. M. (2002) Hybridization of mismatched or partially matched DNA at surfaces. *J. Am. Chem. Soc.* **124,** 14601–14607.
30. Jacobs, K. A., Rudersdorf, R., Neill, S. D., Dougherty, J. P., Brown, E. L., and Fritsch, E. F. (1988) The thermal stability of oligonucleotide duplexes is sequence independent in tetraalkylammonium salt solutions: application to identifying recombinant DNA clones. *Nucleic Acids Res.* **16,** 4637–4650.

31. Wood, W. I., Gitschier, J., Lasky, L. A., and Lawn, R. M. (1985) Base-composition-independent hybridization in tetramethylammonium chloride: a method for oligonucleotide screening of highly complex gene libraries. *Proc. Natl. Acad. Sci. U. S. A.* **82,** 1585–1588.
32. Maskos, U. and Southern, E. M. (1992) Parallel analysis of oligodeoxyribonucleotide (oligonucleotide) interactions. I. Analysis of factors influencing duplex formation. *Nucleic Acids Res.* **20,** 1675–1678.
33. Maskos, U. and Southern, E. M. (1993) A study of oligonucleotide reassociation using large arrays of oligonucleotides synthesized on a large support. *Nucleic Acids Res.* **21,** 4663–4669.
34. Gotoh, M., Hasegawa, Y., Shinohara, Y., Schimizu, M., and Tosu, M. (1995) A new approach to determine the effect of mismatches on kinetic parameters in DNA hybridization using an optical biosensor. *DNA Res.* **2,** 285–293.
35. Dunbar, S. A. and Jacobson, J. W. (2005) Rapid screening for 31 mutations and polymorphisms in the cystic fibrosis transmembrane conductance regulator gene by Luminex xMAP suspension array. *Methods Mol. Med.* **114,** 147–171.
36. Anthony, R. M., Brown, T. J., and French, G. L. (2000) Rapid diagnosis of bacteremia by universal amplification of 23S ribosomal DNA followed by hybridization to an oligonucleotide array. *J. Clin. Microbiol.* **38,** 781–788.
37. Hermanson, G. T. (1996) Zero-length cross-linkers, in *Bioconjugate Techniques*, Academic Press, San Diego, CA, pp. 169–186.
38. Wetmur, J. G. (1991) DNA probes: applications of the principles of nucleic acid hybridization. *Crit. Rev. Biochem. Mol. Biol.* **26,** 227–259.

2

Detection of *Salmonella* by Bacteriophage Felix 01

Jonathan C. Kuhn

Summary

Salmonellae are mammalian pathogens that are transmitted mainly through foodstuffs and their handlers. Rapid detection requires both specificity and sensitivity in samples containing other bacteria. A solution to this problem is the use of the great specificity conferred by bacteriophages. After implanting reporter genes in a phage genome, the reporter gene products can be measured with great sensitivity when a bacterial host is present. Bacteriophage Felix 01 infects almost all *Salmonella* strains and has been manipulated to contain the *lux* genes specifying bacterial luciferase, an enzyme that converts chemical energy to visible light. A widely applicable methodology for preventing the escape of such recombinant phage has also been developed.

Key Words: Bacteriophage; *Salmonella*; detection; foreign genes in phage; locking.

1. Introduction

One of the most serious problems that face the microbiologist is the detection of pathogenic bacteria in samples containing many types of organisms and often very low levels of the pathogen. Classical microbiology developed techniques to enrich a given pathogen and then identify it on the basis of biochemical tests or special indicator plates. Although relatively inexpensive, such methods usually required a number of days. When a pathogen was isolated, it could be further characterized by phage typing, immunology, or metabolic tests.

Salmonella is a genus with a wide diversity of strains that have been classified into innumerable species on the basis of antigenic type. Patients afflicted with salmonellosis have relatively high levels of the organism, and so the detection

From: *Methods in Molecular Biology, vol. 394: Salmonella: Methods and Protocols*
Edited by: H. Schatten and A. Eisenstark © Humana Press Inc., Totowa, NJ

in these individuals is not a major problem. However, checking for the presence of *Salmonella* in food samples is difficult because of the requirement that there be less than one *Salmonella* per 25 g. Food handlers who are *Salmonella* carriers also have very low numbers. Thus, rapid detection becomes very difficult. The two most important parameters in such tests are specificity and sensitivity. Various methods have been used to solve this problem: specific antibodies, DNA–DNA hybridization, polymerase chain reaction (PCR), and phage-based diagnostics ***(1)***. All these are specific and do not require purified cultures.

This chapter will focus on the construction and use of a derivative of Felix 01 (F01) ***(2)***, a *Salmonella*-specific bacteriophage (phage) that carries reporter genes that are expressed after infecting this host ***(3)***. A cartoon illustrating this method of detection is shown in **Fig. 1**. When no *Salmonella* are present, the reporter genes (*lux*A and *lux*B) are not expressed because of a lack of an appropriate host. In the presence of *Salmonella*, the phage injects its genome, the reporter genes carried by it are expressed, and a measurable signal (light) results. One very important advantage of this test is that it only detects living cells in contrast to the other aforementioned test procedures that also detect dead cells. F01 is an unusual phage that infects 96% ***(4)*** to 98% ***(5)*** of all *Salmonella* strains. The methods specific to the development of a phage reagent are given in this chapter, but techniques in common use such as cloning and PCR are not ***(6)***.

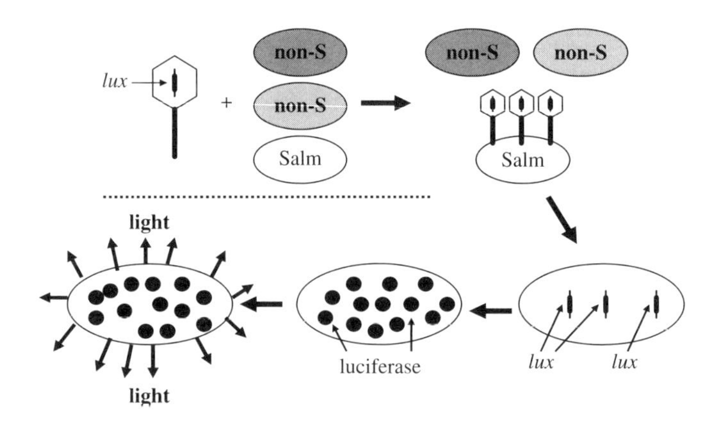

Fig. 1. Detection of bacteria with recombinant phage. Felix 01 with the *lux* genes implanted in its genome (upper left) mixed with a sample containing *Salmonella* (Salm) and other species (non-S). Adsorption of the phage specifically to *Salmonella* cells (upper right). Injection of the phage DNA into *Salmonella* (lower right). Expression of the *lux* genes and synthesis of luciferase (lower middle). Light emission (lower left).

2. Materials

2.1. Media and Reagents

1. Modified Luria-Bertani (LB) broth (per liter H_2O): 10 g tryptone, 5 g yeast extract, 5 g NaCl, 1 g glucose, 5 mL of a 1 M solution of $MgSO_4$, and 5 mL of a 1 M solution of $CaCl_2$. For plating bacteria, add 15 g agar. For bottom-layer agar for phage, add 10 g agar; for top-layer agar, add 6.5 g agar. Sterilize for 15 min at 121 °C. Pour 30–40 mL of LB agar per bacterial plate, and for phage, 30 mL per plate of bottom agar. When required, add 30 mg of ampicillin per liter.
2. Phage buffer: 10 mM $MgCl_2$, 10 mM Tris–HCl, pH 7.6, 10 mM NaCl.
3. TE: 10 mM Tris–HCl, pH 7.9, 1 mM EDTA sodium salt.
4. 3N Sodium acetate.

2.2. Strains

1. *Salmonella* strains: LT2, K772, and as wide a collection of characterized strains as possible which can usually be obtained from local public health services.
2. *Escherichia coli* K12 strains: TG1 *lac*ZΔM15 and W3110 *rec*A (*see* **Note 1**).
3. Phage strains: F01 and λ *red*.

3. Methods

3.1. Propagation and Titration of F01

F01 is a relatively large virulent phage that makes small plaques. To plate the phage for plaques:

1. Make serial dilutions of the stock in LB.
2. Add 0.1 mL phage to 0.1 mL (about 2×10^8 cells) of an overnight culture *Salmonella* strain K772 grown in LB at 37 °C. If this is not available, use an overnight culture of LT2 diluted 1:4 (*see* **Note 2**).
3. Add 4 mL of molten top-layer agar held at 55 °C and mix gently by hand.
4. Pour the contents evenly over the surface of a plate with bottom-layer agar.
5. After the agar hardens, incubate the plate at 37 °C overnight. Count the plaques (*see* **Note 3**).

3.2. Preparation of a Stock of F01

1. Touch an isolated plaque with a sterile needle.
2. Rinse the needle into an early log-phase culture of LT2 or K772 growing in 10 mL LB.
3. Continue incubation at 37 °C until the culture clears.
4. If the culture fails to clear before it stops growing, dilute 1:10 in fresh LB and continue incubation until lysis occurs.
5. Add 15 drops of chloroform, mix, and continue incubation for 10 min. The chloroform kills bacterial and other cells by dissolving their lipid membranes.

6. Centrifuge at 5000 × *g* (6500 rpm in a Sorvall SS34 rotor) in polypropylene Oak Ridge tubes (*see* **Note 4**).
7. Carefully remove and save the supernatant fluid; discard the pellet and chloroform.
8. Titer the lysate by plaque assay as in **Subheading 3.1**.

3.3. Preparation of F01 DNA

F01 DNA is double stranded with a length of 86,155 bp *(7)* and appears to contain neither unusual nor methylated bases *(8)*. To prepare large amounts of F01 DNA:

1. Grow a 1-L culture of LT2 in LB at 37 °C in a 2-L Erlenmeyer to mid-log phase ($A_{600} = 0.5$). The culture has about 2×10^8 cells/mL (2×10^{11} cells total).
2. Add 2×10^{11} F01 particles (about one phage per bacterium).
3. Continue incubation until clearing or until there is a sharp drop in absorbance. The culture can be checked to see whether it is approaching lysis by taking 1 mL of the culture, adding two drops of chloroform, vortexing, and allowing it to stand for 5–10 min. If the sample clears, lysis is near. However, if the culture fails to lyse, dilute it 1:3 with fresh LB and continue incubation until lysis.
4. Add 10 mL chloroform; DNase and RNase to a final concentration of 1 μg/mL.
5. Continue incubation with slow shaking for another 30 min at 37 °C. The chloroform will lyse the remaining cells and release their phages. DNase and RNase degrade nucleic acids that tend to trap free phage particles.
6. Spin the culture at 16,000 × *g* (10,000 rpm in a Sorvall GSA rotor) to remove bacterial debris.
7. Save the supernatant solution; discard the pellet.
8. Precipitate the phage: add 58 g of NaCl per liter lysate with gentle stirring. The lysate can be used immediately or stored overnight in a refrigerator. To the lysate at room temperature, slowly add 100 g of polyethylene glycol 6000 with gentle stirring. When dissolved, cool on ice and wait for several hours during which time the precipitate forms.
9. Spin at 16,000 × *g* (10,000 rpm in a GSA rotor) for 30 min to collect the precipitate.
10. Carefully decant and discard the supernatant fluid and invert the empty centrifuge bottles on a paper towel to remove as much fluid as possible. When drainage has stopped, wipe away any remaining fluid around the centrifuge bottle's mouth.
11. Resuspend the pellet with 10 mL phage buffer. The use of a vortex is alright because the DNA is protected within the phage capsid (*see* **Note 5**). Transfer to an Oak Ridge centrifuge tube and centrifuge the solution at 12,000 × *g* (10,000 rpm in a SS34 rotor) to remove any remaining debris.
12. Add 0.8 g of CsCl per milliliter of concentrated lysate that will give the solution a density of about 1.5 g/mL. Place the solution in small ultracentrifuge tubes. If more fluid is needed to fill the tubes, add phage buffer containing 0.8 g CsCl/mL.
13. Spin at 8 °C at a speed in excess of 30,000 rpm (e.g., 34,000 rpm in an SW 50.1 Beckman rotor, 108,000 × *g*) overnight in an ultracentrifuge (*see* **Note 6**).

The CsCl will form an equilibrium density gradient. The phage will be seen as a grayish band near the middle of the tube; any remaining nucleic acid will be at the bottom of the tube, whereas proteins and lipids will be at the top (*see* **Note 7**).

14. Place a small piece of tape on the side of the tube slightly below the band. Puncture the tube with a 2-mL hypodermic needle through the tape (prevents leakage) and then slowly draw out the fluid until the band disappears.
15. Put the phage solution in dialysis tubing sealed at one end that has been previously boiled in 1 mM EDTA (removes heavy metals and the glycerol preservative), seal the other end with a clip, and dialyze for 12 h against 1 L phage buffer at 4 °C to remove the CsCl. Repeat the dialysis.
16. Extract the phage DNA by adding an equal volume of phenol saturated with 100 mM Tris–HCl buffer (pH 7.6). Mix by inverting a number of times (*see* **Note 8**) and allow the tube to stand for 10 min. A protein precipitate may form at the interface. Separate the phases by centrifugation and carefully remove and save the upper DNA-containing layer. Try to avoid taking phenol that will tend to jump through the interface. Phenol causes burns, so wear gloves.
17. Add an equal volume of chloroform to extract phenol from the aqueous layer. Mix by inverting. Centrifuge for 1 min. Remove and save the upper layer. Repeat the procedure.
18. Precipitate the phage DNA by adding a one-tenth volume of 3 N sodium acetate and 2.5 volumes of cold ethanol. Mix and put at −20 °C for several hours. Spin in a microfuge at full speed for 15 min, carefully draw off the fluid and discard it, add cold 70 % ethanol to the pellet, and spin again. The ethanol steps will remove any remaining phenol, and the second ethanol wash removes salts. Decant the fluid but do not dry, as large DNA is very hard to dissolve after drying.
19. Add 1 mL TE to dissolve the DNA. Expected yield is 1–2 mg DNA.

3.4. Cloning Segments of F01 DNA

F01 DNA is cleaved much less than expected by many restriction enzymes *(8)*. Therefore, *Hae*III (GG↓CC) digestion was performed to generate fragments, and these fragments were cloned into pHG165 (a low copy version of pUC8) *(9)* cut with *Sma*I (CCC↓GGG) using T4 ligase to connect the elements. Transformants with inserts were detected by a lack of lacα complementation in strain TG1 on MacConkey Agar plates. Several clones were isolated that contained F01 inserts, and these inserts were DNA sequenced. Among the clones isolated, several segments were contiguous. Only a small part of the F01 genome was isolated. However, relatively short segments ($\geq$3 kb) are sufficient for the construction of a recombinant F01, and this should be true for other phages into which one wishes to implant foreign reporter genes (*see* **Note 9**).

3.5. Introduction of Foreign DNA Into F01

There are three main ways to introduce reporter genes into the genome of a phage: (1) transposition, (2) direct cloning, and (3) recombination ***(10)***. Transposition is very efficient with temperate phages but much less so with virulent ones. For temperate phages, the reporter genes can be coupled with an antibiotic resistance and cloned between the target sequences of a transposable element. When the phage is grown on a strain with such a transposable element, some (usually about 1 per 10^5–10^6) of the progeny will have acquired it. After infection of a strain sensitive to the same antibiotic, some bacterial cells become lysogenic and contain the phage genome. Transposition events are then detected by the formation of lysogenic bacteria that have become resistant to the antibiotic whose resistance is carried within the transposon. Transposition is much more difficult to use with virulent phages because the direct selection of transposition events seemingly cannot take advantage of antibiotic resistance. However, antibiotic resistance might be able to be used if one looks for a "window of opportunity." This would entail making a population of phages that have been subjected to transposition by an element containing the reporter gene and, for example, kanamycin resistance. A few minutes after the infection when the resistance gene has been already expressed by the rare phage carrying it, kanamycin is added which prevents the phages that have not been transposed from developing but allows those that have to continue until lysis. Further cycles of enrichment can then be performed. Although not tested, the use of chromosomal genes for selection of recombinant phage might be efficacious. For example, the piece to be transferred might be combined with a tryptophan pathway (*trp*) gene, and only a phage receiving this gene will be able to multiply on a bacterial cell whose chromosome is mutated in this gene in a medium lacking tryptophan.

A more general method for isolating virulent phages with reporter genes after transposition employs amber mutants of the phage ***(3)***. Both transposable elements and recombination (detailed below in this section and 3.8) can be used to insert the reporter genes in the phage genome when the reporter gene is adjacent to a selective gene (*sup*$^+$) that encodes a tRNA that is suppressor of amber mutations. Only transposed or recombinant phages are able to multiply and form plaques on a strain that is *sup*$^-$ (i.e., lacks an amber suppressor gene). Transposition of the luciferase genes without any selection has also been successfully used for a few virulent phages by screening plates with about 10^5 plaques for those plaques that have luminescent halos ***(10)***.

A second method for introducing reporter genes into a phage genome is simply direct cloning. One needs to know the DNA sequence of the phage and the location of open reading frames (ORFs) that most probably encode proteins ***(11)***. The sequence imparts the location of all restriction endonuclease cleavage sites and may allow one to attempt direct cloning after isolating phage DNA. The construct must then be introduced into the host by transformation.

The method used to isolate a derivative of F01 carrying the *luxA* and *luxB* genes employed a mixture of cloning, DNA sequencing, and recombination. The steps were (1) isolation of amber mutants, (2) isolating phage with two amber mutations, (3) construction of genetic elements by DNA cloning procedures, and (4) homologous recombination and selection of phage able to form plaques on *sup*$^-$ hosts to isolate the desired phage type.

3.6. Isolation of Amber Mutations of F01

1. Transform K772 or LT2 with a plasmid containing an amber suppressor. A number of such plasmids based on pBR322 are available ***(3)***.
2. Mutagenesis of the phage with *N*-nitroso-*N*′-nitro-*N*-methylguanidine (NG): To a log-phase culture of K772 *sup*$^+$ growing at 37 °C in LB, add NG to a final concentration of 50μg/mL, and add five phages per bacterium (*see* **Note 10**). After 20–30 min, centrifuge the culture in the cold at 16,000 × g and resuspend the pellet in 10 mM $MgSO_4$. Repeat the centrifugation and resuspension steps. This removes the NG. Then, centrifuge again and resuspend the cells in LB and continue incubation until lysis. Alternatively, subject the phage lysate to UV light so that survival is 10^{-3}–10^{-4} and then infect the *sup*$^+$ strain. With UV treatment, the phage can be directly plated for plaques. The use of a *sup*$^+$ strain permits the growth of the amber mutants.
3. Add chloroform (15 drops/10 mL) and centrifuge to remove debris.
4. Titer the lysate on K772 *sup*$^+$ or LT2 *sup*$^+$ to obtain isolated plaques (100–300 per plate).
5. Pick plaques with sterile toothpicks to K772 *sup*$^-$ and K772 *sup*$^+$ (LT2 *sup*$^-$ and LT2 *sup*$^+$). About 1 per 1000 should be amber mutants that fail to grow on *sup*$^-$ strains.
6. From the amber mutant plaques, make small lysates and test them to ensure that they are indeed amber mutants.

3.7. Isolating Double-Amber Mutants

The reversion index of lysates of single-amber mutants on LT2 *sup*$^-$ or K772 *sup*$^-$ is about 1 per 10^4–10^5 phages. Double ambers revert at 10^{-8}–10^{-9}, and therefore, reversion is negligible. Double-amber mutants are made by crossing single-amber mutants with each other. Crosses to accomplish this are done in the following way:

1. Infect 0.1 mL of a late log-phase culture of LT2 *sup*$^+$ with about five particles of each mutant ($A_{600} = 0.8$, approximately 3×10^8 cells/mL; 1.5×10^9 phages of each mutant).
2. Continue incubation at 37 °C. After 20 min, dilute the culture 100-fold in LB and continue incubation for 120 min. Add chloroform, centrifuge, and save the supernatant fluid.
3. Titer on LT2 *sup*$^-$ for wildtype recombinants and on LT2 *sup*$^+$ for total phages.
4. Crosses that give at least 5 % wildtype recombinants (plaques on *sup*$^-$ per total phages on *sup*$^+$) are desired. This indicates that about 5 % of the phage progeny will be double-amber recombinants, which arise by the reciprocal crossing-over event.
5. Take 50 individual plaques from the plate with LT2 *sup*$^+$ using a Pasteur pipette. Put each agar plug in 1 mL 10 mM $MgSO_4$ at 4 °C to allow the phages to elute from the plug overnight. A single plaque usually has between 10^6 and 10^7 phage particles.
6. Prepare a 100 mL early log-phase culture of LT2 *sup*$^-$, and distribute 2-mL aliquots into 50 tubes. To each tube, add 0.5 mL from the individually eluted plaques. Continue incubation at 37 °C. Within 2 h, those plaques that were from wildtype phages will lyse their cultures. After 3–5 h, those plaques that come from phages with single-amber mutations will lead to lysis because there were a few revertants in the inoculum. Double-amber mutants will not cause lysis because the inoculum contained no wildtype revertant particles.
7. Several tubes will show no lysis, and these are from phages that are putative double-amber mutants. Return to the $MgSO_4$ tubes of these phages and plate 0.1 mL on LT2 *sup*$^-$. Double ambers should show no reversion. Dilute and titer the $MgSO_4$ suspension on LT2 *sup*$^+$ to ascertain the total number of particles.
8. From those that look to be double ambers, make a lysate on LT2 *sup*$^+$ and then titer on both LT2 *sup*$^+$ and LT2 *sup*$^-$ to confirm their genotype.

3.8. Isolation of Recombinant Phages Carrying lux*AB and* sup$^+$

The *red* gene of F01 ***(11)*** encodes a protein of 244 amino acids that is non-essential for phage growth. Clones of this F01 gene allow the growth of λ *red* mutants on W3110 *rec*A on which λ *red* mutants do not normally grow. Downstream of the F01 *red* gene are three small ORFs ***(7)***. It is not known whether these are essential for F01 growth. Further downstream is a large ORF (ORF4) encoding a protein of 533 amino acids followed by ORF5 that specifies a protein of 488-amino-acid residues (*see* **Fig. 2**). Two *Hae*III fragments were joined and the resultant clone contains *red*, ORF4, and part of ORF5. There is a *Bst*XI site in the *red* gene. A piece of DNA containing the *lux* and *sup*$^+$ genes

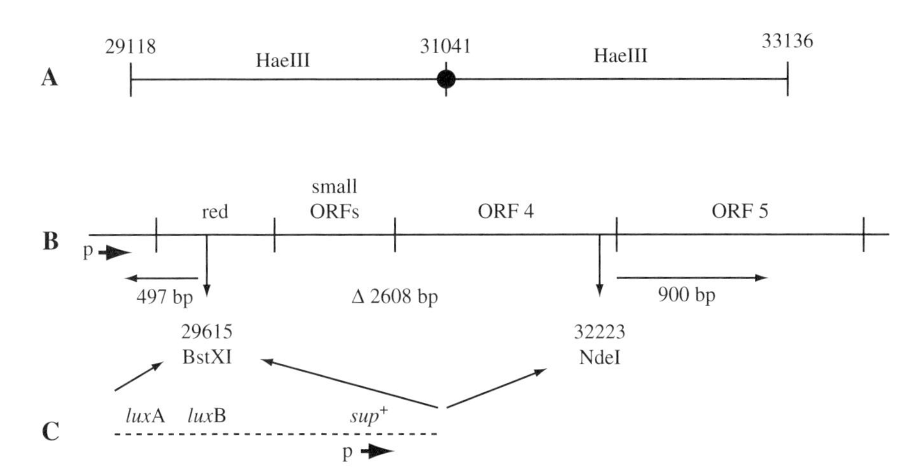

Fig. 2. Diagrammatic representation of the sites and genes mentioned in the text. **(A)** The two cloned *Hae*III fragments connected by ligation (the joint is represented by a filled circle). **(B)** The genes of Felix 01 in the relevant segment. A promoter (p) is upstream of the *red* gene, and all genes in the segment are in the same orientation as this promoter. The two sites for insertion (*Bst*XI and *Nde*I) are shown as are the flanking regions of Felix 01 used for recombination. Replacement of the *Bst*XI–*Nde*I segment deletes (Δ) 2608 bp. **(C)** The foreign DNA for insertion that contains the promoterless *lux*A and *lux*B genes and a suppressor gene with its associated promoter (p).

can be directly inserted into the *Bst*XI site (at bp 29615; *see* **ref.** *7*) and the inserted piece should be checked to ensure that it is in the same orientation as *red* with regard to transcription/translation. No promoters need be associated with the *lux* piece because the *red* gene has an upstream promoter within the phage genome and the *lux* genes will be read from this after their transfer into the F01 genome. The sup^+ gene has its own promoter. The plasmid with this piece of F01 DNA and containing the *lux-*sup^+ insertion should be transferred to LT2 that make LT2 sup^+. The double-amber phage is grown on this strain and the progeny phage plated on LT2 sup^-. The amber mutants should be in genes outside of the cloned piece (*see* **Note 11**). Plaques that form on the sup^- strain will be either rare revertants or those that have picked up the *lux-*sup^+ piece by recombination. The plaques that form can be examined for light emission as described in **subheading 3.11**. A better way of detecting the recombinants is to plate on a strain of LT2 sup^- into which has been introduced an F′ *lac*Z kan^R plasmid that carries a *lac*Z amber mutation (*see* **Note 12**). When this strain is used, add 1 mg of X-gal (5-bromo-4-chloro-3-indoyl β-D-galactopyranoside dissolved in 100 μL dimethyl sulfoxide) and 10 μL isopropyl β-D-1-thiogalactopyranoside

(IPTG, 0.1 M in H_2O) into the top agar at the time of plating (*see* **Note 13**). Phages carrying the *sup*$^+$ gene will give blue plaques because they suppress the *lacZ* mutation and allow the production of an active β-galactosidase, which cleaves X-gal to give a colored product. Phage that are revertants to wild type (reversion of the amber mutations) will give colorless plaques.

3.9. Locked Phages

The recombinant phage constructed in the previous section might escape from the laboratory because it is able to grow on wildtype, *sup*$^-$ *Salmonella* strains. It is, however, unlikely that such a phage would be able to survive for any length of time because even the suppression of amber mutations does not lead to an organism as fit as the original wild type. Suppression of amber mutations is never 100 % because the best suppressors only have an efficiency of about 60 %. However, the Federal Drug Administration of the United States prefers that a biological reagent have built-in features that prevent its spread. This was accomplished by removing a segment of F01 DNA that contains an essential gene(s) and replacing the segment with the *lux-sup*$^+$ fragment. Rather than singly cleave the F01 segment, a region between the *red* gene and the end of ORF4 was removed by cleavage with *Bst*XI (in the *red* gene at bp 29615 of the F01 sequence) and *Nde*I (at the distal end of ORF4 at bp 32223) followed by the insertion of the *lux-sup*$^+$ genes in place of the deleted segment. In such a clone, the regions of F01 DNA that remain are 497 bp to the *red* side and about 900 bp after ORF4. These flanking regions provide homology with F01 and allow recombination to occur between the plasmid and the F01 phage carrying two amber mutations. Such recombinants will be rare.

Although these rare recombinant phages carry *lux-sup*$^+$ that suppresses their amber mutations, they will still be unable to grow because they are deleted for the small ORFs and ORF4. At least one of these genes is essential for growth. A clone was therefore made that contains exactly this region under the control of the *plac* promoter of *E. coli*. As ORF4 runs from base 30238 to 32241, a cloned *Bst*XI-to-*Nde*I (32223) segment will lack the last six amino acids of ORF4. To solve this, a piece of synthetic DNA was attached to the *Nde*I end of the isolated fragment by ligation before cloning the *Bst*XI–*Nde*I piece. The synthetic DNA encodes the same amino acids as the end of ORF4, but the codon usage of the sequence is entirely different than that used by F01 and there is no homology between this synthetic DNA and F01 DNA. When this fragment is added to the cloning vector, the resultant plasmid encodes the small ORFs and ORF4 and will permit the growth of recombinant phage missing these genes. This plasmid was introduced into strain LT2 *sup*$^-$. The resultant strain allows the growth of the desired F01 recombinant by supplying the products of the missing genes and selects for the recombinant phage carrying the sup^+ gene. Rare revertants of the amber mutations can grow on LT2 sup^- without the plasmid with the F01 segment. Phages containing *lux*-sup^+ were recovered, and these grow only on the strain containing the plasmid expressing the missing genes. The recombinant phage has absolutely no DNA homology to the plasmid in whose presence it can grow. Should this mutant F01 phage escape from the laboratory, it cannot grow because *Salmonella* strains will neither have nor express the missing phage genes. Recombination with a wildtype F01 phage to regain these genes will, by necessity, delete the *lux*-sup^+ genes. As a result, the phage is locked and good for a single round of bacterial detection but can neither escape into nature nor be propagated by commercial competitors.

3.10. Large Amounts of the Phage Reagent

1. Follow the same protocol, steps 1–7, as that for DNA preparation.
2. Spin the lysate at 43, 000 × *g* (19,000 rpm in an SS34 rotor) for 60 min. The phages form a pellet at this speed.
3. Gently resuspend the pellet with 10 mL phage buffer.
4. Spin the mixture at 12,000 × g to remove any remaining debris and decant and save the supernatant fluid. This will provide a stock of about $1–3 \times 10^{12}$ phages/mL of the recombinant phage producing luciferase. This protocol damages the viability of the phage particles much less than that for DNA purification in which phage viability is irrelevant.

3.11. Determination of Light Emission in Samples

Bacterial luciferase is a heterodimer, αβ, encoded by *lux*A(α) and *lux*B(β). Its substrates are $FMNH_2$, a long-chain aldehyde, and O_2 ***(12)***. The reaction catalyzed by luciferase generates FMN, a fatty acid, and produces light in the visible range. All eubacteria contain $FMNH_2$. Externally supplied aldehyde rapidly enters the cells by diffusion. In luminescent bacteria, the long-chain aldehyde has 14 carbons, but dodecyl aldehyde is more soluble in water and hence superior as an addition.

1. To a 1-mL sample of growing bacteria in LB, add 10^9–10^{10} F01-*lux*. Use a glass scintillation vial.
2. Incubate for 30 min at 28 °C.
3. Add 20 μL dodecyl aldehyde (0.1 % v/v in ethanol) and mix. To liquefy pure aldehyde, melt it in a 60 °C incubator or place the bottle in water and heat.
4. After 4 min, read the light emitted in a Packard Tri-Carb (1600 TR) scintillation counter with the coincidence off. Read for 12 s. An average counter will take 30 s between vials when the reading time is 12 s. Usually, the first eight vials can be

empty and aldehyde is added to the first sample when the first empty vial descends into the counter or begins reading. Then, aldehyde is added sequentially to each additional sample as the next empty vial moves into the reading position within the counter and so forth. This protocol ensures that a vial will be read exactly after 4 min.
5. Many other luminometers have different formats but can be used for this purpose.
6. Maximum light emission takes place after 4 min and continues for several minutes. Charge-coupled devices have been developed for the detection of very low levels of light, and their use should permit the detection of even lower numbers of *Salmonella* than is possible with a scintillation counter in which as few as 100 luciferase-containing *Salmonella* cells/mL can be detected.

3.12. Enrichment of Samples Before Detection

Food samples usually contain only very low levels of *Salmonella*, if at all. The food sample should be placed in peptone water (10 g peptone/L) to increase the numbers of *Salmonella* (most other bacterial species also grow) and incubated with shaking at 37 °C overnight. Further enrichment can be accomplished, if necessary, by a second round of growth in selenite broth (Difco Corp.) or tetrathionate broth (Difco Corp.). These procedures might be coupled at any stage to a method that concentrates *Salmonella*. An example of such a procedure is the use of magnetic beads linked to anti-*Salmonella* antibodies, and such beads are commercially available.

3.13. Resistance and Restriction-Modification

Resistance is defined as the inability of a given phage to form plaques on a lawn of a particular host. In many cases, resistance results from a lack of adsorption by the phage to a particular host strain. That is, the host does not have appropriate receptors on its external surface and is not recognized as a host by the phage. While many phages choose host receptor molecules that are abundant and important to the cell, at the same time to avoid phage infection either by not synthesizing the receptor at all or by modifying it through mutation so that it still carries out its biological function without being recognized by the phage. Besides the loss of receptors, there are also other ways by which a bacterium can be resistant to a given phage: (1) the lack of a host factor essential for the growth of a particular phage, (2) slow adsorption, and (3) the presence of restriction-modification systems. With regard to F01, our experience shows that the first of these ways seems unlikely and no instance of the last of these has yet been documented. However, a lowered rate of adsorption can occur, and a detailed study ***(13)*** found that only about 50 % of *Salmonella* strains allow F01 to form plaques when they are used as hosts rather than the 96–98 % that are sensitive by spot tests. When adsorption rate is much reduced, the phage grows on the strain, albeit slowly, but fails to form plaques because plaque formation is essentially a race between the growth of the bacterial lawn and the rate of propagation of the phage. Phage growth usually ceases after 6–7 h at 37 °C when the host ceases to grow. To form a plaque, a phage must go through at least four cycles of infection during this time and the cycle time is much slower than that seen in liquid because diffusion through the top-layer agar is necessary to reach new hosts. In contrast, in a spot test, the area is small and there are about 10^8 or more phages. Even if adsorption is slow, all the bacteria in the spot will be killed, becauses the phage are in great excess. This accounts for the disparity between plaque formation and spot tests. To overcome the problem of slow adsorption, it is recommended that a high concentration of phage be used: the constant of adsorption (k) cannot be changed, but the rate of adsorption can change because it depends on the concentration of phage (P) according to the equation: $-dN/dt = kNP$, where dN/dt reflects the number of uninfected bacteria, t is time, N is the concentration of *Salmonella* cells, and P the concentration of free phage ***(14)***.

3.14. Future Improvements

3.14.1. Coverage of Resistant Strains

F01 fails to infect most of the E group strains of *Salmonella*. A phage, OE1, that does infect this group was isolated. OE1 was made manipulated to carry the lux genes by transposition. When a phage such as F01 has been successfully engineered to carry lux or other reporter genes, it may be worthwhile to look for host range mutants or other phage types related to F01 rather than repeat the genetic manipulations with a second phage type such as OE1 that is not in the F01 group. To avoid duplicating the effort to make recombinant F01 with OE1, perhaps it is worthwhile to look for host-range mutants of F01 or phages that are related to F01 that can infect these resistant strains. Host-range mutants can occasionally be isolated by plating large numbers of mutagenized phage on a resistant strain and searching for rare plaques that are host-range mutants. However, it is not always possible to obtain such mutants. Several phages closely related to F01 have already been isolated by us from nature ***(8)***. It seems that a good approach for covering resistant strains might be to search for phages from the environment that grow on a given resistant strain and examine whether some of them are related to F01 by plaque hybridization. Phages are found where their hosts are. For *Salmonella*, the most common sources of phage are sewage and effluents from locales or farms where fowl are raised. Such a related phage could, in all probability, be made to carry the *lux* genes by recombination with the constructed F01-*lux* in a host that can be infected by both of them. That is, an exchange of tail and tail fiber genes should create a recombinant of the desired type that carries both *lux* and also grows on the strain originally resistant to F01.

3.14.2. Reporter Genes

Fluorescent proteins as reporters are based on the detection of emitted light of a wavelength distinct from that used for excitation. The instrumentation developed for light detection is again applicable for these. In addition, with the development of very sensitive confocal microscopes, visual examination of a sample has become possible. Although phages carrying genes encoding these proteins have not yet been used for the detection of *Salmonella*, it would seem that fluorescent proteins will prove to be superior to all other reporter genes. The genes for various fluorescent proteins have been cloned and can be manipulated to allow their insertion into phage genomes as readily as for the luciferase genes used previously.

Several other reporter systems should be mentioned here. The ice nucleation gene from *Pseudomonas* has been used with F01 for the detection of *Salmonella* ***(15)*** and gives extreme sensitivity because even a single molecule of this protein is enough to lead to rapid ice formation that can be detected by a change in color. The main drawback to this reporter system is that crude samples often have contaminants that can also act as nucleation centers for ice formation, and this leads to an increase in false-positive tests. A second, but untried, type of reporter gene might be the use of genes from extreme thermophiles. Any such protein synthesized from such genes will be active at up to 100 °C, and this will allow samples to be heated to boiling after phage infection and then assayed for the presence of the thermo-resistant protein synthesized by the phage-borne gene. This technique should entirely eliminate background. Another, but as yet untried, technique might take advantage of the rapid expansion of fluorescent reagents that bind proteins or nucleic acids. If one could label a particular phage without affecting its adsorption, then the techniques for fluorescent proteins could be used without the necessity of the genetic manipulations outlined earlier.

3.15. Conclusions

The techniques detailed in this chapter should be generally applicable ***(1)*** for the creation of phage reagents for the detection of almost any pathogen, as there are few, if any, bacterial species that lack phage that infects them.

4. Notes

1. Because the sequence of the F01 genome and that of its *red* gene are now known, *rec*A mutant strains and the λ *red* strain can be dispensed with.
2. Large phages such as F01, P1, and T4 tend to make small plaques, whereas some relatively small phages (T1, T3, and T7) form large plaques. Plaque size depends mainly on the rate of diffusion of the phage particles through the soft agar and the time necessary for an infective cycle. When plaques are larger than desired, one can use a higher number of host cells for plating to reduce the time between plating and that when bacterial growth ceases. Increasing the concentration of agar in the top-layer agar helps somewhat in this regard. When plaques are small, one can (1) lower the amount of agar in the top-layer agar, (2) use less host cells to increase the length of the growth period on the plates, (3) use a mutant host strain that grows more slowly to allow more time for phage diffusion, and/or (4) try a variety of different host strains in the hope of finding a better plating strain.
3. The variance of the experimental error in plating for colonies or plaques is approximately equal to the number appearing. The standard deviation is the square root of the variance. One should try to keep the number of bacterial colonies per plate between 50 and 500 and the number of plaques between 70 and 700 to minimize the relative error. The higher numbers for both are dictated by the appearance of overlapping colonies or plaques above these limits.
4. Polycarbonate tubes are sensitive to many organic compounds, and chloroform is one of these compounds.
5. Vortexing often shears off the tail and tail fibers of phage. If one is following the titer of the phage to measure purification and its associated loss in total phage number, one should take this into account.
6. Although a gradient will be formed more quickly at higher speeds, it is desirable to spin at a speed of around 30,000 rpm (approximately $100,000 \times g$). At equilibrium, the steepness of the gradient is a function of rotor velocity, and a better separation of biological macromolecules is achieved when the difference in specific gravity between the top and bottom of the gradient is less.
7. The basis of this purification step is that phage particles are a complex of nucleic acid and protein, which imparts a density that is unique except for ribosomes that are destroyed in the protocol by RNase. Phages usually have densities of about 1.5 g/mL, whereas proteins, polysaccharides, and lipids are 1.3 g or less and nucleic acids above 1.6 g/mL. The phage band is therefore free of these other macromolecules.
8. The phage DNA is now free and very susceptible to breakage. Therefore, the use of a vortex mixer or other treatments that cause shearing should be avoided.
9. Now that DNA sequencing can be outsourced quite cheaply, some of these steps can be eliminated or modified when large regions of the sequence become available.
10. NG is highly mutagenic and carcinogenic. Take appropriate precautions.

11. Although it is probable that in most cases the mutations will lie outside the particular segment used for cloning, this can be checked by plating the single-amber mutants on a strain that carries the segment and is also *sup*⁻. If a mutation lies within the segment, then "rescue" (i.e. recombination) will occur and there will be a large increase in plaque number in comparison with a *sup*⁻ strain without the segment.
12. The F fertility factor of *E. coli* can promote its own transfer from *E. coli* to *Salmonella*. F-factor derivatives carrying the *lac* operon of *E. coli* are available, and some of these have an introduced *lac*Z amber mutation in the gene specifying β-galactosidase. Resistance to kanamycin (*kan*^{R}) is used to allow the easy identification of *Salmonella* recombinant cells carrying this F-factor derivative. On minimal synthetic medium containing both citrate as the sole carbon source and kanamycin, only recombinant *Salmonella* can form colonies.
13. X-gal is a colorless substrate of β-galactosidase that enters the cell by diffusion. IPTG causes induction of the *lac* operon genes of *E. coli* and induction is necessary for good color formation.

References

1. Ulitzur, S. and Kuhn, J. (1989) Detection and/or identification of microorganisms in a test sample using bioluminescence or other exogenous genetically introduced marker. US Patent No. 4,861,709.
2. Felix, A. and Callow, B. R. (1943) Typing of paratyphoid B bacilli by means of Vi bacteriophage. *Br. Med. J.* **2,** 127–130.
3. Kuhn, J., Suissa, M., Wyse, J., et al. (2002) Detection of bacteria using foreign DNA: the development of a bacteriophage reagent for *Salmonella*. *Int. J. Food Microbiol.* **74,** 229–238.
4. Fey, H., Burgi, E., Margadant, A., and Boller, E. (1978) An economic and rapid diagnostic procedure for the detection of *Salmonella/Shigella* using the polyvalent *Salmonella* phage 0-1. *Zent. Bakteriol. Mikrobiol. Hyg. Abt. I Orig. A* **240,** 7–15.
5. Kallings, L. O. and Lindberg, A. A. (1967) Resistance to Felix 0-1 phage in *Salmonella* bacteria. *Acta Pathol. Microbiol. Scand.* **70,** 455–460.
6. Maniatis, T., Fritsch, E. F., and Sambrook, J. (1982) *Molecular Cloning*. Cold Spring Harbor Laboratory, Cold Spring Harbor, NY.
7. Sriranganathan, N., Whichard, J. M., Pierson, F. W., Kapur, V., and Weigt, L. A. Bacteriophage Felix 01, complete genome. NCBI accession number: AF320576.
8. Kuhn, J., Suissa, M., Chiswell, D., et al. (2002) A bacteriophage reagent for *Salmonella*: molecular studies on Felix 01. *Int. J. Food Microbiol.* **74,** 217–227.
9. Stewart, G. S. A. B., Lubinsky-Mink, S., Jackson, C. G., Cassel, A., and Kuhn, J. (1986) pHG165: a pBR322 copy number derivative of pUC8 for cloning and expression. *Plasmid* **15,** 172–181.

10. Ulizur, S. and Kuhn, J. (2000) Construction of *lux* bacteriophages and the determination of specific bacteria and their antibiotic sensitivities. *Methods Enzymol.* **305,** 543–557.
11. Kuhn, J. C., Suissa, M., Chiswell, D., Ulitzur, S., Bar-On, T., and Wyse, J. (1998) Bacteriophage Felix 01 and some of its genes. GenBank accession number AF071201.
12. Meighen, E. A. (1991) Molecular biology of bacterial bioluminescence. *Microbiol. Rev.* **55,** 123–142.
13. Kallings, L. O. (1967) Sensitivity of various salmonella strains to Felix 0-1 phage. *Acta Pathol. Microbiol. Scand.* **70,** 446–454.
14. Stent, G. S. (1963) *Molecular Biology of Bacterial Viruses.* W.H. Freeman & Co., San Francisco and London, pp. 89–96.
15. Wolber, P. K. and Green, R. L. (1990) New rapid method for the detection of *Salmonella* in foods. *Trends Food Sci. Technol.* **1,** 80–82.

3

Physical Mapping of *Salmonella* Genomes

Shu-Lin Liu

Summary

Physical mapping is a key methodology for determining the genome structure of *Salmonella* and revealing genomic differences among different strains, especially regarding phylogenetic relationships and evolution of these bacteria. In fact, physical mapping is the only practical approach to genomic comparisons among *Salmonella* involving large numbers of strains to document their insertions, deletions, and rearrangements that may be related to pathogenesis and host specificity. The core technique in physical mapping is pulsed field gel electrophoresis (PFGE), which can separate DNA fragments ranging from less than one kilobase to several thousand kilobases. After genomic DNA has been cleaved by an endonuclease and the DNA fragments have been separated on PFGE, a number of techniques will be employed to arrange the separated DNA fragments back to the original order as in the genome. These techniques include Southern hybridization with known DNA as the probe to identify the DNA fragments, Tn10 insertion inactivation to locate genes and identify the fragments that contain these genes, double cleavage to determine the physical distances of cleavage sites between different endonucleases for further refining the physical map, and I-*Ceu*I partial cleavage to lay out the overall genome structure of the bacteria. The combination of these mapping techniques makes it possible to construct a *Salmonella* genome map of high resolution, sufficient for comparisons among different *Salmonella* lineages or among strains of the same lineage.

Key Words: *Salmonella*; physical map; genome comparison; PFGE; I-*Ceu*I; Southern hybridization; Tn*10* insertion; double cleavage.

1. Introduction

Salmonella, with its currently recognized over 2500 lineages, is a good model for studies of bacterial genome divergence and evolution, because these closely related bacteria may have quite different biological properties, especially

From: *Methods in Molecular Biology, vol. 394: Salmonella: Methods and Protocols*
Edited by: H. Schatten and A. Eisenstark © Humana Press Inc., Totowa, NJ

regarding host range and pathogenicity. For systematic studies of the genomic evolutionary processes that have led to different pathogens, it is necessary to know the levels of genomic variation both among and within the *Salmonella* lineages. Additionally, molecular markers are needed for rapid and reliable detection or differentiation of *Salmonella* pathogens, because serotyping, which has been used for over 70 years ***(1)*** and is still the main method of *Salmonella* typing, is tedious and expensive and can be done only at a few reference laboratories. Local research institutions or hospitals often have available only a limited number of commercial antisera that may not be quality assured. Useful molecular markers are to be found through genome analysis.

Whole genome sequencing and comparison between pairs of *Salmonella* strains would provide the ultimate answers regarding their similarities and differences at the single-nucleotide resolution, as is in the cases of two *S. typhi* strains ***(2,3)*** and a representative strain each of some other *Salmonella* lineages ***(4–6)***. However, revelation of genome features common to all members of a *Salmonella* lineage would require large numbers of strains in the comparison, which makes it unrealistic to employ whole genome sequencing as the main methodology because of the enormous cost and time involved.

Physical mapping, on the contrary, can provide genome data on populations of bacteria within relatively short times at low cost, and the resolution is sufficient for most comparative studies. The first *Salmonella* physical maps were published in 1992 ***(7,8)***. Over the years, the mapping techniques have been greatly optimized and the accuracy has been significantly improved ***(9,10)***. Genomic features, including insertions, deletions, inversions, and translocations, revealed first by mapping in *S. typhimurium* LT2 ***(11)***, *S. typhi* Ty2 ***(12)***, and *S. paratyphi* A ATCC9150 ***(13)***, were all confirmed later by sequencing ***(3–5)***, and their genome sizes estimated by physical mapping differed by only 1, 0.25, and 0.22 %, respectively, from those determined by sequencing.

2. Materials (See Note 1)

2.1. Bacterial Culture, Cell Lysis, and Genomic DNA Isolation

1. Luria-Bertani (LB) broth: 10 g tryptone, 5 g yeast extract, 10 g NaCl, and water to 1 L. Autoclave. LB plates also contain 15 g agar. Tetracycline, when needed, is added to a final concentration of 20 μg/mL, and kanamycin at 50 μg/mL.
2. 0.5 M Ethylenediaminetetraacetic acid (EDTA), pH 8.0: 186.12 g EDTA (Sodium salt), ca. 600 mL H_2O, and ca. 30 g NaOH. When EDTA and NaOH are completely dissolved, adjust pH to 8.0 with concentrated HCl or 10 N NaOH. In the case of EDTA free acid, use 142.12 g of EDTA and about 60 g of NaOH, adjust pH to

8.0 with concentrated HCl or 10 N NaOH. After autoclaving, adjust the volume to 1 L with sterile water (*see* **Note 2**).

3. 1 M Tris–HCl, pH 8.0: 121.14 g Tris base, ca. 600 mL H_2O, and ca. 60 mL concentrated HCl. Stir with a magnetic rod in a glass beaker on a magnetic stirrer until Tris base is completely dissolved. Adjust pH with concentrated HCl or 10 N NaOH. After autoclaving, adjust the volume to 1 L with sterile water.
4. 10 % (w/v) Sodium dodecyl sulfate (SDS): 100 g SDS and ca. 600 mL H_2O (sterile). Stir with a magnetic rod in a glass beaker on a magnetic stirrer with the temperature adjusted to about 68 °C, until SDS is completely dissolved. Adjust the volume to 1 L with sterile water. No autoclaving is needed.
5. Tuberculin syringe with the end cut off; one syringe is for one DNA sample. Syringes are kept in 70 % ethanol before and after use.
6. 1.4 % Agarose: For 10 mL (the actual volume can be brought up or down according to the needs), add water first to a container (plastic or glass), then add 140 mg agarose on top of water. Briefly mix with hand. Boil the liquid completely and keep it at 72 °C in a water bath.
7. Cell suspension solution: 10 mL 1 M Tris–HCl (pH 8.0), 4 mL 5 M NaCl, 200 mL 0.5 M EDTA, and ca. 400 mL H_2O. After autoclaving, adjust the volume to 1 L with sterile water.
8. Cell lysis buffer: 10 mL 1 M Tris–HCl (pH 8.0), 10 mL 5 M NaCl, 10 g *N*-lauryl-sarcosine sodium, 200 mL 0.5 M EDTA, and ca. 400 mL H_2O. After autoclaving, add 20 mL 10 % SDS and adjust the volume to 1 L with sterile water. *N*-Lauryl-sarcosine sodium is toxic, so wear mask when weighing.
9. Wash solution: 10 mL 1 M Tris–HCl (pH 8.0), 200 mL 0.5 M EDTA, ca. 600 mL H_2O. After autoclaving, adjust the volume to 1 L with sterile water.
10. Proteinase K solution: 200 mL 0.5 M EDTA, ca. 700 mL H_2O, and 10 g *N*-lauryl-sarcosine sodium. After autoclaving, add 20 mL 10 % SDS and adjust the volume to 1 L with sterile water. *N*-Lauryl-sarcosine sodium is toxic, so wear mask when weighing. Powder of proteinase K is added to proteinase K solution to make the required concentration, which is usually 1 mg/mL, prior to use.
11. 100 mM Phenylmethylsulfonylfluoride (PMSF): 870 mg PMSF and 50 mL isopropanol. Store at −20 °C. Prior to use, dissolve the crystallized PMSF in the solution in a warm water bath (e.g., 37–50 °C) for 1–2 min.
12. Storage solution: 1 mL 1 M Tris–HCl (pH 8.0), 20 mL 0.5 M EDTA, and ca. 600 mL H_2O. After autoclaving, adjust the volume to 1 L with sterile water.

2.2. Endonuclease Cleavage and Pulsed Field Gel Electrophoresis (PFGE)

1. Endonucleases: For *Salmonella* genomes, the most frequently used endonucleases include I-*Ceu*I (NEBiolabs), *Xba*I (Roche), *Avr*II (NEBiolabs), and *Spe*I (NEBiolabs).

2. Electrophoresis buffer (5× TBE): 54 g Tris base, 27.5 g boric acid, 20 mL 0.5 M EDTA, and ca. 900 mL H_2O. After autoclaving, adjust the volume to 1 L with sterile water. Prior to use, dilute the 5× TBE by 10-fold in water into 0.5× TBE.
3. Ethidium bromide stock solution (10 mg/mL): 1 g ethidium bromide and 100 mL H_2O. Stir on a magnetic stirrer until ethidium bromide is completely dissolved. Wrap the container in aluminum foil or store the solution in a dark bottle at room temperature.
4. 0.5 % Agarose gel: For a gel of 400 cm^2 with a thickness of 1 cm, add 2 g agarose to 200 mL water, boil for 10 s, add 200 mL water and 12 μL ethidium bromide stock solution, shake, and pour the liquid to a casting tray.

2.3. Southern Hybridization

1. Immobilon-P transfer membrane (Millipore).
2. 0.4 N NaOH.
3. 0.25 N HCl.
4. 20× SSC solution: 175.3 g NaCl, 88.2 g sodium citrate, and ca. 800 mL H_2O. After autoclaving, add sterile water to 1 L.
5. Random priming DNA labeling kit (Pharmacia).
6. 50× Denhardt's solution (in H_2O): 1 % (w/v) Ficoll 400, 1 % (w/v) polyvinylpyrrolidone, and 1 % (w/v) bovine serum albumin (Fraction V). Sterilize by filtering and store at −20 °C.
7. Hybridization solution (in H_2O): 45 % formamide, 5× SSC, 1× Denhardt's solution, 20 mM sodium phosphate (pH 6.5), 0.02 % denatured sheared salmon sperm DNA, and 5 % dextran sulfate. Sterilize the components except Denhardt's solution and formamide by autoclaving. Denhardt's solution has been sterilized by filtering, and formamide does not need to be sterilized by filtering or autoclaving.
8. Membrane washing solutions: Solution 1, 2× SSC and 0.5 % SDS; solution 2, 2× SSC and 0.1 % SDS; solution 3, 0.1× SSC and 0.1 % SDS; solution 4, 0.1× SSC.

*2.4. Tn*10 *Insertion by Bacteriophage P22-Mediated Transduction*

1. *S. typhimurium* LT2 with Tn10 inserted in known genes (*Salmonella* Genetic Stock Center, SGSC; to search and request: http://www.ucalgary.ca/~kesander).
2. Bacteriophage P22 (SGSC; to search and request: http://www.ucalgary.ca/~ kesander).

2.5. Phenotype Testing

1. 5× M9 salts: 64 g $Na_2HPO_4 \cdot 7H_2O$, 15 g KH_2PO_4, 2.5 g NaCl, 5 g NH_4Cl, and 1 L H_2O. Autoclave.
2. M9 minimal medium: 200 mL 5× M9 salts, ca. 600 mL sterile water, 2 mL 1 M $MgSO_4$, 20 mL 20 % solution of a carbon source, and 0.1 mL 1 M $CaCl_2$. Add sterile water to 1 L. 1 M $MgSO_4$ and 1 M $CaCl_2$ should be made and autoclaved separately and added to the bottle after 5× M9 salt stock solution has been diluted.

3. Amino acids or other nutrients for phenotype testing to confirm the inactivation of the gene by Tn10 insertion.
4. Semisolid LB agar plates for motility tests: To LB broth, add agar to final concentration of 0.3 % and gelatin to final concentration of 2 %. Autoclave, mix well, and pour the medium to Petri dishes on the bench. After this, do not move the plates. To test the bacterial motility, inoculate the bacteria in the center and observe the migration of the bacteria.
5. Soft nutrient gelatin agar (NGA) plates for motility tests: Into nutrient broth (Difco Laboratories), add 8.0 % gelatin and 0.3 % agar.

2.6. Double Endonuclease Cleavages

1. Selected endonucleases (for *Salmonella*, the most commonly used ones are I-*Ceu*I, *Xba*I, *Avr*II, and *Spe*I).
2. Razor blades.
3. End-labeling solution (using a Pharmacia Random Priming Oligo-synthesis kit): 3 μL dNTP (reagent mix), 1 μL α-^{32}P dCTP, 1 mL H_2O, and 1 μL Klenow fragment. Add Klenow fragment immediately prior to the labeling reaction. Mix the solution by gently inverting the tube two to three times; avoid vortexing or pipetting.

3. Methods

3.1. Preparation of Intact Genomic DNA Embedded in Agarose Blocks From Cultured Bacteria

1. Start a culture in 5 mL LB broth from a freshly grown single colony and shake the culture at 37 °C overnight (15–18 h).
2. Spin down the bacteria and resuspend the bacteria in 0.5 mL cell suspension solution and mix well by pipetting (*see* **Note 3**).
3. Alternatively, streak a freshly grown single colony to make a lawn of bacteria on one small area of an LB plate. Four to eight strains can be grown on the same LB plate, but extreme care has to be taken to ensure that these different bacterial strains do not contaminate one another. When the lawn is grown up, take a half inoculating loop of bacteria from a lawn on the LB plate and suspend the bacteria in 0.5 mL cell suspension solution, mix well by pipetting (*see* **Note 3**).
4. Prepare 1.4 % agarose in water for the required volume, keep it at 72 °C after boiling, and take 0.5 mL to mix with the 0.5 mL bacterial suspension in a 2-mL container (*see* **Note 4**).
5. Immediately, draw the mixture into a tuberculin syringe (with the end cut off); avoid bubbles.
6. Leave the syringe in room temperature for at least 10 min, and slice the agarose rod with a razor blade into 1-mm thick disks on a piece of parafilm.
7. Transfer the agarose disks into a 12-mL plastic tube that has a tight-fitting lid and contains 3 mL cell lysis buffer. Shake the tube at 20 rpm for 90 min in a 72 °C

water bath. Ensure that marker or paper labels on the tubes are not erased or peal off at this stage.

8. Decant the liquid, ensuring that the agarose disks are not damaged (*see* **Note 5**). Add 3 mL wash solution and shake the tube at 20 rpm at room temperature for 15 min. Repeat once.
9. Decant the liquid and add 3 mL freshly made proteinase K solution. Incubate with shaking at 12 rpm in a 42 °C water bath for 72 h (*see* **Note 6**).
10. Decant the proteinase K solution, add 3 mL wash solution and shake the tube at 20 rpm at room temperature for 15 min. Repeat once.
11. Add 3 mL wash solution and then 0.03 mL PMSF stock solution to the same tube (*see* **Note 7**) and shake the tube at 20 rpm at room temperature for 2 h.
12. Decant the PMSF solution, add 3 mL wash solution and shake the tube at 20 rpm at room temperature for 15 min. Repeat once.
13. Decant the wash solution, add 3 mL storage solution and shake the tube at 20 rpm at room temperature for 15 min. Repeat once.
14. Decant the storage solution, add 3 mL fresh storage solution and keep the tube at 4 °C. The DNA under such conditions will be stable for at least 10 years (*see* **Note 8**).

3.2. Titration of Endonucleases, Cleavage with Titrated Endonucleases, and PFGE

1. Pulse centrifuge all unopened endonuclease tubes to collect all liquid to the bottom of the tubes.
2. Find out the appropriate amounts of the endonucleases by serial twofold dilutions using genomic DNA of a batch that has been known to be of good quality as a control (*see* **Note 9**). Record the most appropriate concentration and incubation time for each of the endonucleases that will be used currently or in the near future.
3. Incubate the DNA with the selected endonuclease at the determined concentration for the required length of time. Although there is no general rule for all enzymes regarding lengths of incubation for optimal cleavages, an incubation time of 6 h is appropriate for *Xba*I, *Avr*II, and *Spe*I from any supplier. The optimal length of incubation time for complete cleavage with I-*Ceu*I is 2 h.
4. While waiting for completion of endonuclease cleavage, cast the gel for PFGE. Although some particular experiments may require higher or lower agarose concentrations for separating very small or very large DNA fragments, 0.5 % agarose gels are almost universal for fragments ranging from as small as 1 kb without diffusion to as large as over 2000 kb without migration difficulty. Gels harden within half an hour at room temperature, but it is often important to have the gel made at least 2 h prior to use for straight and sharp PFGE bandings.
5. Load the DNA samples when the endonuclease cleavage terminates, seal the loading wells with melted agarose, and carefully slide the gel into the PFGE buffer box. Set the PFGE conditions so that the first cycle of electrophoresis aims at the whole

range of fragment sizes, for example, from a few kilobases to over 1000 kb. This is conveniently done by running the gel at 10 s ramping to 100 s over 12–16 h. The second cycle usually aims at small fragments, for example, from a few kilobases to about 100 kb, which can be achieved by running the gel at 3–12 s over 6–8 h. Two or more additional cycles may be needed to separate middle-sized (200–500 kb) and large fragments (>1000 kb), which require running at 20 s ramping to 30 s and 60 s ramping to 100 s, respectively. Ramping for a certain range of lengths rather than a fixed time is advantageous to separate fragments crowded over a range of sizes. The pulse voltage is usually set at 6 V/cm. For very large DNA fragments, such as those larger than 2–3 Mb, the pulse voltage has to be set much lower, for example, 2–3 V/cm, and the running will continue for a week or longer. The buffer temperature is usually set at about 16 °C; PFGE runs faster at higher temperatures, but the resolution may not be optimal. Make sure that cooler is turned on.
6. Visualize and photograph the gel using a gel doc system, Polaroid camera, or other imaging systems (*see* **Note 10**).

3.3. Southern Blotting, Probe Preparation and Hybridization to Identify PFGE Bandings or Locate Genes

1. Cut the Immobilon-P membrane to the size of the gel. Do not touch the membrane with fingers (use forceps or wear gloves).
2. Wet the membrane with methanol. After washing briefly in water, keep the membrane in 0.4 M NaOH until use.
3. Trim the gel on a UV light box to remove the edge parts of the gel that do not contain DNA. Then, soak the gel in 0.25 M HCl for 15 min to depurinate the DNA for more convenient transfer of DNA from the gel to the membrane.
4. Pour off the 0.25 M HCl, wash the gel briefly in water, and soak the gel in 0.4 M NaOH to denature the DNA.
5. In the center of a glass or plastic container that is 2–3 cm wider and 4–5 cm longer than the gel, place a firm support that is no bigger than the gel and is 4–8 cm thick. On the support, place a thin (3–5 mm thick) glass or plastic plate that is of the same size as, or a little larger than, the gel. On the plate, place a piece of filter paper that has the same width as the plate but is longer than the plate, with the two ends hanging to touch the bottom of the container. Fill the container with 0.4 M NaOH to a level that covers about two-thirds of the height of the support.
6. Rinse the filter paper with 0.4 M NaOH and place the gel bottom up on the filter paper. Eliminate all bubbles between the filter paper and the plate and those between the gel and the filter paper.
7. Mark the orientation of the Immobilon-P membrane. For example, cut off the left top corner of the membrane. Place the membrane onto the gel, so that the left top corner now becomes right top corner. On top of the membrane, place three pieces of filter paper of the same size as the membrane.

8. Place a stack (about 10 cm) of absorbent paper (paper towels) of the same size as or a little larger than the membrane on top of the filter paper.
9. Finally, place a glass or plastic plate of similar size as the absorbent paper on top of the stack of the absorbent paper, then put a small weight of about 500 g on top of the plate.
10. Seal the container with Saran Wrap to avoid evaporation of the 0.4 M NaOH solution. Leave the blotting device for 18–24 h to allow the 0.4 M NaOH solution to flow through the membrane by capillary action.
11. After the DNA transfer from the gel to the membrane, carefully peal the membrane off the gel, face (the side that directly contacted the gel) up on a piece of absorbent paper to air dry it for at least 30 min.
12. Bake the membrane in an 80 °C vacuum oven or gel dryer for 2 h.
13. Roll the membrane and place it in a roller bottle. Add hybridization solution and place the roller bottle in hybridization oven adjusted to 42 °C for prehybridization. Several membranes can be put in the same roller bottle if the same probe is to be used for them. In such cases, the membranes should be separated from one another by nylon cloth. The optimal prehybridization time is for about 6–12 h.
14. To denature the probe DNA in liquid, boil the tube for 5 min, transfer the tube into ice water and keep it there for 5 min.
15. To denature the probe DNA in low-melting-point agarose gel, which is usually as large as several hundred kilobases or even larger, melt the gel at about 80 °C for a few minutes, sonicate the sample to shear DNA into small pieces, boil the tube for 5 min, transfer the tube into a 37 °C water bath, and immediately and swiftly add the components at step 16.
16. Add reagent mix containing dNTP, α-^{32}P-dCTP, and Klenow fragment to the denatured DNA using the amounts as suggested in the manufacturer's instruction of the kit (*see* **Subheading 2.3., item 5**) and incubate at 37 °C for 30 min.
17. Boil the tube for 5 min and immediately add the labeled probe to about 5 mL hybridization solution, pour the whole content to the roller bottle that contains the membranes for prehybridization, and place the roller bottle in hybridization oven adjusted to 42 °C for hybridization. Keep the roller bottle rolling in hybridization oven for 2 days.
18. Into a container that is at least 2 cm larger than the membrane in all dimensions, add an appropriate volume of solution 1 that can cover the whole membrane (about 1.5 mL for every square centimeter of membrane). Stop the hybridization oven and transfer the membrane from the roller bottle directly to solution 1. Shake the container gently for 15 min. Save the hybridization solution in a bottle and record the probe in it. Store the recycled hybridization solution at −20 °C (*see* **Note 11**).
19. Into another container of similar size, add a similar volume of solution 1. Transfer the membrane from the first container directly to the second, as swiftly as possible. Shake the container gently for 15 min.

20. Continue washing the membrane between two containerswith solutions 2, 3, and 4 (*see* **Subheading 2.3., item 8**) twice, 15 min for each. When the washing is finished, place the membrane face up on a piece of absorbent paper to decant the solution, and then transfer the membrane to another piece of absorbent paper to air-dry it.
21. Place the membrane on a piece of a thin paper board or filter paper, wrap the membrane to the paper board or filter paper with Saran Wrap, place it in an X-ray film cassette, and load an X-ray film on top of it. Expose the X-ray film overnight and record the exact exposure time.
22. Develop the X-ray film and determine the exposure times for the next few X-ray films. Ideally, three or more exposures should be obtained: at least one optimally exposed for overall analysis, publication, or other presentation purposes; one underexposed for analyzing areas that may have multiple bands showing the signals crowded together, which might be obscured as just a single thick band with "normal" exposures; and one overexposed to reveal the signals that might be real but too weak to be revealed with "normal" exposures. **Fig. 1** shows examples of Southern hybridization.

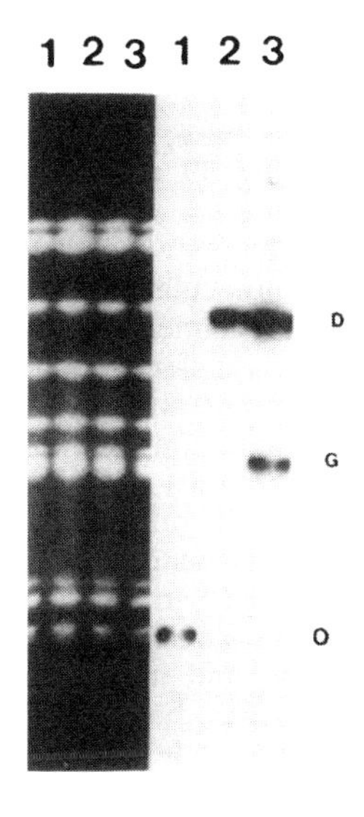

Fig. 1. Pulsed field gel electrophoresis (PFGE) patterns and Southern blotting analysis of *Xba*I digests of the genome of *Salmonella typhimurium*. Left: PFGE pattern of *Xba*I digests of genomic DNA. Right: Southern blot and hybridization. The gel from the left was blotted to membrane strips, and lanes were probed as follows: lane 1, plasmid pJN13 as the probe, which contains the gene *pyrE* and hybridizes with fragment O; lane 2, *pyrF* as the probe hybridizing with fragment D; lane 3, TT15246 *pyrF*::MudP as the probe, which has an *Xba*I cleavage site and hybridizes with two *Xba*I fragments D and G. Adapted from **ref.** *7* with modifications.

3.4. Localization of Genes by Tn10 Transferred from S. typhimurium LT2

1. Choose a batch of bacteriophage P22 that has a high titer (10^{11} pfu/mL or higher) by making a serial 10-fold dilutions and testing on a lawn of *S. typhimurium* LT2.
2. On an LB plate containing tetracycline, grow the selected *S. typhimurium* LT2 mutant that has a Tn*10* insertion in a known gene.
3. Pick up a single colony, inoculate it into 1 mL LB broth, and shake the culture at 37 °C overnight.
4. Add 3 mL fresh LB broth into the tube and continue shaking for 3 h.
5. Add 10 μL P22 lysate (10^{11} pfu/mL or higher) and incubate the bacteria–phage culture at 37 °C for 6 h.
6. To release and harvest the phage, add 100 μL chloroform into the bacteria–phage culture, vortex the tube for 20 s, centrifuge, and transfer the supernatant to a fresh tube. Before this is done, make sure that the plastic tube to be used is resistant to chloroform. The lysate will be stable for decades at 4 °C. Such lysate will contain rare particles that have the Tn*10* insertion and the surrounding regions of the *S. typhimurium* LT2 genomic DNA instead of phage DNA, altogether about 45 kb, packed within a protein head of the P22 phage. Such pseudo phage can attach to *Salmonella* cells that express the O12 antigen and inject the DNA into the cells. In many cases, the DNA in such pseudo phage particles will be incorporated into the host bacterial genome at the homologous site via the *Salmonella* DNA surrounding the Tn*10* sequence. In this way, the location of that gene in the target *Salmonella* genome will be determined on the map when analyzed by PFGE with the genomic DNA being cleaved by *Xba*I or *Avr*II.
7. The Tn*10* sequence contains cleavage sites for both *Xba*I and *Avr*II, so the location of the extra *Xba*I or *Avr*II sites will be where the Tn*10* is inserted or where the target gene resides. **Fig. 2** shows examples of Tn*10* insertions to determine gene locations.
8. To locate a gene on the genome of a *Salmonella* strain, which expresses the O12 antigen, by Tn*10* insertion transferred from *S. typhimurium* LT2 via P22-mediated transduction, grow a single colony in 1 mL LB at 37 °C overnight. The next morning, place 100 μL of the bacterial culture in the center of an LB plate containing tetracycline and 1–2 μL of a lysate from a Tn*10* mutant of *S. typhimurium* LT2 on top of the bacterial culture. Spread the bacteria–phage mixture with a T-shaped spreader on the LB plate.
9. To locate a gene on the genome of a *Salmonella* strain, which does not express the O12 antigen, by Tn*10* insertion transferred from *S. typhimurium* LT2 via P22-mediated transduction, first make the host *Salmonella* strain to be sensitive to P22 infection with a cosmid, pPR1347 ***(14)***, which contains the genes encoding the O12 antigen. This cosmid can be mobilized by F plasmid through conjugation, which requires the *Salmonella* strain to be made resistant to an antibiotic like

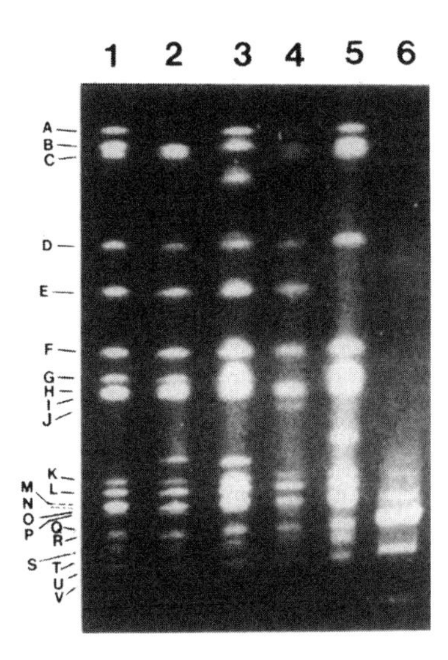

Fig. 2. Pulsed field gel electrophoresis patterns of *Xba*I-cleaved genomic DNA of *Salmonella typhimurium* LT2 and its Tn*10* insertion derivatives. Lanes: 1, wild-type LT2; lane 2, Tn*10* insertion in *zbf* is located on fragment A; lane 3, Tn*10* insertion in *srl* is located on fragment C; lane 4, Tn*10* insertion in *ompD* is located on fragment G; lane 5, Tn*10* insertion in *mel* is located on fragment E. Adapted from **ref.** *7* with modifications.

streptomycin for counterselection, as can be done by growing the bacteria in increasing concentrations of streptomycin, starting from 5 μg/mL (*see* **Note 12**). On the contrary, pPR1347 contains kanamycin resistance gene.

10. When the streptomycin resistance has reached 1000 μg/mL, combine cultures of this strain and the *E. coli* strain that harbors both F plasmid and pPR1347 (request the strain from SGSC) at a ratio of 5:1, for example, 5 mL of the streptomycin-resistant *Salmonella* strain and 1 mL of the *E. coli* strain, centrifuge the bacteria, and resuspend the bacteria in 2 mL of fresh LB broth. Incubate the bacteria at 37 °C for 3–6 h without shaking and spread the bacteria on an LB plate that contains both streptomycin (500–1000 μg/mL) and kanamycin (50 μg/mL). Test the colonies appearing on the plates (transconjugants) for P22 sensitivity. Grow a single colony of a P22-sensitive transconjugant in 1 mL LB at 37 °C overnight. The next morning, place 100 μL of the bacterial culture in the center of an LB plate containing tetracycline and 1–2 μL of lysate from a Tn*10* mutant of *S. typhimurium* LT2 on top of the bacterial culture. Spread the bacteria–phage mixture with a T-shaped spreader on the LB plate.
11. Pick up three to six colonies from the LB plate containing tetracycline for phenotype testing.

3.5. Phenotype Testing

1. For nutrient requirement tests, grow the tetracycline-resistant transconjugant on an M9 minimal plate and M9 minimal plate supplemented with the nutrient supposed to be needed by the bacteria as a result of Tn*10* insertion into a gene for the biosynthesis of that nutrient (*see* **Note 13**).
2. For motility testing, grow the tetracycline-resistant transconjugant on an LB plate, pick up a colony, and inoculate it to a semisolid LB agar plate (*see* **Subheading 2.5., item 4**) or a soft NGA plate (*see* **Subheading 2.5., item 5**) or both. Judgment of nonflagellate (*fla*⁻), nonmotile (*mot*⁻), or nonchemotactic (*che*⁻) phenotypes is shown in **Fig. 3** by examples with *S. typhi*.

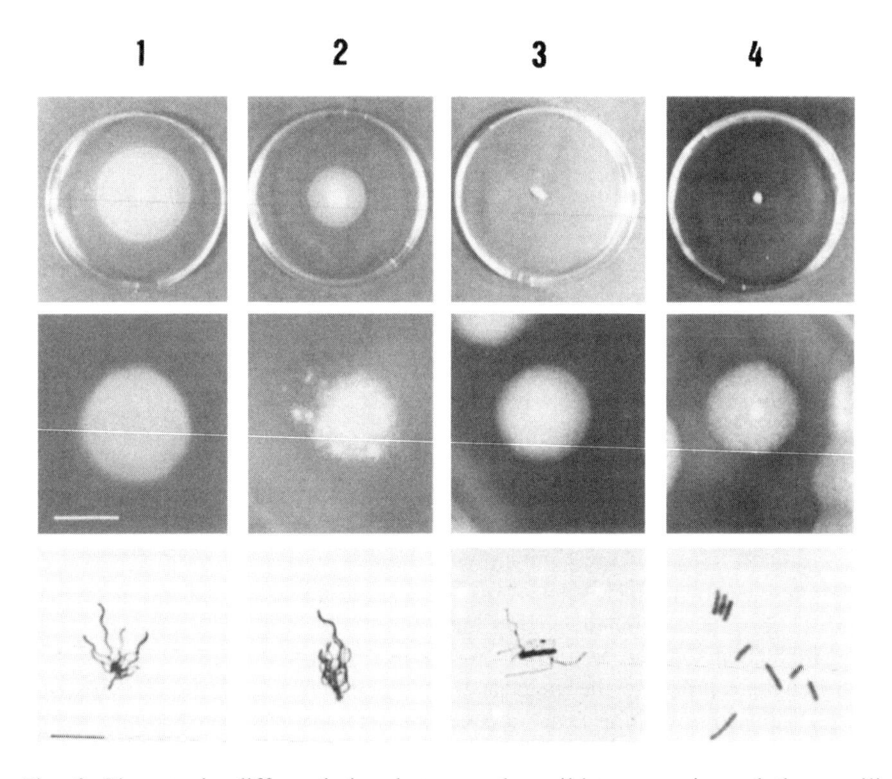

Fig. 3. Phenotypic differentiation between the wild-type strain and the motility mutants. Top row, motility observed on an LB semisolid agar plate; middle row, colony morphology on a nutrient gelatin agar plate (bar, 1 mm); bottom row, flagellation visualized by flagella staining (bar, 5 μm). Strains: 1, wild-type *S. typhi*; 2, a *che*⁻ mutant; 3, a *mot*⁻ mutant; 4, a *fla*⁻ mutant. Adapted from **ref. *15*** with modifications.

3.6. Refinement of the Salmonella *Genome Maps by Double Endonuclease Cleavages*

1. For the purposes of double endonuclease cleavages on DNA fragments produced by the cleavage of one endonuclease, usually three to five lanes of PFGE separated DNA are needed. For example, I-*Ceu*I cleavage of any *Salmonella* genome produces seven fragments. To determine the physical distances of these seven cleavage sites to those of other endonucleases, such as *Xba*I, *Avr*II, and *Spe*I, three lanes of DNA on the PFGE are necessary, one for each enzyme. It is a good practice, however, to run two or more extra lanes, in case some of the excised agarose blocks are damaged, lost, or in poor quality; running another gel just for one missing DNA band is not practical. Excised DNA in agarose blocks is stable for up to 1 week.
2. On the photographed image of the PFGE gel, mark the bands to excise.
3. Mark Eppendorf tubes for the DNA bands to be excised from the PFGE gel, add 0.5 mL storage solution to each tube.
4. On a UV light box, identify the marked bands, excise them with a razor blade, trim off all agarose that does not contain DNA, and transfer the agarose blocks containing DNA into the Eppendorf tubes that have been marked for the individual bands. Be extremely careful not to expose the DNA for more than 10 s and not to place the DNA into wrong tubes.
5. Cleave the DNA of the first cleavage with a second endonuclease, which usually requires less enzyme and shorter cleavage time, because the purity of DNA is much higher than the whole genome DNA first isolated when the bacteria were embedded in agarose and treated with proteinase K.
6. Following the second cleavage, much of the DNA disappears as a result of random breakage during PFGE running and the many manipulations afterward. Typically, up to 20 % of the DNA remains at this stage. Therefore, some small fragment will be difficult to visualize on a UV light box and to be photographed with ethidium bromide straining. In such cases, end-labeling of the DNA fragments with radioisotopes becomes necessary.
7. To end-label the DNA fragments, remove the storage solution from the excised DNA blocks, add 0.5 mL water, wait for 5 min, remove the water, add 50 μL end-labeling solution, and leave the labeling reaction at 37 °C for 30 min.
8. Load the samples to the wells of a PFGE gel and start the PFGE. Because of the radioactivity of the DNA sample, the gel cannot be conveniently viewed on a UV light box to determine the running conditions from one cycle to another. Therefore, all conditions will have to be determined and programmed to run the gel from the very beginning. For most *Salmonella* double cleavage experiments (**Fig. 4**), the following conditions for running a radioactive gel would work, although adjustments may be needed for certain *Salmonella* strains based on a first run: (1) 10 s ramping to 80 s over 12 h; (2) 3 s ramping to 18 s over 3 h; and

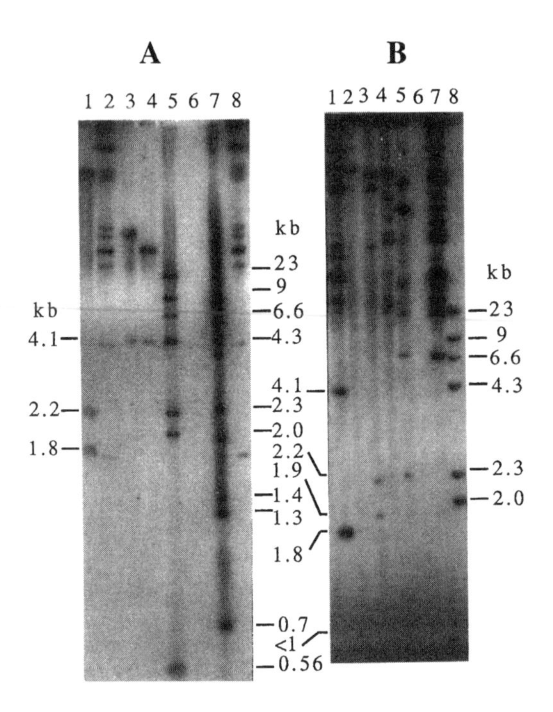

Fig. 4. Double cleavage of *Salmonella typhimurium* LT2 genomic DNA and end-labeling with ^{32}P. **(A)** Set 1. Lanes: 1, I-*Ceu*I fragment C recleaved with *Avr*II; 2 and 8, genomic DNA of LT2 cleaved with *Avr*II only; 3, I-*Ceu*I fragment E, recleaved with *Avr*II; 4, I-*Ceu*I fragment D recleaved with *Avr*II; 5, λ *Hin*dIII standard; 6, λ DNA concatemer (not end-labeled with ^{32}P so no signals are shown here); 7, λ BstEII standard. The sizes of some of the fragments of the standards are shown on the right side; the sizes of some of the small fragments from the digests are on the left. **(B)** Set 2. Lanes: 1, genomic DNA of LT2 cleaved with *Avr*II only; 2, I-*Ceu*I fragment A, recleaved with *Avr*II; 3, I-*Ceu*I fragments B and G (these fragments have very similar sizes so cannot be fully separated by pulsed field gel electrophoresis), recleaved with *Avr*II; 4, I-*Ceu*I fragment A, recleaved with *Xba*I; 5, I-*Ceu*I fragments B and G, recleaved with *Xba*I; 6, λ concatemer (not end-labeled); 7, genomic DNA of LT2 cleaved with *Xba*I only; 8, λ *Hin*dIII standard. The sizes of some of the fragments of the standards are on the right side; the sizes of some of the small fragments from the digests are on the left. Adapted from **ref. *11*** with modifications.

(3) 20 s ramping to 30 s over 6 h. Set the pulse voltage at 6 V/cm and the buffer temperature at 16 °C. Make sure that the cooler is turned on.

9. Carefully clean the area for the labeling process and properly dispose of solid and liquid radioactive wastes.

10. During the PFGE run, the radioisotope that is not incorporated will be "washed" off the samples into the whole buffer box. These free radioisotope molecules will have to be eliminated, otherwise the gel, full of the free radioisotope molecules, will have a very dark background when exposed to the X-ray film. This can be done early next morning after the PFGE run was started the previous afternoon or evening by pausing the PFGE, draining off all buffer in the buffer box into the liquid radio waste container (*see* **Note 14**), washing the buffer box with ample crude distilled water (about 3–4 L) with the pump being turned on for 15 min, and repeating the washing process once. The run can then be resumed for the remaining scheduled time with fresh 0.5× TBE buffer.
11. When the PFGE run is complete, drain off the buffer, wash the buffer box twice as in step 10, and then take out the gel. On the metal mesh plate of a Bio-Rad gel dryer, place a piece of filter paper that is about 1 cm larger than the gel at all four edges, and put the gel on the filter paper. Cover the whole gel with a piece of Saran Wrap, which is a little larger than the filter paper for all dimensions. The gel on the filter paper and the Saran Wrap on the gel should both be as smooth as possible. Place the soft sheet of the gel dryer on the gel and start the vacuum. When the soft sheet becomes tightly stuck to the metal mesh plate around the gel and the liquid starts to drain into the vacuum bottle, close the hard lid of the dryer and start the heating. Set the temperature at 80 °C and time at 2 h. After 2 h, heating is automatically turned off, but do not turn off the vacuum until the metal mesh plate is cooled to about room temperature.
12. On the back of the filter paper, record the gel number with pencil if it was not done before the drying process. Trim off the edges of the filter paper and Saran Wrap that do not contain the dried gel to fit the gel to the X-ray cassette. Load a sheet of X-ray film. Expose the X-ray film for 3 h and develop it. Determine the exposure time for the next X-ray films. Have at least one X-ray film optimally exposed, and others overexposed or underexposed for reasons stated in **Subheading 3.3., step 22**.

3.7. Determination of Overall Genome Structure by I-CeuI Partial Cleavage

1. Cleave the genomic DNA with the concentration of I-*Ceu*I that is appropriate as determined by titration (*see* **Subheading 3.2.**).
2. Determine the sizes of the complete and incomplete cleavage products by comparing to DNA size markers (λ DNA concatemers are among the most frequently used ones).
3. Try to figure out the neighboring relationships of the completely cleaved fragments by sizes of individual completely cleaved fragments and the incompletely cleaved fragments. As shown in **Fig. 5**, if the sizes of completely cleaved fragments are significantly different, that is, at least 50 kb, it is not difficult to order the fragments on to a map.

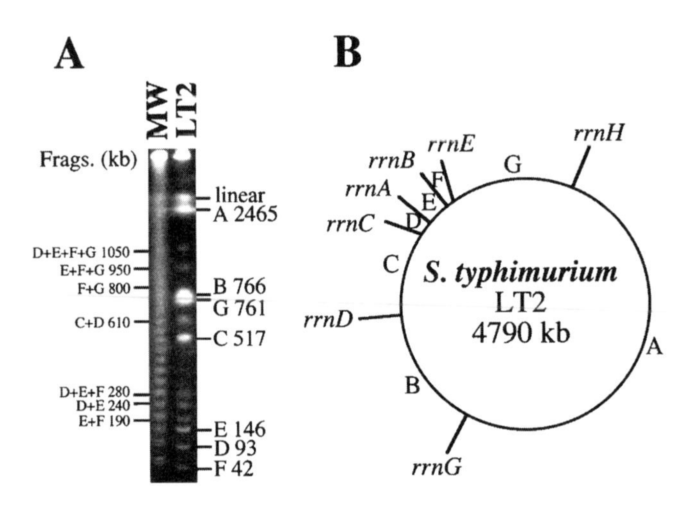

Fig. 5. Determination of overall genome structure by I-*Ceu*I partial cleavage exemplified by *Salmonella typhimurium* LT2. **(A)** I-*Ceu*I cleavage pattern of genomic DNA on pulsed field gel electrophoresis, with DNA fragments and their sizes of complete cleavage indicated on right with larger letters and DNA fragments and their sizes of incomplete cleavage indicated on left with smaller letters. Lane MW, λ concatemer as molecular size markers. **(B)** Circular genome map based on complete as well as incomplete cleavage data in panel A. Adapted from **ref. *16*** with modifications.

4. If the identity of an incompletely cleaved fragment is in question, that fragment can be excised from a PFGE gel and recleaved with I-*Ceu*I. A comparison of this double cleavage with the same endonuclease with the single and complete cleavage of the genomic DNA will unambiguously clarify the situation.

3.8. Construction of the Physical Map

1. For *Salmonella* genomes, the easiest way to construct a physical map is to establish an I-*Ceu*I map first, then align *Xba*I, *Avr*II, or *Spe*I maps to the I-*Ceu*I map.
2. Although circular maps will eventually be constructed for the *Salmonella* genomes, it is more convenient to construct linear maps first. An efficient order of the mapping process is (1) construct an I-*Ceu*I map based on complete and incomplete cleavage data, (2) roughly assign *Xba*I, *Avr*II, or *Spe*I fragments to regions of the I-*Ceu*I map based on I-*Ceu*I/*Xba*I, I-*Ceu*I/*Avr*II, or I-*Ceu*I/*Spe*I double cleavage data, (3) and adjust the relative positions of the *Xba*I, *Avr*II, or *Spe*I fragments based on *Xba*I/*Avr*II, *Xba*I/*Spe*I, *Avr*II/*Xba*I, *Avr*II/*Spe*I, *Spe*I/*Xba*I, and *Spe*I/*Avr*II double cleavage data, and, occasionally, *Xba*I/I-*Ceu*I, *Avr*II/I-*Ceu*I, and *Spe*I/I-*Ceu*I double cleavage data.

3. Further refine the cleavage maps and locate genes based on Tn*10* insertion data. At this stage, the map usually becomes fairly accurate.
4. Some *Salmonella* genomes are more difficult to map than others. For example, the map of *S. enteritidis* ***(17)*** took 3 months and the map of *S. paratyphi* B ***(18)*** took 1 month to construct, but it was not possible to finish the map of *S. typhi* Ty2 until Tn*10* data on *Spe*I fragments were analyzed, although all other published *Salmonella* maps were finished before analysis of Tn*10* data on *Spe*I cleavage

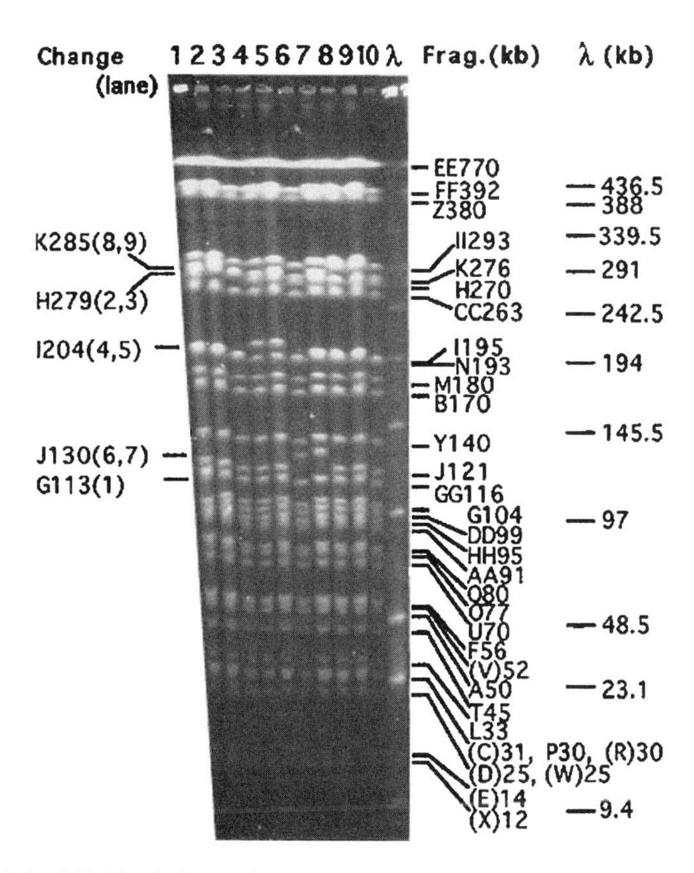

Fig. 6. Pulsed field gel electrophoresis of *Spe*I-cleaved genomic DNA of *Salmonella typhi* Ty2 Tn*10* insertion mutants. Lanes: 1, *proBA*::Tn*10*; 2, *purE*::Tn*10*; 3, *ahp*::Tn*10*; 4, *nadA*::Tn*10*; 5, *bio*::Tn*10*; 6, *aspC*::Tn*10*; 7, *pncB*::Tn*10*; 8, *flgL*::Tn*10*; 9, *trp*::Tn*10*; 10, Ty2 wild-type strain; 11, λ molecular size markers. The changed fragments with the Tn*10* insertions are shown on the left, and the lanes are indicated in brackets; the *Spe*I fragments and their sizes of the wild-type strain are indicated on the right, along with the λ molecular size markers. Adapted from **ref.** ***12*** with modifications.

became necessary. Tn*10* does not have an *Spe*I cleavage site, but it will increase the fragment size by 9.3 kb, which is sometimes very useful for estimating relative or even exact distances between two genes if they are on different *Xba*I and *Avr*II fragments but on the same *Spe*I fragment. The use of *Spe*I cleavage on Tn*10* insertion strains is shown in **Fig. 6** with *S. typhi* Ty2 as the example.

4. Notes

1. Unless specified, no special sources of chemicals or plastic/glassware are necessary.
2. Throughout this chapter, water (sometimes written as H_2O) means pure water with a resistance of 18.2 MΩ.
3. Vortexing often cannot completely suspend the bacteria, leaving clusters of bacteria in the agarose blocks.
4. The container could be plastic or glass plates with flat-bottomed wells or tubes with a flat bottom of about 1.5–2 mL volume. The regular Eppendorf tubes are not recommended, because a tuberculin syringe cannot reach the narrow bottom.
5. A kind of mesh, like that used for screen windows, can be used to cover the tubes when decanting the liquid.
6. For many *Salmonella* strains, proteinase K treatment for 3 h at 42 °C is enough, but 72 h is optimal for most *Salmonella* and other enteric bacterial strains.
7. PMSF should not come into contact with water until use.
8. Do not freeze the DNA samples. If they are frozen accidentally at no lower than −3 °C for no more than 6 h, the DNA is often still more or less usable, although good-quality data are not guaranteed.
9. Enzyme activities of the same endonuclease from different suppliers vary significantly, although activities among different batches of the same endonuclease from the same supplier are often fairly consistent. It is good practice to use an endonuclease from a single supplier; once a batch is titrated, titration of other batches of the same endonuclease from the same supplier is usually not needed.
10. Many authors stain the PFGE gel after the run, but putting ethidium bromide in the gel usually results in cleaner and sharper banding patterns.
11. The recycled hybridization solution can be reused when the radioisotope ^{32}P has nearly completely decayed.
12. Start the streptomycin resistance selection with 5 μg/mL streptomycin in LB broth, spread 100 μL overnight culture in this broth to LB plates containing 10 μg/mL streptomycin. Pick up a colony from such a plate and inoculate it in LB broth containing 20 μg/mL streptomycin. This process is repeated with increasing concentrations of streptomycin until the bacteria form colonies on LB plates containing 1000 μg/mL streptomycin.
13. Many *Salmonella* lineages are natural auxotrophs for certain nutrients. For example, *S. typhi* requires cysteine to grow on M9 plates. Such nutrients have to be added to the M9 plates for such tests.
14. The radioactive wastes have to be disposed of according to the rules of the institute.

Acknowledgments

The author thanks Gui-Rong Liu for technical assistance. This work was supported by a 985 Project grant of Peking University Health Science Center, a National Natural Science Foundation of China grant (30370774), and a Discovery Grant from Natural Sciences and Engineering Research Council (NSERC) of Canada.

References

1. Kauffmann, F. and Edwards, P. R. (1957) A revised, simplified Kauffmann-White schema. *Acta Pathol. Microbiol. Scand.* **41,** 242–246.
2. Parkhill, J., Dougan, G., James, K. D., et al. (2001) Complete genome sequence of a multiple drug resistant *Salmonella enterica* serovar Typhi CT18. *Nature* **413,** 848–852.
3. Deng, W., Liou, S. R., Plunkett, G., 3rd, et al. (2003) Comparative genomics of Salmonella enterica serovar Typhi strains Ty2 and CT18. *J. Bacteriol.* **185,** 2330–2337.
4. McClelland, M., Sanderson, K. E., Spieth, J., et al. (2001) Complete genome sequence of *Salmonella enterica* serovar Typhimurium LT2. *Nature* **413,** 852–856.
5. McClelland, M., Sanderson, K. E., Clifton, S. W., et al. (2004) Comparison of genome degradation in Paratyphi A and Typhi, human-restricted serovars of *Salmonella enterica* that cause typhoid. *Nat. Genet.* **36,** 1268–1274.
6. Chiu, C. H., Tang, P., Chu, C., et al. (2005) The genome sequence of *Salmonella enterica* serovar Choleraesuis, a highly invasive and resistant zoonotic pathogen. *Nucleic Acids Res.* **33,** 1690–1698.
7. Liu, S. L. and Sanderson, K. E. (1992) A physical map of the *Salmonella typhimurium* LT2 genome made by using XbaI analysis. *J. Bacteriol.* **174,** 1662–1672.
8. Wong, K. K. and McClelland, M. (1992) A BlnI restriction map of the *Salmonella typhimurium* LT2 genome. *J. Bacteriol.* **174,** 1656–1661.
9. Liu, G. R., Rahn, A., Liu, W. Q., Sanderson, K. E., Johnston, R. N., and Liu, S. L. (2002) The evolving genome of *Salmonella enterica* serovar Pullorum. *J. Bacteriol.* **184,** 2626–2633.
10. Liu, G. R., Edwards, K., Eisenstark, A., et al. (2003) Genomic diversification among archival strains of *Salmonella* enterica serovar typhimurium LT7. *J. Bacteriol.* **185,** 2131–2142.
11. Liu, S. L., Hessel, A., and Sanderson, K. E. (1993) The XbaI-BlnI-CeuI genomic cleavage map of *Salmonella typhimurium* LT2 determined by double digestion, end labelling, and pulsed-field gel electrophoresis. *J. Bacteriol.* **175,** 4104–4120.
12. Liu, S. L. and Sanderson, K. E. (1995) Genomic cleavage map of *Salmonella typhi* Ty2. *J. Bacteriol.* **177,** 5099–5107.

13. Liu, S. L. and Sanderson, K. E. (1995) The chromosome of *Salmonella paratyphi* A is inverted by recombination between *rrnH* and *rrnG*. *J. Bacteriol.* **177,** 6585–6592.
14. Neal, B. L., Brown, P. K., and Reeves, P. R. (1993) Use of *Salmonella* phage P22 for transduction in *Escherichia coli*. *J. Bacteriol.* **175,** 7115–7118.
15. Liu, S. L., Ezaki, T., Miura, H., Matsui, K., and Yabuuchi, E. (1988) Intact motility as a *Salmonella typhi* invasion-related factor. *Infect. Immun.* **56,** 1967–1973.
16. Liu, S. L., Schryvers, A. B., Sanderson, K. E., and Johnston, R. N. (1999) Bacterial phylogenetic clusters revealed by genome structure. *J. Bacteriol.* **181,** 6747–6755.
17. Liu, S. L., Hessel, A., and Sanderson, K. E. (1993) The XbaI-BlnI-CeuI genomic cleavage map of *Salmonella enteritidis* shows an inversion relative to *Salmonella typhimurium* LT2. *Mol. Microbiol.* **10,** 655–664.
18. Liu, S. L., Hessel, A., Cheng, H. Y., and Sanderson, K. E. (1994) The XbaI-BlnI-CeuI genomic cleavage map of *Salmonella paratyphi* B. *J. Bacteriol.* **176,** 1014–1024.

4

Cloth-Based Hybridization Array System for the Identification of Antibiotic Resistance Genes in *Salmonella*

Burton W. Blais and Martine Gauthier

Summary

A simple macroarray system based on the use of polyester cloth as the solid phase for DNA hybridization has been developed for the identification and characterization of bacteria on the basis of the presence of various virulence and toxin genes. In this approach, a multiplex polymerase chain reaction (PCR) incorporating digoxigenin-dUTP is used to simultaneously amplify different marker genes, with subsequent rapid detection of the amplicons by hybridization with an array of probes immobilized on polyester cloth and immunoenzymatic assay of the bound label. As an example of the applicability of this cloth-based hybridization array system (CHAS) in the characterization of foodborne pathogens, a method has been developed enabling the detection of antibiotic resistance and other marker genes associated with the multidrug-resistant food pathogen *Salmonella enterica* subsp. *enterica* serotype Typhimurium DT104. The CHAS is a simple, cost-effective tool for the simultaneous detection of amplicons generated in a multiplex PCR, and the concept is broadly applicable to the identification of key pathogen-specific marker genes in bacterial isolates.

Key Words: DNA hybridization; macroarray; polyester cloth; antibiotic resistance; *Salmonella*.

1. Introduction

An important element in the delivery of an effective risk-based food microbiology inspection system is the operation of analytical testing programs capable of elucidating the maximum amount of information regarding the nature of microbial hazards identified in foods (and related agricultural commodities).

From: *Methods in Molecular Biology, vol. 394: Salmonella: Methods and Protocols*
Edited by: H. Schatten and A. Eisenstark © Humana Press Inc., Totowa, NJ

Food microbiology testing laboratories worldwide are capable of identifying the presence of bacterial pathogens such as *Salmonella* in foods at the species level, but most have little capacity to provide information on the potential severity of human health impact. Most laboratories have limited capability to confirm the identity of isolates and rely on the services of reference laboratories for definitive identification and detailed characterization (e.g., serotyping and phage typing). The ability to detect certain genes, such as species-specific markers, virulence factors, toxin and antibiotic resistance genes, in the food microbiology testing laboratory would provide more timely confirmation of isolates, and possibly, some key information for consideration in the development of appropriate responses to incidents of contamination.

The polymerase chain reaction (PCR) technique offers a potentially sensitive, specific, and robust approach for the detection of key target genes in bacterial isolates. Conventional PCR approaches utilize a pair of primers targeting unique DNA sequences in the genome of the bacterial species of interest, with postamplification detection of the amplicons on the basis of their molecular size as determined by agarose gel electrophoresis. A refinement of the PCR concept has been the development of multiplex approaches in which several target DNA sequences are coamplified in a single reaction, thus increasing the amount of information that can be rapidly generated from isolates in order to assist in investigations of contamination incidents. The analysis of amplicons generated in multiplex PCR systems is commonly achieved on the basis of their differential mobility by agarose gel electrophoresis, as previously described for multiplex PCR systems targeting bacterial toxin genes ***(1,2)***. Agarose gel electrophoretic analysis can be time-consuming, and complex electrophoretic patterns may be difficult to interpret accurately, especially when spurious amplicons arise. Although the specific nature of the amplicons can in many instances be confirmed by restriction endonuclease digestion, the resulting fragment mixtures can compound the complexity of electrophoretic analysis. Ideally, the nature of the amplicons generated by multiplex PCR should be confirmed by hybridization with arrays of amplicon-specific probes. High-density microarray systems for the detection and characterization of multiple gene markers in bacteria have been developed ***(3,4)***, but these systems rely on the use of highly sophisticated instruments to prepare and process the arrays and are not generally within the scope of analytical capabilities of the basic food microbiology laboratory.

As a practical alternative for the analysis of multiplex PCR products, we have developed a simple low-density array technique based on the use of a macroporous, hydrophobic polyester cloth as a solid phase for the detection of multiplex PCR products by their hybridization with an array of immobilized DNA probes. The advantages of polyester cloth as a DNA adsorbent for nucleic acid hybridization assays have previously been demonstrated *(5)*. Polyester cloth is a cost-effective support yielding improved reaction kinetics because of a large and readily accessible surface and is easy to wash between reaction steps to remove unbound reagents. Strips of polyester cloth bearing immobilized arrays of capture probes have recently been used in a reverse dot blot hybridization format for the detection of multiplex PCR products in a variety of food microbiology applications, including the identification of antibiotic resistance and other marker genes associated with *Salmonella enterica* subsp. *enterica* serotype Typhimurium DT104 *(6)*, the detection of *Clostridium botulinum* type A, B, E, and F neurotoxin genes *(7)*, and the detection of toxin genes associated with major foodborne pathogenic bacteria *(8)*. In this approach, termed cloth-based hybridization array system (CHAS), an array of target-specific DNA probes is immobilized on a polyester cloth strip and hybridized with amplicons bearing a label (e.g., digoxigenin [DIG]) generated in a multiplex PCR targeting various gene markers, followed by immunoenzymatic detection of the bound label (*see* **Fig. 1**).

The incorporation of a detectable label such as DIG in the amplicons constitutes a critical step in enabling subsequent detection of hybridization events on the array using a simple colorimetric immunoassay approach. DIG labeling can be achieved during the PCR process either through the use of 5′-DIG-labeled oligonucleotide primers or by *Taq* polymerase-mediated incorporation of DIG-dUTP into the amplicon DNA *(9)*. The capture probes constituting the probe arrays can be either PCR-generated DNA fragments, which must be converted to single-strand form by heat denaturation prior to immobilization on the polyester cloth surface *(6,8)*, or they can be synthetic oligonucleotides *(7)*, both of which are fixed to the polyester surface through a simple process involving adsorption and cross-linking with UV light. The present method also optionally incorporates a simple internal amplification control (IAC) system to gauge PCR inhibition by sample matrix components. This IAC is based on the incorporation of a primer pair with complementary 3' ends, resulting in the generation of a unique "primer-dimer" detectable by hybridization with a specific capture probe immobilized on polyester cloth.

This chapter will focus on the detection of antibiotic resistance genes in the multidrug-resistant food pathogen *S*. Typhimurium DT104. The emergence of multidrug-resistant food pathogens such as *S*. Typhimurium DT104, commonly exhibiting resistance to the antibiotics ampicillin (A), chloramphenicol (C), streptomycin (S), sulfonamides (Su), and tetracycline (T), has become a major public health issue for the food industry and regulatory agencies worldwide.

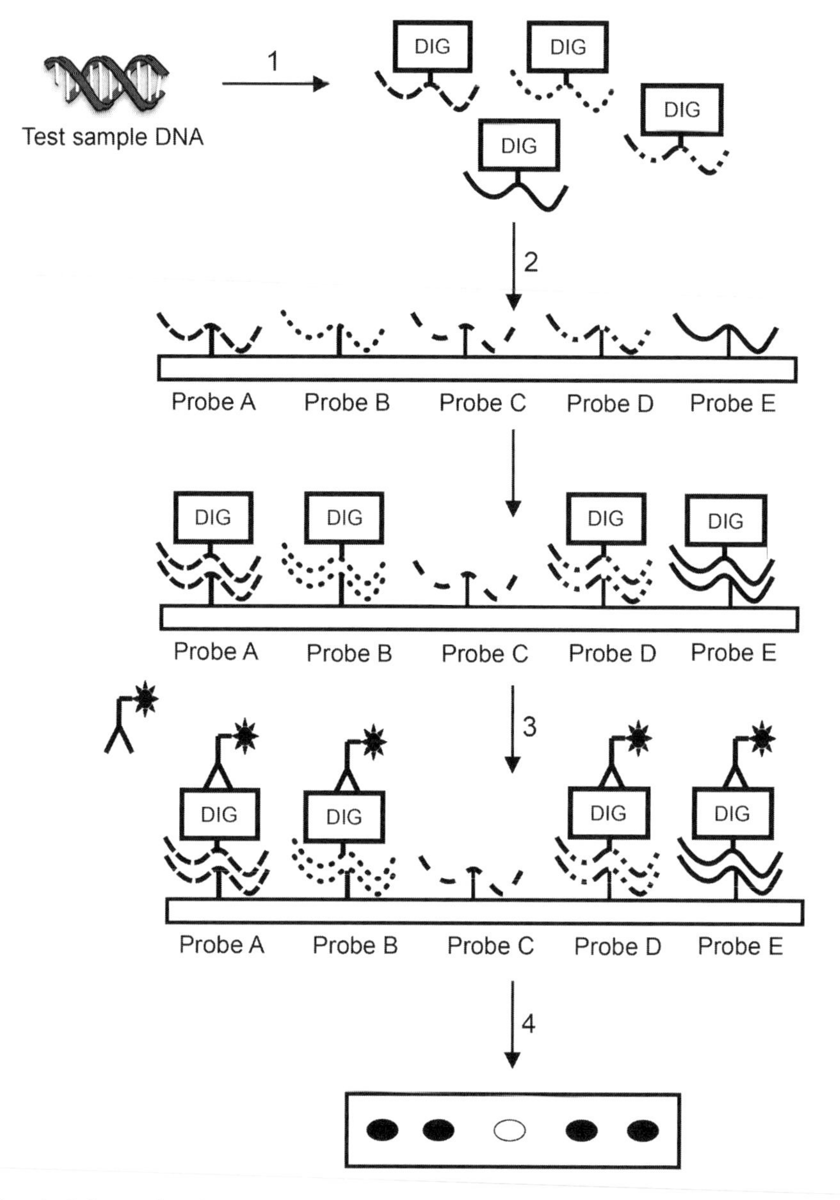

Fig. 1. Scheme for the cloth-based hybridization array system (CHAS). Amplicons incorporating digoxigenin (DIG)-labeled dUTP generated in a multiplex PCR **(1)** are detected by hybridization with an array of amplicon-specific DNA probes immobilized in discrete spots on a polyester cloth strip **(2)**, followed by immunoenzymatic assay of the bound label using anti-DIG antibody–peroxidase conjugate **(3)**, and colorimetric detection using tetramethylbenzidine membrane peroxidase substrate solution **(4)**.

Rapid detection methods specific to this pathogen constitute an important element in any strategy to control its spread in the food supply, particularly in poultry products where there is a high incidence of *Salmonella* contamination.

Here, we describe a simple low-density array technique based on the use of polyester cloth as a solid phase for the detection of amplicons from a multiplex PCR targeting key *S.* Typhimurium DT104 genes by their hybridization with immobilized DNA probes. In the present assay, suspect colonies isolated on plating media using standard culture techniques are subjected to a multiplex PCR-amplifying DNA sequences from a variety of *S.* Typhimurium DT104 marker genes. This is followed by hybridization of the DIG-labeled PCR products with an array of amplicon-specific DNA probes immobilized in discrete spots on a polyester cloth strip, and subsequent immunoenzymatic assay of the bound label using anti-DIG antibody–peroxidase conjugate (*see* **Fig. 2**). In the present application, the genes of interest include those conferring

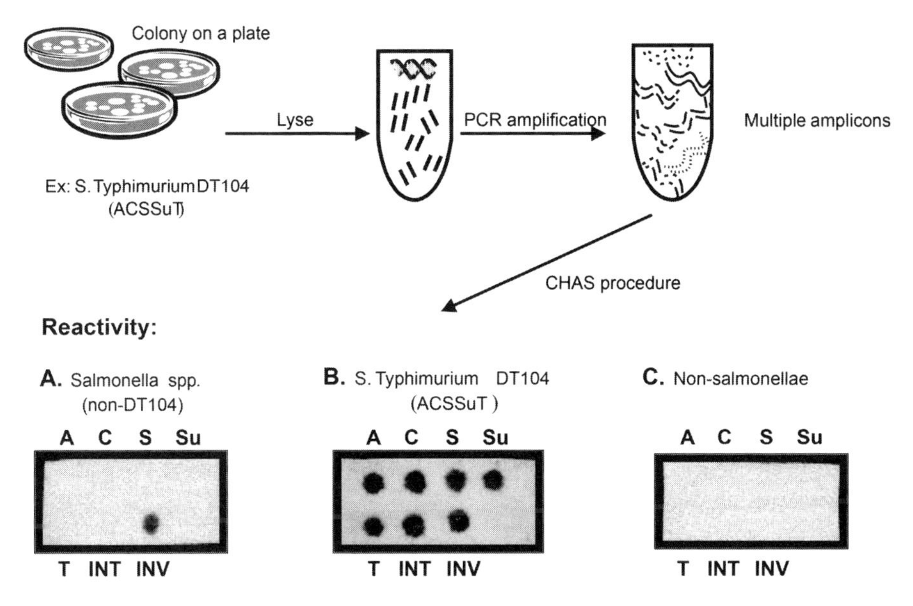

Fig. 2. General scheme for the identification of *Salmonella enterica* subsp. *enterica* serotype Typhimurium DT104 using the multiplex polymerase chain reaction (PCR) and cloth-based hybridization array system (CHAS) procedures. Expected patterns of reactivity for different types of bacteria are shown. Probe spots on the array are as follows: A, ampicillin; C, chloramphenicol; S, streptomycin; Su, sulfonamide; T, tetracycline; INT, integrase; INV, *invA*.

the classical penta-resistance phenotype (ACSSuT), the gene encoding integrase associated with the DT104 integrons ***(10)***, and the *Salmonella invA* virulence gene that is a key genus-specific marker ***(11)***. Thus, the identity of isolates obtained by standard enrichment culture techniques is inferred by the detection of various marker genes most commonly associated with *S*. Typhimurium DT104 strains.

2. Materials

2.1. Bacterial Strains and Media

For the purposes of this assay system, the test sample consists of a suspect *Salmonella* colony isolated on plating media using standard culture techniques. Typical plating media for this purpose include tryptone soya agar (TSA), bismuth sulfite agar (BS), brilliant green sulfa agar (BGS), and MacConkey agar.

An isolated colony of any *S*. Typhimurium DT104 laboratory strain exhibiting the penta-resistant phenotype (ACSSuT) grown on TSA (or any suitable plating media) can serve as a positive control in the test and as a source of genomic DNA for probe preparation (*see* **Subheading 3.3.1., step 1**).

2.2. Chemicals and Biologicals

The following is a list of chemicals and biologicals required to carry out the PCR and CHAS procedures. Chemicals can be purchased from any reputable supplier. However, where a specific supplier is indicated, it is highly recommended that the item be purchased from that source. Molecular biology-grade chemicals should be used wherever possible.

1. Agar.
2. Ammonium acetate.
3. Anti-DIG antibody–peroxidase conjugate (Roche, Mannheim, Germany).
4. Blotting grade blocker (Bio-Rad, Hercules, CA).
5. Bovine serum albumin.
6. Deoxyribonucleoide triphosphates (dNTP): dATP, dGTP, dCTP, and dTTP.
7. DIG-11-dUTP (Roche).
8. Ethanol, 95 and 70 %.
9. PCR buffer, 10× (Qiagen, Mississauga, Ontario).
10. *Taq* DNA polymerase (HotStarTaq, Qiagen).
11. Tetramethylbenzidine (TMB) membrane peroxidase substrate (Kirkegaard and Perry Laboratories, Gaithersburg, MD).
12. Tryptone Soya Broth (TSB).
13. Wizard Genomic DNA Purification Kit (Promega, Madison, WI).

2.3. Supplies and Equipment

1. Aluminum foil.
2. Analytical balance.
3. Centrifuge tubes, 50-mL capacity.
4. Crushed ice.
5. Dry heating block.
6. Filtration apparatus (*see* **Note 1**).
7. Fluorometer or spectrophotometer, for DNA quantification.
8. Ice bucket.
9. Incubators.
10. Microcentrifuge.
11. Microcentrifuge tubes, 0.5-mL capacity.
12. Microcentrifuge tubes, 1.5-mL capacity.
13. Petri dishes, square and round.
14. Pipettes of various sizes (2, 5, and 10 mL) and pipette bulb.
15. Pipettors covering volume range of 0.5–1000 μL, with appropriate filtered tips.
16. Polyester cloth (Sontara No. 8100, E I DuPont Nemours & Co., Wilmington, DE) (*see* **Note 2**).
17. Racks, to accommodate microcentrifuge tubes and centrifuge tubes.
18. Ruler.
19. Thermal cycler.
20. Tweezers (forceps).
21. UV cross-linker.
22. Vortex mixer.
23. Wash bottles.

2.4. CHAS Reagents

1. 1 M Tris–HCl (pH 8.0). Store at room temperature.
2. 1 M $MgCl_2$. Store at room temperature.
3. 5 M NaCl. Store at room temperature.
4. High salt buffer (HSB): 0.1 M Tris–HCl (pH 8.0), 0.01 M $MgCl_2$, and 0.15 M NaCl. Store at room temperature.
5. 20× SSC: 3 M NaCl and 0.3 M sodium citrate (pH 7.0). Store at room temperature.
6. 10 % (w/v) Sodium dodecyl sulfate. Store at room temperature.
7. 10 % (w/v) *N*-Lauroyl sarcosine. Store at room temperature.
8. Hybridization working solution (2× HS): 10× SSC; 2 % (w/v) blotting grade blocker; 0.2 % (v/v) *N*-lauroyl sarcosine, and 0.04 % (v/v) sodium dodecyl sulfate. Store in aliquots at −20 °C (*see* **Note 3**).
9. PBST: 0.01 M phosphate (pH 7.2), 0.15 M NaCl, and 0.05 % (v/v) Tween-20.
10. PBST-B: PBST containing 0.5 % (w/v) blotting grade blocker.

2.5. Oligonucleotides

Satisfactory synthetic oligonucleotides can be obtained from commercial sources. They should be provided desalted and preferably gel-purified or HPLC-purified. Oligonucleotides should be rehydrated at a concentration of 100 μM in 10 mM Tris–HCl (pH 8.0) containing 1 mM EDTA, then stored at −20 °C until use.

2.5.1. Primers for Multiplex PCR

Cell lysates are subjected to a multiplex PCR incorporating seven primer pairs targeting antibiotic resistance and marker genes (*see* **Table 1**). These primer pairs target the genes associated with the following DT104 markers: ampicillin (A-1 and A-2), chloramphenicol (C-1 and C-2), streptomycin (S-1 and S-2), sulfonamide (Su-1 and Su-2), tetracycline (T-1 and T-2), integrase (INT-1 and INT-2), and *invA* (INV-1 and INV-2). In addition to the sample target-specific primers, the multiplex PCR can also optionally include a pair of IAC primers (IAC-1 and IAC-2) (*see* **Note 4**).

2.5.2. Primers for Preparation of DNA Capture Probes

DNA capture probes for immobilization on the arrays can be prepared by PCR (*see* **Subheading 3.3.1., step 1**) using primer pairs specific to each of the genes associated with the following DT104 markers: ampicillin (A-3 and A-4), chloramphenicol (C-3 and C-4), streptomycin (S-3 and S-4), sulfonamide (Su-3 and Su-4), and tetracycline (T-3 and T-4), integrase (INT-3 and INT-4), and *invA* (INV-3 and INV-4) (*see* **Table 1**). These probes are designed to be internal to the amplicons from the multiplex PCR system (*see* **Note 5**).

3. Methods

3.1. Sample Preparation

3.1.1. Purification of Genomic DNA

Genomic DNA is extracted from *S.* Typhimurium DT104 for the preparation of DNA capture probes by PCR. The following is a brief outline of the procedure for the extraction of genomic DNA using the Wizard Genomic DNA Purification Kit (more detailed instructions are provided in the kit).

1. Inoculate 5 mL TSB with one loopful of *S.* Typhimurium DT104, and grow overnight at 37 °C.
2. Transfer 2× 1 mL portions of the cells to 1.5-mL microcentrifuge tubes. Pellet the cells by centrifugation at 15,000 g for 2 min.

Table 1
Oligonucleotide Primers for PCR

Primer	Sequence (5′ to 3′)	Amplicon size (bp)	Source,[a] accession number
Multiplex PCR			
A-1	CAA GTA GGG CAG GCA ATC ACA CTC G	216	NCBI, AF261825
A-2	GAG TTG TCG TAT CCC TCA AAT CAC C		
C-1	CAT TGA TCG GCG AGT TCT TGG GAT G	236	NCBI, AF261825
C-2	AGC CGT CGA GAA GAA GAC GAA GAA G		
S-1	TCA TTG AGC GCC ATC TGG AAT CAA C	210	NCBI, AF261825
S-2	GTG ACT TCT ATA GCG CGG AGC GTC T		
Su-1	CCG ATG AGA TCA GAC GTA TTG CGC C	209	NCBI, AF261825
Su-2	GCG CTG AGT GCA TAA CCA CCA GCC T		
T-1	CTC TAT ATC GGC CGA CTC GTG TCC G	204	NCBI, AF261825
T-2	CGG CGG CGA TAA ACG GGG CAT GAG C		
INT-1	CGC ACG ATG ATC GTG CCG TGA TCG A	235	NCBI, AF261825
INT-2	TAC GGC AAG GTG CTG TGC ACG GAT C		
INV-1	GTG AAA TTA TCG CCA CGT TCG GGC A	284	NCBI, M90846
INV-2	TCA TCG CAC CGT CAA AGG AAC CGT A		**Ref.** ***11***
IAC-1[b]	CAT AAT ATC ACT CGC GTC CGT TGA AGC TTA		
IAC-2[b]	GAC GAA ATC GTA AGC TTC AA		
Multiplex PCR Probe preparation			
A-3	ACT ATG ACT ACA AGT GAT AA	146	NCBI, AF261825
A-4	GAG CTT ACC TTC ATT TAA AT		

(*Continued*)

Table 1
(Continued)

Primer	Sequence (5′ to 3′)	Amplicon size (bp)	Source,[a] accession number
C-3	GCA GGC GAT ATT CAT TAC TT	177	NCBI, AF261825
C-4	GTG CCC ATA CCG GCG CTA AA		
S-3	TTG CTG GCC GTG CAT TTG TA	146	NCBI, AF261825
S-4	GAA AGC CGA AGC CTC CAT AA		
Su-3	TCT TAG ACG CCC TGT CCG AT	144	NCBI, AF261825
Su-4	GCA ATA TCG GGA TAG AGC GC		
T-3	TCA CGG GCG CAA CCG GAG CT	141	NCBI, AF261825
T-4	CCG AGC ATG CCA CCA AGT GC		
INT-3	GAT CCT TGA CCC GCA GTT GC	174	NCBI, AF261825
INT-4	TGG CTT CAG GAG ATC GGA AG		
INV-3	TTA TTG GCG ATA GCC TGG CG	219	NCBI, M90846
INV-4	TCC CTT TCC AGT ACG CTT CG		**Ref. *11***
IAC-P	CAT AAT ATC ACT CGC GTC CGT TGA AGC TTA CGA TTT CGT C	40-mer	

[a]NCBI, National Center for Biotechnology Information, http://www.ncbi.nlm.nih.gov/.
[b]Complementary regions in the IAC primers are underscored.

3. Resuspend the cell pellets in 600 μL "Nuclei Lysis Solution."
4. Incubate at 80 °C for 10 min to lyse the cells and cool to room temperature.
5. Add 3 μL of "RNase Solution" and mix by inversion.
6. Incubate at 37 °C for 60 min and cool to room temperature.
7. Add 200 μL of "Protein Precipitation Solution" to the RNase-treated cell lysate. Vortex vigorously at high speed for 20 s to mix. Incubate on ice for 5 min.
8. Centrifuge at 15,000 × *g* for 3 min.
9. Transfer supernatant to a new 1.5-mL microcentrifuge tube containing 600 μL room temperature isopropanol. Gently mix by inversion until the thread-like strands of DNA form a visible mass.
10. Centrifuge at 15,000 × *g* for 10 min.
11. Remove supernatant and drain tube with absorbent paper. Add 600 μL of room temperature 70 % ethanol and gently invert tube several times to wash DNA pellet.
12. Centrifuge at 15,000 × *g* for 10 min. Carefully aspirate all ethanol. Drain tube and let air-dry.
13. Add 100 μL of "DNA Rehydration Solution" to tube and rehydrate DNA by storing it overnight at 4 °C.
14. Once rehydrated, DNA can be stored at −20 °C for long-term storage.

3.1.2. Preparation of DNA Template for Multiplex PCR

Bacterial colonies are suspended in 50 μL 1× PCR buffer and then lysed by the addition of an equal volume of 2 % (w/v) Triton X-100, followed by heating at 100 °C for 10 min. Cell lysates are then subjected to PCR.

3.2. Multiplex PCR

1. In a 0.5-mL PCR tube, combine 10 μL of cell lysate (*see* **Subheading 3.1.2.**) with 90 μL PCR mixture (2.5 U HotStarTaq and 1.5× PCR buffer containing 2.25 mM $MgCl_2$, plus 200 μM each dNTP, 5 μM DIG-11-dUTP, 0.1 μM each sample target-specific primer [*see* **Table 1**], and 2 μg BSA/mL). For quality control purposes, the multiplex PCR can optionally incorporate a set of IAC primers (*see* **Note 4**), IAC-1 and IAC-2 (*see* **Table 1**), which are added to the PCR mixture at 0.04 μM each.
2. Carry out the PCR in a thermal cycler using the following program: initial heating at 94 °C for 15.5 min, followed by 35 cycles of denaturation at 94 °C for 30 s, primer annealing at 52 °C for 30 s, and primer extension at 72 °C for 1.5 min, with an additional 2 min at 72 °C following the last cycle.

3.3. Detection of Amplicons

3.3.1. Detection of Amplicons from DNA Targets

In **Subheading 3.2.**, target sequences originating from DNA molecules were amplified in the presence of DIG-11-dUTP, resulting in the incorporation of this label in the amplicons. The procedure outlined in this subheading describes a cloth-based hybridization assay for the detection of the DIG-labeled PCR amplicons by hybridization with an immobilized DNA probe. The hybridized amplicons are then detected by sequential reactions of the cloth strip with anti-DIG antibody–peroxidase conjugate and TMB substrate solution.

The DNA probe-coated cloth strips can be prepared ahead of time and stored refrigerated until use. In the present application, the DNA probe is conveniently made by amplification of a portion of the target sequence using the PCR technique.

3.3.1.1. Preparation of DNA Capture Probes

DNA capture probes can be prepared by PCR amplification of the individual target genes using purified *S*. Typhimurium DT104 genomic DNA as template. Each gene target is amplified individually using its corresponding primer pair (*see* **Subheading 2.5.2.**).

1. In a 0.5-mL PCR tube, combine 10 μL distilled H_2O containing 10 ng of *S*. Typhimurium DT104 purified genomic DNA with 90 μL PCR mixture (2.5 U HotStarTaq DNA polymerase and 1× PCR buffer with 1.5 mM $MgCl_2$, 200 μM each dNTP, 0.5 μM each primer [*see* **Table 1**], and 2 μg BSA/mL).
2. Carry out the PCR in a thermal cycler using the following program: initial heating at 94 °C for 15.5 min, followed by 35 cycles of denaturation at 94 °C for 30 s, primer annealing at 55 °C for 30 s, and primer extension at 72 °C for 1.5 min, with an additional 2 min at 72 °C following the last cycle.
3. Remove all of the PCR mixture to a new 1.5-mL microcentrifuge tube. Add 0.5 volume of 7.5 M ammonium acetate, followed by 2.5 volume of 95 % ethanol, and place at −20 °C for at least 1 h.
4. Centrifuge at 15,000 × *g* for 30 min.
5. Carefully decant the supernatant. Add 750 μL 70 % ethanol to wash pellet. Centrifuge at 15,000 × *g* for 5 min.
6. Carefully decant the supernatant and allow pellet to air-dry.
7. Resuspend DNA in Tris–EDTA buffer.

3.3.1.2. Immobilization of DNA Probes on Polyester Cloth

1. Cut polyester cloth into 2 cm × 5 cm strips (*see* **Note 6**). Mark cloth strips with pencil to guide spotting and to identify each strip.
2. Place the cloth strips flat on the filtration apparatus (*see* **Note 1**), and rinse briefly with 95 % (v/v) ethanol. Wash the cloth strips seven times with deionized distilled water (*see* **Note 7**).
3. With the vacuum still on, remove the cloth strips using forceps, place in an uncovered Petri dish and allow to air-dry overnight at room temperature (*see* **Note 8**).

4. Dilute probes in HSB containing 30 % (v/v) ethanol to a final coating concentration of 10 ng/μL. Denature probes by placing in a heat block for 10 min at 100 °C followed by chilling probes on ice water (ensure that tubes are tightly sealed and secured with lid locks) (*see* **Note 9**). If the multiplex PCR includes IAC primers (*see* **Note 4**), an IAC capture oligonucleotide probe should also be immobilized on the strips. The IAC probe stock is diluted to 10 μM in HSB containing 30 % (v/v) ethanol, then heated and chilled on ice water.
5. Spot 5 μL of the denatured probe in discrete spots on each cloth strip. A sample spotting pattern is shown in **Fig. 2** (*see also* **Note 6**). Allow the cloth strips to dry for 30 min at 37 °C.
6. Cross-link the probes to the cloth by exposing the strips to UV light for ca. 1 min (254 nm, 100 mJ/cm^2) using a UV cross-linker (*see* **Note 10**).
7. Block the cloth strips with 1× HS for at least 1 h at 37 °C. Wash the strips five times with PBST. The probe-coated cloth strips can be air-dried and stored in a sealed container at 4 °C until use (*see* **Note 11**).

3.3.1.3. Hybridization and Colorimetric Detection of Multiplex PCR Amplicons

1. Denature DIG-labeled PCR products (100 μL) in a heat block at 100 °C for 10 min and chill in ice water.
2. Transfer denatured DIG-labeled PCR products (100 μL) to 900 μL ice-cold hybridization solution containing 50 % formamide (1 volume 2× HS plus 1 volume formamide).
3. Saturate the cloth strips with the solution from step 2 and incubate at 45 °C for 30 min. Wash the strips five times with PBST (*see* **Note 12**).
4. Add 1 mL anti-DIG antibody–peroxidase conjugate diluted 1:2000 in PBST containing 0.5 % (w/v) blocking reagent (PBST-B). Incubate 10 min at room temperature. Wash the strips five times with PBST.
5. Add 1 mL TMB membrane peroxidase substrate solution to each strip. Let the color develop for 10 min. Reactions are graded qualitatively as follows: positive (blue spot) and negative (no spot). An example of the expected reactivity patterns obtained with different *Salmonella* and non-*Salmonella* bacteria is shown in **Fig. 2**. Strips can be photographed to keep a permanent record of results.

4. Notes

1. The filtration apparatus is composed of three components: a detachable macroporous filter, a filtering flask, and rubber tubing connecting the flask to a vacuum source. A typical filtration apparatus setup is diagrammed in **Fig. 3**. Before washing cloth strips on synthetic polymeric filter surfaces, which tend to have water-repelling properties (resulting in interference with uniform suction across the surface), it is important to prewet the filter by rinsing with 95 % (v/v) ethanol and then distilled water.

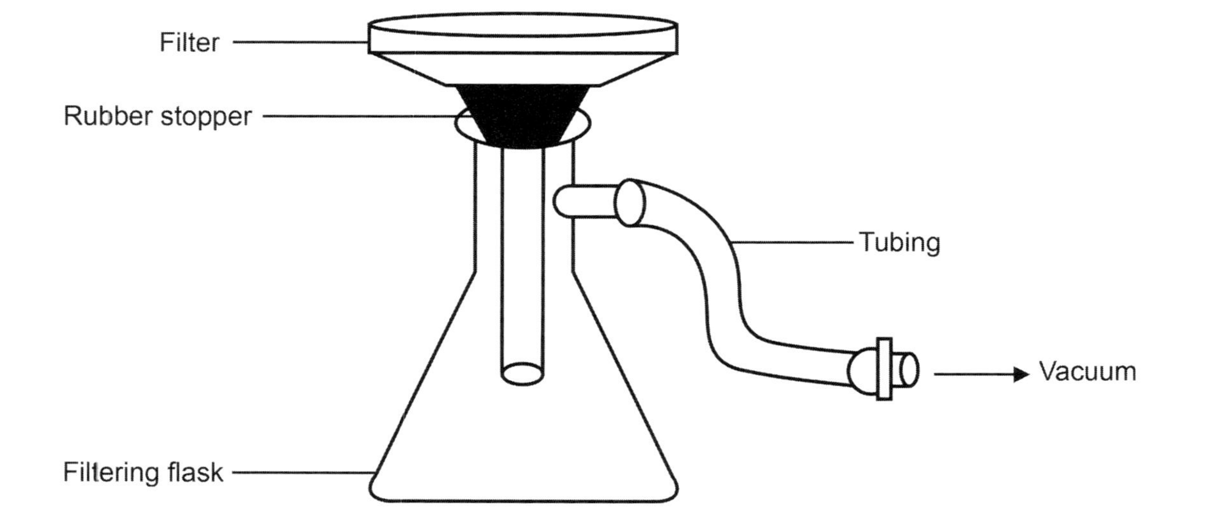

Fig. 3. Filtration apparatus consisting of three components: a detachable filter head, a filtering flask, and flexible tubing connecting the flask to a vacuum source.

2. The DuPont Sontara 8100 brand of nonwoven polyester cloth has been extensively tested and found to be highly suitable in terms of uniformity, thickness, and absorptive properties as a solid phase for DNA hybridization procedures. If this particular brand of cloth cannot be obtained, a suitable alternative nonwoven cloth fabric of uniform-density polyester fibers may be substituted. Care must be taken to ensure that the fabric source is pure polyester with no surface treatments or impurities. DuPont Sontara 8100 polyester cloth has a uniform fiber density and a weight of ca. 12 mg/cm^2 with a thickness of ca. 1 mm.
3. When preparing the working hybridization solution (2× HS), mix all reagents together, adding the blotting grade blocker last. Any precipitates formed can be dissolved by warming the solution in a 37 °C water bath. The HS can be prepared in large volumes (e.g., 200 mL) and stored in 25-mL aliquots at −20 °C. Prehybridization solution (1× HS) or hybridization solution containing 50 % (v/v) formamide is readily prepared using the 2× HS by dilution with equivalent volumes deionized distilled water (for 1× HS) or formamide (for HS with 50 % formamide).
4. It is a well-known fact that in some instances PCR can suffer inhibition because of the presence of certain sample matrix components. The IAC provides a simple internal control system for the PCR based on the incorporation of a primer pair with complementary 3' ends, resulting in the generation of a unique "primer-dimer" detectable by hybridization with a specific capture probe immobilized on polyester cloth as part of the array of amplicon-specific probes. The inclusion of a low concentration of IAC primers in the reaction mixture does not adversely affect the amplification and subsequent detection of target gene sequences by hybridization with immobilized probes in either single-gene amplification or multiplex PCR systems. Failure to amplify target gene sequences because of the presence of inhibitors is mirrored by a failure to amplify the internal control primer-dimer.
5. Oligonucleotide primers for the production of DNA capture probes by PCR should be designed to avoid complementarity with the primers used to amplify the corresponding target amplicons in the multiplex PCR. Otherwise, false-positive signals may arise in the CHAS in the event that "primer-dimer" formation occurs in the multiplex PCR.
6. The size of the cloth strips will depend on the number of probes comprising the array. It is important to ensure that probes are not spotted too close to the margins of the strip. Depositing probes at least 0.5 cm from the edges of the cloth strip and about 1 cm apart provides good results.
7. Saturating the cloth strips with 95 % (v/v) ethanol serves to remove any impurities from the strips and render them more hydrophilic. To wash, stop the vacuum (e.g., by pinching tubing on filtration apparatus), saturate the strips with water, and open the vacuum to remove excess water. Repeat until strips have been washed seven times.

8. It is critical that the strips be completely dry before proceeding with the probe application. Ensure that strips are not piled on top of each other to permit adequate drying. Allow the strips to air-dry in the Petri dishes with the lids removed.
9. Alternatively, to avoid evaporation of ethanol during heating, dilute the appropriate amount of probe stock into ca. 0.7 volume of HSB, heat denature (*see* **Subheading 3.3.1.2.**), then cool and add ca. 0.3 volume of 95 % ethanol (i.e., final 30 % [v/v] ethanol). The addition of ethanol to the HSB allows penetration of the probe coating solution into the hydrophobic cloth matrix.
10. Cross-linking using UV light covalently links the probes to the polyester cloth fibers. This is a critical step in the assay as the probes will not remain bound to the strips without UV cross-linking.
11. Prehybridization of the cloth strips in HS is critical to block any nonspecific binding sites available on the cloth strips that would otherwise cause nonspecific adsorption of assay reagents resulting in background staining. After blocking and washing, the strips are ready for the hybridization assay. Alternatively, after blocking and washing, the strips can be air-dried and stored in a sealed container at 4 °C for up to 6 months.
12. Washing with PBST between each step ensures that unbound reagents are washed away to prevent the formation of background. It is important to manipulate cloth strips using clean forceps. Rinse forceps and filter with deionized distilled water between each treatment.

References

1. Nagano, I., Kunishima, M., Itoh, Y., Wu, Z., and Takahashi, Y. (1998) Detection of verotoxin-producing *Escherichia coli* O157:H7 by multiplex polymerase chain reaction. *Microbiol. Immunol.* **42,** 371–376.
2. Osek, J. (2001) Multiplex polymerase chain reaction assay for identification of enterotoxigenic *Escherichia coli* strains. *J. Vet. Diagn. Investig.* **13,** 308–311.
3. Strizhkov, B. N., Drobyshev, A. L., Mikhailovich, V. M., and Mirzabekov, A. D. (2000) PCR amplification on a microarray of gel-immobilized oligonucleotides: detection of bacterial toxin- and drug-resistant genes and their mutations. *Biotechniques* **29,** 844–857.
4. Chizhikov, V., Rasooly, A., Chumakov, K., and Levy, D. D. (2001) Microarray analysis of microbial virulence factors. *Appl. Environ. Microbiol.* **67,** 3258–3263.
5. Blais, B. W. and Phillippe, L. M. (1995) Macroporous hydrophobic cloth (Polymacron) as a solid phase for nucleic acid probe hybridizations. *Biotechnol. Tech.* **9,** 377–382.
6. Gauthier, M. and Blais, B. W. (2004) Cloth-based hybridization array system for the detection of multiple antibiotic resistance genes in *Salmonella enterica* subsp. *enterica* serotype Typhimurium DT104. *Lett. Appl. Microbiol.* **38,** 265–270.

7. Gauthier, M., Cadieux, B., Austin, J. W., and Blais, B. W. (2005) Cloth-based hybridization array system for the detection of *Clostridium botulinum* type A, B, E and F neurotoxin genes. *J. Food Protect.* **68,** 1477–1483.
8. Gauthier, M. and Blais, B. W. (2005) Cloth-based hybridization array system for detection of toxin genes associated with major foodborne pathogenic bacteria. *J. Rapid Methods Autom. Microbiol.* **13,** 243–256.
9. Gauthier, M. and Blais, B. W. (2003) Comparison of different approaches for the incorporation of non-radioactive labels into polymerase chain reaction products. *Biotechnol. Lett.* **25,** 1369–1374.
10. Ng, L.-K., Mulvey, M. R., Martin, I., Peters, G. A., and Johnson, W. (1999) Genetic characterization of antimicrobial resistance in Canadian isolates of *Salmonella* serovar Typhimurium DT104. *Antimicrob. Agents Chemother.* **43,** 3018–3021.
11. Rahn, K., De Grandis, S. A., Clarke, R. C., et al. (1992) Amplification of an *invA* gene sequence of *Salmonella typhimurium* by polymerase chain reaction as a specific method of detection of *Salmonella. Mol. Cell. Probes* **6,** 271–279.

5

Genome-Based Identification and Molecular Analyses of Pathogenicity Islands and Genomic Islands in *Salmonella enterica*

Michael Hensel

Summary

Pathogenicity islands and genomic islands (GI) are key elements in the evolution of bacterial virulence and environmental adaptation. In *Salmonella enterica*, *Salmonella* pathogenicity islands (SPI) confer important virulence traits; however, many of these loci have not been characterized in molecular detail. In this chapter, procedures for the identification and molecular characterization of SPI and GI are described. Based on genome sequence data, bioinformatics approaches allow the identification of putative SPI and GI. The role of these loci can be analyzed after the generation of deletion mutant strains using the Red recombination approach. For further analyses, cosmid libraries of *S. enterica* genomic DNA are screened for clones harboring entire SPI or GI. Such cosmid clones are then used for complementation of SPI or GI deletions as well as for the transfer of these loci to other bacterial species and subsequent functional assays. This set of methods allows the rapid and efficient analyses of the functions of SPI and GI.

Key Words: Pathogenicity islands; genomic islands; horizontal gene transfer; virulence evolution; cosmid cloning.

1. Introduction

1.1. Pathogenicity Islands and Genomic Islands in Salmonella

The horizontal transfer of genetic material is considered as the major force of bacterial evolution *(1)*. For pathogenic bacteria such as *S. enterica*, the horizontal gene transfer allows the acquisition of the complex repertoire of virulence determinants. This large number of virulence factors appears to be

From: *Methods in Molecular Biology, vol. 394: Salmonella: Methods and Protocols*
Edited by: H. Schatten and A. Eisenstark © Humana Press Inc., Totowa, NJ

necessary for the successful colonization of very different habitats within the host *(2)*. Among the various genetic elements contributing to horizontal gene transfer, pathogenicity islands (PAI) are of specific interest, because acquisition of PAI can transfer rather complex virulence functions in a single step *(3)*. PAI are defined by the following criteria:

1. Large contiguous genomic regions (often more than 20 kb)
2. Presence of one or more virulence genes
3. Genetic instability and deletion of the locus
4. Presence of DNA sequences associated with DNA mobility (genes for transposases, integrases, bacteriophages genes, and direct repeats)
5. Base composition different from that of the core genome (higher or lower G+C content of the DNA and different codon usage)
6. Association with genes encoding tRNA

It is important to note that these conditions are not stringent and only very few PAI meet all the criteria listed. There is a wide range in the size of PAI, and the genetic elements responsible for DNA mobility may have been deleted resulting in PAI that became stable constituents of the genome of the pathogen (for recent review, *see* **ref. *4***). However, the association of PAI with genes for tRNAs and/or the differences in base composition appear characteristic for the majority of PAI. Therefore, these features will have a central role in the methods described in this chapter.

Horizontal gene transfer of large regions of genomic DNA is not restricted to pathogenic bacteria. It has also been observed that metabolic traits were transferred between various bacteria by a similar mechanism. The term "genomic islands" (GI) was created as a broader definition for large genomic regions acquired by horizontal gene transfer *(5)*.

Many of the virulence determinants of *S. enterica* are encoded by genes on *Salmonella* pathogenicity islands (SPI) (reviewed in **ref. *6***). The invasion of eukaryotic host cells and the intracellular survival and replication—the two major virulence traits of *Salmonella*— are functions encoded by genes on SPI1 and SPI2, respectively. SPI1 and SPI2 genes encode type III secretion systems for the translocation of virulence proteins into the host. In addition to these well-characterized SPI, there is a large number of SPI for which the contribution to virulence is less well understood. Depending on the *S. enterica* serotype analyzed, up to 11 SPI were defined, but the total number of SPI is not known. The available *S. enterica* genome sequences indicate that the distribution of SPI is variable *(7–9)*.

The identification and characterization of unknown SPI and GI in *Salmonella* is important for the integrated understanding of the virulence of this pathogen.

Here, I describe methods for the identification of SPI and GI, the mutational analysis, and the cloning and transfer of these loci. The procedures are based on my group's work on genome-based approaches to identify putative SPI and GI *(10)*, for the deletion of the entire loci, and for the cloning and experimental transfer of cloned loci to heterologous bacterial species *(10,11)*. Although the methods were developed for and applied to *S. enterica* serovar Typhimurium, it is likely that the techniques are also applicable to other *Salmonella* serotypes and other enterobacteria.

1.2. Genome-Based Identification of PAI and GI in Salmonella

Many PAI have been identified by functional screens, such as random mutagenesis and screening for mutant strains with virulence defects. Other PAI were identified by the detailed characterization of the genomic context of a known virulence gene. The large number of genome sequences of bacterial species now allows genome-based identification of putative PAI and GI without an initial functional screen. The typical characteristics of PAI can be used for an inspection of complete genome sequences as well as partial genome sequences from ongoing sequencing projects. Using the search criteria defined in **Subheading 3.3.**, many known SPI as well as new GI were identified in the genome sequence of *S. enterica* serovar Typhi *(10)*. Similar search criteria have also been incorporated into search algorithms *(12)*. There are also very useful Internet sites that list putative PAI and GI predicted from genome sequences of various bacterial species *(13)*.

1.3. Deletion of SPI and GI

The deletion of an entire SPI may be a useful approach to investigate the role of the locus in virulence and an initial step followed by mutational analyses of individual genes within the SPI. The availability of the one-step inactivation approach, initially developed for *Escherichia coli* *(14)*, allows the rapid and precise deletion of loci of almost any size in *S. enterica*. My group has successfully used the one-step inactivation approach to delete entire SPI2, SPI3, SPI4, SPI9, or part of these SPI as well as various GI *(10)*. So far, no failure of this approach has been observed. In addition to *S. enterica* serotype Typhimurium, the deletion protocol could also be successfully applied to clinical isolates and reference strains of *S. enterica* serotype Enteritidis.

1.4. Cloning and Transfer of SPI and GI

The experimental transfer of cloned SPI or GI to bacterial strains lacking the locus of interest could be of interest for the analysis of the virulence function of the SPI, their regulation, and their interaction with other virulence determinants. The large size of most SPI restricts the use of PCR-based approaches. In our hands, a conventional λ library-based approach worked well. This approach involved the generation of a cosmid library, ordering library clones, and identifying SPI-harboring clones by hybridization. Although generation of a cosmid library requires a significant investment of hands-on time, an existing cosmid library of sufficient complexity can be used for repeated screening and isolation of SPI. We have successfully used a cosmid library of *S. enterica* serovar Typhimurium genomic DNA generated with pSuperCos1 to obtain clones that harbor entire functional SPI2, SPI4, or SPI9.

1.5. Functional Analyses

The most challenging task will be to identify a phenotype that mediated the newly identified SPI or GI. The initial characterization of mutant strains or hosts harboring the cloned locus of interest may include analyses of adherence, invasion, and intracellular pathogenesis using conventional murine and human cell lines. In the case of GI, the analyses should be extended toward a physiological characterization of mutant strains. Because SPI are likely to contribute to the host specificity of *S. enterica* serovars, it could be of interest to analyze phenotypes in models with cells lines of additional host species.

2. Materials

2.1. Identification of Putative New SPI and GI

2.1.1. Hardware

1. Personal computer (PC): Standard PC equipment will work, but analysis of entire genomes is more efficient if sufficient RAM is available (at least 1 Gb). Viewing graphical analyses of genome sequences using software such as Artemis benefits from large monitors (19”).
2. Internet access.

2.1.2. Software

1. Software for genome analysis, for example, Artemis version 7.0, available free of charge at http://www.sanger.ac.uk/software/artemis.
2. Software for genome comparison, for example, ACT version 4, available free of charge at http://www.sanger.ac.uk/software/act.

   Various commercial standard software packages for sequence are useful, for example:

3. Gene construction kit (GCK) (http://www.textco.com).

4. MacVector for Apple Mac or DSGene for PC or GCG for various platforms (http://www.accelrys.com).
5. Vector NTI (http://www.invitrogen.com).

2.1.3. Useful Internet Resources

1. http://colibase.bham.ac.uk (comprehensive database and tool collection for sequence analyses of bacterial genome sequences).
2. http://www.pathogenomics.sfu.ca/islandpath (lists of putative PAI and PI in sequenced genomes).
3. http://selab.janelia.org/software.html (identification of tRNA genes).
4. http://www.sanger.ac.uk/Software (further sequence analysis software).

2.2. Deletion of Putative SPI and GI

1. Plasmids for Red-mediated recombination: pKD3, pKD4, pKD46, and pCP20 ***(14)***.
2. Oligonucleotides for PCR-mediated generation of targeting constructs, usually 60-mer with 40 bases complementary to a given target region (xyz) and 20 bases complementary to pKD3 or pKD4, for example,
   a. xyz-del-for: 5′-40 bases-GTGTAGGCTGGAGCTGCTTC-3′.
   b. xyz-del-rev: 5′-40 bases-CATATGAATATCCTCCTTAG-3′.
3. Oligonucleotides for confirmation of the integration of targeting constructs and subsequent deletion, for example,
   a. forward primer, xyz-specific, 20-mer.
   b. K1-red-del: 5′-CAGTCATAGCCGAATCGCCT-3′ (for kanamycin resistance cassette).
   c. C1-red-del: 5′-TTATACGCAAGGCGACAAGG-3′ (for chloramphenicol resistance cassette).
4. Incubators at 30, 37, and 42 °C.
5. Luria broth (LB).
6. LB agar plates containing antibiotics for selection.
7. Arabinose stock solution (500 mM, filter sterilized, and stored in aliquots at −20 °C).
8. SOC media.
9. Electroporation instrument and electroporation cuvettes.

2.3. Cloning and Transfer of Putative SPI

1. High-molecular-weight genomic DNA of the *Salmonella* isolate of interest, use standard commercial kits for preparation or the CTAB method ***(15)***.
2. Restriction enzymes *Sau*IIIA, *Bam*HI, and others.
3. Cosmid vector for cloning (e.g., pSuperCos1 and pBeloBac11).
4. In vitro packaging kit (e.g., GigaPack III, Stratagenc).

5. *E. coli* host strain (XL1-Blue MR, Stratagene).
6. Sterile microtiter plates for growth and storage of clones.
7. 2× YT broth.
8. Dimethyl sulfoxide (DMSO).
9. Device for replica plating (e.g., Sigma R2383).
10. Membranes for hybridization (e.g., ZetaPlus GT, BioRad).
11. For plating of 96-well plate, 15-cm^2 Petri dishes with LB agar and selective antibiotics.
12. Material for colony hybridization, dependent on the hybridization method used (radioactive or nonradioactive).

3. Methods

3.1. Genome-Based Identification of SPI and GI

Genomic loci acquired by horizontal gene transfer can be identified by using user-defined criteria for identification of SPI and GI.

Approach A:

1. Scan genome sequences for regions of altered base composition. Use Artemis to scan entire genomes or partial genome data for regions with G+C content higher or lower than the genome average. Artemis plots the average G+C content of the sequence file, and regions with significantly higher or lower G+C content are indicated. *See* **Fig. 1A** for an example.

Approach B:

1. Scan the vicinity of genes for tRNAs. This approach requires the comparison to the genome data of a related nonpathogenic species. The genome organization in *S. enterica* is largely parallel to that of *E. coli* K-12; thus, the *E. coli* genome is useful for comparison. For complete genome sequences, the positions of tRNA genes are usually listed in the corresponding publications.
2. Analyze the vicinity of the tRNA gene in the genome for parallel or divergent gene organization. This includes prediction of open reading frames and database searches with the predicted amino acid sequences. Putative SPI or GI will appear as regions specific for the pathogenic species that are inserted between parallel arranged genes present in both genomes. *See* **Fig. 1B** for an example.

Both approaches are likely to reveal a large number of candidate loci. Additional criterion such as the presence of genetic elements associated with DNA mobility, the presence of genes with similarity to known virulence genes, or the presence of genes for metabolic functions should be considered to shortlist those loci that are of interest for the subsequent functional analyses.

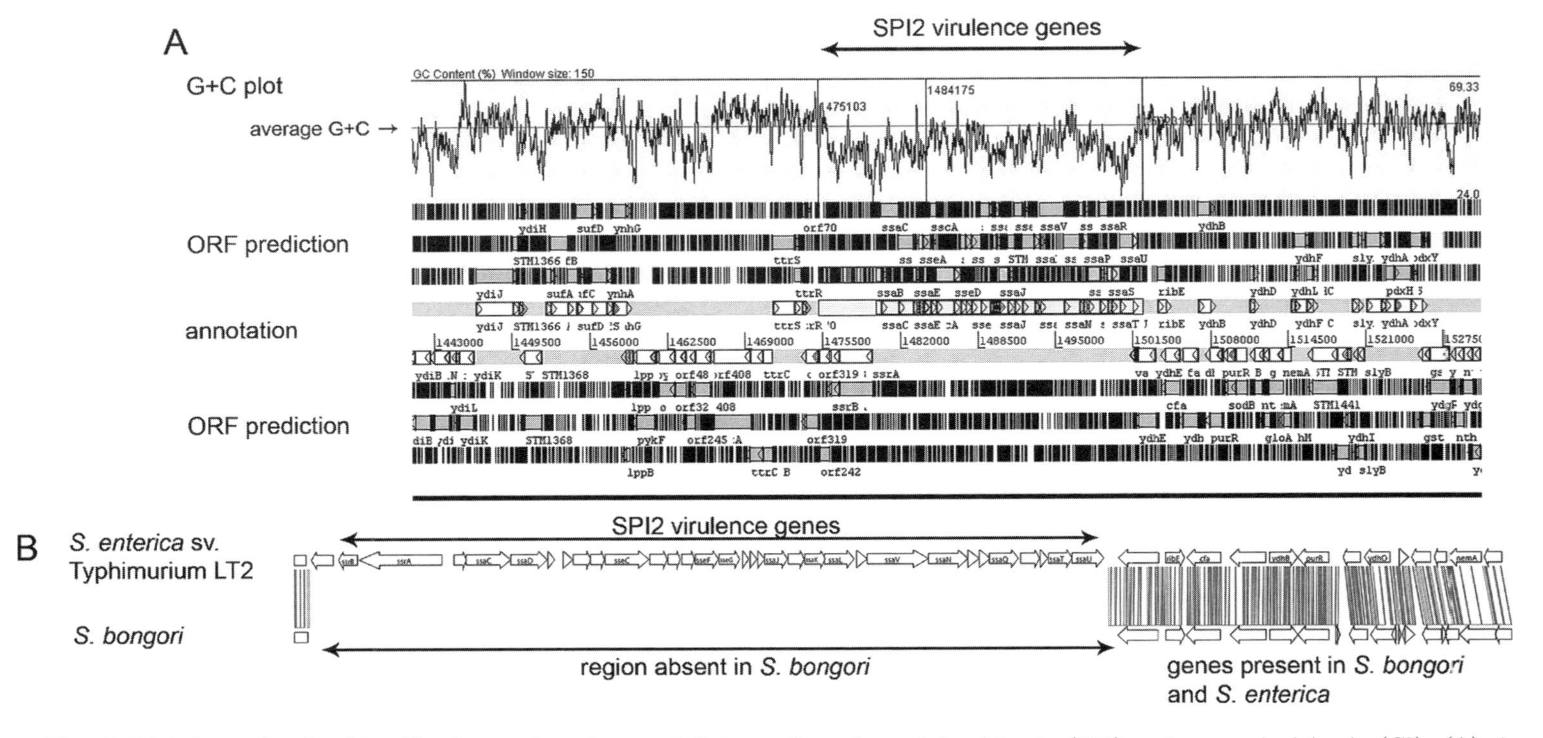

Fig. 1. Bioinformatics for identification and analyses of *Salmonella* pathogenicity islands (SPI) and genomic islands (GI). **(A)** A screenshot of the analysis of the *Salmonella enterica* serovar Typhimurium LT2 genome sequence in Artemis 7.0. A segment of the genome containing SPI2 is shown as an example of the program application. The upper part of the figure shows the G+C plot. SPI and GI are often characterized by a base composition different from that of the core genome. Such alterations can be identified by G+C plots. **(B)** The result a genome comparison using the Internet resource ColiBase (http://colibase.bham.ac.uk/) is shown. This example shows a MUMmer alignment of the region of *S. enterica* sv. Typhimurium LT2 genome that contains SPI2, and the equivalent regions from *S. bongori*. The vertical lines between the gene symbols indicate sequence similarity between the genomic regions. SPI and GI can be identified as insertions in regions containing genes with a parallel organization between the compared genomes. ColiBase allows the comparison of various complete and partial genome sequences.

SPI and GI may also be identified using available databases and published algorithms. Of specific interest are the following sites:

1. http://colibase.bham.ac.uk. This site can be used for genome comparison and contains various tools for sequence analyses.
2. http://www.pathogenomics.sfu.ca/islandpath. This site lists predicted PAI of GI from various bacterial genome sequences.

3.2. Deletion of SPI

The experimental analysis of putative new SPI or GI in *S. enterica* may require the deletion of the locus. In our hands, the Red-mediated recombination approach ***(14)*** works very reliably. For *S. enterica* serovar Typhimurium strains ATCC14028, we routinely apply the following protocols.

3.2.1. Generation of Targeting DNA

1. Use pKD3 or pKD4 as template for PCR amplification of targeting constructs with chloramphenicol or kanamycin resistance genes, respectively, with 60-mer primer pair specific to the target locus. Optimize PCR for a low number of amplification cycles (usually 25) and a high amount of template plasmid in order to minimize PCR-generated sequence errors in the product. Use proofreading polymerase mixes. A typical program is 95 °C, 10 min (95 °C, 45 s; 58 °C, 45 s; 72 °C, 2 min), 25×; 72 °C, 10 min.
2. Perform clean-up of the PCR product (e.g., PCR purification on spin columns, Qiagen).
3. Optional: *DpnI* digest to remove residual template plasmid—28 μL PCR product, 20 U *DpnI* (New England Biolabs), and 3.2 μL buffer; incubate for 4 h (overnight) at 37 °C, purify DNA using Qiagen nucleotide removal column, and elute with 30 μL elution buffer (EB).

3.2.2. Preparation of Competent Cells

1. Grow *Salmonella* strains harboring pKD46 overnight at 30 °C in LB + 50 μg/mL carbenicillin.
2. Inoculate 50 mL culture LB + 50 μg/mL carbenicillin + 10 mM arabinose with 500 μL overnight culture (use flasks with baffles).
3. Incubate in shaking water bath at 30 °C to OD_{600} of about 0.6 (0.5–0.8).
4. For the following steps, it is important to keep cells on ice and to use prechilled solutions, tubes, and centrifuges. Transfer the culture to a 50-mL tube, incubate on ice for 20 min, and centrifuge for 10 min at 7000 × *g* at 4 °C.
5. Resuspend pellet in 50 mL ice-cold ddH_2O, incubate on ice for 20 min, and centrifuge for 10 min at 7000 × *g*.
6. Resuspend pellet in 25 mL ice-cold ddH_2O, incubate on ice for 20 min, and centrifuge for 10 min at 7000 × *g*.

7. Resuspend pellet in 2 mL ice-cold 10 % glycerol, incubate on ice for 20 min, and centrifuge for 10 min at 7000 × *g*.
8. Resuspend pellet in 500 μL ice-cold 10 % glycerol.
9. Use cells directly for electroporation (keep on ice) or freeze aliquots of 45 μL (or multiples) at −70 °C. Frozen competent cells are about 10-fold in transformation efficiency.

3.2.3. Transformation and Selection

1. Transfer 40 μL competent cells into electroporation cuvette (2 mm gap, prechilled on ice), add 2 μL (1–4 μL) of purified PCR product, and mix by pipetting.
2. Perform electroporation at 2.5 kV, 200 Ω, 25 μF, recover cells immediately in 950 μL prewarmed SOC or LB, and incubate with agitation for 1 h at 37 °C.
3. Plate 100 μL bacteria suspension on LB plates containing 50 μg/mL kanamycin or 10 μg/mL chloramphenicol, recover bacteria from the remaining 900 μL suspension by centrifugation for 5 min at 8000 × *g*, remove 800 μL supernatant, resuspend pellet in remaining 100 μL SOC or LB and plate as described on selective plates, and incubate overnight at 37 °C.
4. Pick colonies resistant to kanamycin or chloramphenicol, purify on LB plates containing antibiotics, and incubate overnight at 42–44 °C.
5. Perform colony PCR to confirm the proper insertion of the cassette using a reverse primer within the resistance cassette and a forward primer upstream of the deletion.
6. Streak confirmed positive clones on new LB plates containing selective antibiotics (kanamycin or chloramphenicol), in parallel streak on LB plates containing carbenicillin, and incubate overnight at 37 °C.
7. Select carbenicillin-sensitive and kanamycin-resistant (or chloramphenicol-resistant) clones. Pick a single colony and inoculate in 3 mL LB medium with selective antibiotics, and incubate overnight at 37 °C.
8. Prepare frozen stocks by adding DMSO to 7 % final concentration.
9. Confirm sensitivity to carbenicillin by plating 100 μL of the overnight culture on LB agar plates with carbenicillin, and incubate overnight at 37 °C (no single colony should grow).

3.2.4. Deletion of Resistance Markers Using FLP Recombinase (Optional)

1. Prepare competent cells of confirmed mutant strains as described in **Subheading 3.2.2.**, except growing the culture at 37 °C.
2. Introduce pCP20 by electroporation as described in **Subheading 3.2.3.**
3. Select transformed clones by selection on LB agar plates with carbenicillin at 30 °C.
4. Purify clones on LB agar plates without antibiotics and grow at 37 °C.
5. Check individual clones for sensitivity to carbenicillin and kanamycin or chloramphenicol.
6. Repeat steps 4 and 5 if necessary to obtain antibiotic-sensitive clones.

7. With selected clones, perform colony PCR to confirm the deletion of the locus and the resistance cassette. Use forward and reverse primers flanking the deleted locus of interest.
8. Prepare stock cultures as described in **Subheading 3.2.3., step 7**.

3.3. Generation of Cosmid Libraries and Cloning of SPI

1. Prepare high-molecular-weight genomic DNA of the *S. enterica* strain to be analyzed. We use the CTAB method ***(15)*** for large-scale preparations. A critical parameter is the integrity of the chromosomal DNA. Specific care should be taken to avoid fragmentation caused by shearing.
2. Perform partial digestion with *Sau*IIIA. Digestion should be optimized in order to obtain highest amounts of restriction fragments of 30–50 kb. Conditions for partial digestion have to be established empirically by altering the units of the restriction enzyme (RE) at a fixed incubation time or by varying the incubation time with a fixed amount of RE units. In our hands, the first condition was more reproducible.
3. Optionally, perform size fractionation to enrich DNA fragments within the range of the cloning capacity of the cosmid according to standard procedures ***(15)***. This step is not essential as in vitro packaging selects for cosmids that harbor 35–45 kb inserts; however, this step may increase cloning efficiency.
4. Purified, size-fractionated, *Sau*IIIA-digested fragments are ligated to *Bam*HI-digested cosmid pSuperCos1 (Stratagene) or another cosmid vector of choice. For packaging of cosmids in phage λ heads, use an in vitro packaging kit, for example GigaPack III Gold Packaging Kit (Stratagene). Infect a suitable *E. coli* host strain (e.g., XL1-Blue MR) with the cosmid library in phage λ.
5. Store a number of clones that cover the genome of the host strain with sufficient redundancy (*see* **Note 1**). After infection of *E. coli* with the λ library, pick well-isolated clones, inoculate 100 mL of 2× YT medium in wells of a 96-well plate with a single colony and grow bacteria at 37 °C to stationary phase (about 24 h).
6. Prepare colony blots of the library by spotting cultures on nitrocellulose or nylon membranes (*see* **Note 2** for further considerations). Place membrane on top of LB plates containing antibiotics. Inoculate with strains from the master plate using a sterile replicator. Grow bacteria to sufficient colony size and perform lysis and further treatment of the membrane according to the instructions of the subsequent hybridization protocol.
7. As hybridization probes, use DNA fragments generated by PCR or restriction digests of cloned DNA fragments for screening the library for clones harboring the SPI or GI.
8. For storage of the library, add 50 μL medium containing 21 % DMSO to each well, mix by tapping the plates, and store at −70 °C. If selected clones have to be regrown, use a sterile toothpick to streak over the frozen surface of the respective well and use the toothpick to streak out on an agar plate. Avoid thawing of the plates, as this reduces the viability of stock cultures.

9. For confirmation of positive clones, isolate cosmid DNA using standard protocols. Perform restriction analyses and Southern hybridizations with probes derived from the most 5′ and 3′ regions of the locus of interest. Alternatively, the presence of the most 5′ and 3′ regions can be checked by colony PCR of the positive clones.
10. Use clones harboring the entire locus of interest for further analyses.

3.4. Transfer of SPI and Functional Analysis

1. Prepare cosmid DNA from confirmed clones using DNA purification kits for large plasmids or cosmids (Qiagen). For the efficient transformation, the cosmid DNA should be in supercoiled conformation.
2. Introduce the cosmid into the strain of interest by electroporation.
3. Perform functional assays depending on the phenotype that is of relevance to the locus under investigation (the copy number of the cosmid may be critical, *see* **Note 3**).

4. Notes

1. A cosmid library of 1000 independent clones was sufficient to clone various SPI of *S. enterica* serovar Typhimurium ATCC 14028.
2. Conventional radioactive as well as nonradioactive methods for screening clone libraries are available. In our hands, the nonradioactive DIG system (Roche) worked efficiently. For nonradioactive detection, Zeta-Plus GT (Bio-Rad) nylon membranes worked best.
3. The copy number of the cosmid vector is a critical parameter. The cosmid pSuperCos1 used for the construction of the library is present in 7–10 copies per cell and the presence of virulence genes in multicopy can be detrimental to the host or affect the regulation. For analyses with a single copy of clones SPI or GI, we successfully used pBeloBac11, a BAC vector suitable for cloning large inserts *(16)*. Inserts from pSuperCos1 clones can be recovered as *Not*I fragments and subcloned into pBeloBac11.

Acknowledgments

Work in my group on SPI was supported by grants of the Deutsche Forschungsgemeinsschaft. The approach described here would not have been possible without the provision of DNA sequences from ongoing sequencing projects and excellent bioinformatics software by the Sanger Institute (Hinxton, UK) and ColiBase (University of Birmingham, UK). I especially thank Dr. Imke Hansen-Wester for her inspired work on the identification, molecular analyses, and transfer of SPI.

References

1. Ochman, H., Lawrence, J. G., and Groisman, E. A. (2000) Lateral gene transfer and the nature of bacterial innovation. *Nature* **405,** 299–304.
2. Groisman, E. A. and Ochman, H. (1997) How *Salmonella* became a pathogen. *Trends Microbiol.* **5,** 343–349.
3. Hacker, J. and Kaper, J. B. (2000) Pathogenicity islands and the evolution of microbes. *Annu. Rev. Microbiol.* **54,** 641–679.
4. Schmidt, H. and Hensel, M. (2004) Pathogenicity islands in bacterial pathogenesis. *Clin. Microbiol. Rev.* **17,** 14–56.
5. Dobrindt, U., Hochhut, B., Hentschel, U., and Hacker, J. (2004) Genomic islands in pathogenic and environmental microorganisms. *Nat. Rev. Microbiol.* **2,** 414–424.
6. Hensel, M. (2004) Evolution of pathogenicity islands of *Salmonella enterica. Int. J. Med. Microbiol.* **294,** 95–102.
7. Parkhill, J., Dougan, G., James, K. D., et al. (2001) Complete genome sequence of a multiple drug resistant *Salmonella enterica* serovar Typhi CT18. *Nature* **413,** 848–852.
8. McClelland, M., Sanderson, K. E., Spieth, J., et al. (2001) Complete genome sequence of *Salmonella enterica* serovar Typhimurium LT2. *Nature* **413,** 852–856.
9. McClelland, M., Sanderson, K. E., Clifton, S. W., et al. (2004) Comparison of genome degradation in Paratyphi A and Typhi, human-restricted serovars of *Salmonella enterica* that cause typhoid. *Nat. Genet.* **36,** 1268–1274.
10. Hansen-Wester, I. and Hensel, M. (2002) Genome-based identification of chromosomal regions specific for *Salmonella* spp. *Infect. Immun.* **70,** 2351–2360.
11. Hansen-Wester, I., Chakravortty, D., and Hensel, M. (2004) Functional Transfer of *Salmonella* Pathogenicity Island 2 to *Salmonellae bongori* and *Escherichia coli. Infect. Immun.* **72,** 2879–2888.
12. Yoon, S. H., Hur, C. G., Kang, H. Y., Kim, Y. H., Oh, T. K., and Kim, J. F. (2005) A computational approach for identifying pathogenicity islands in prokaryotic genomes. *BMC Bioinformatics* **6,** 184.
13. Hsiao, W., Wan, I., Jones, S. J., and Brinkman, F. S. (2003) IslandPath: aiding detection of genomic islands in prokaryotes. *Bioinformatics* **19,** 418–420.
14. Datsenko, K. A. and Wanner, B. L. (2000) One-step inactivation of chromosomal genes in *Escherichia coli* K-12 using PCR products. *Proc. Natl. Acad. Sci. U. S. A.* **97,** 6640–6645.
15. Ausubel, F. M., Brent, R., Kingston, R. E., et al. (1987) *Current Protocols in Molecular Biology*. Wiley, New York.
16. Kim, U. J., Birren, B. W., Slepak, T., et al. (1996) Construction and characterization of a human bacterial artificial chromosome library. *Genomics* **34,** 213–218.

6

Determination of the Gene Content of *Salmonella* Genomes by Microarray Analysis

Steffen Porwollik and Michael McClelland

Summary

Microarray technology provides a convenient and relatively inexpensive way of investigating the genetic content of bacterial genomes by comparative genomic hybridization. In this method, genomic DNA of an unknown bacterial strain of interest and that of a closely related sequenced isolate are hybridized to the same array. Hybridization signals are subsequently translated into gene absence and presence predictions for the experimental strain. Our nonredundant microarray of PCR products representing almost all genes from a number of the sequenced *Salmonella enterica* serovars (including Typhimurium, Typhi, Paratyphi A, and Enteritidis) allows accurate predictions of gene presence and absence in hundreds of *Salmonella* isolates on whole genome scale, for a fraction of the cost of complete genome sequencing, or resequencing using tiled oligo-arrays.

Key Words: Comparative genomic hybridization; labeling; genome variability; genome evolution; microarray.

1. Introduction

The ongoing international sequencing efforts have revealed the base composition of complete genomes of many bacterial isolates in several hundred species (http://wit.integratedgenomics.com/ERGO_supplement/genomes.html). These advances permitted the development of comparative genomic hybridization on DNA microarrays, a method that reveals differences in sequence content at single-gene, or higher, resolution. The arrays are a collection of DNA probes, generated either by PCR amplification or by oligonucleotide synthesis, that

From: *Methods in Molecular Biology, vol. 394: Salmonella: Methods and Protocols*
Edited by: H. Schatten and A. Eisenstark © Humana Press Inc., Totowa, NJ

have been immobilized on a solid support (usually a glass slide). The technique involves the use of a laser scanner that is capable of scanning glass slides at 10 μm resolution (or higher) and can produce excitation wavelengths of around 543 and 633 nm, with emission filters of around 570 and 670 nm (for the most common fluorophores, Cy5 and Cy3). It also requires an appropriate software program that can calculate intensities of these signals. The signal strengths obtained from a comparative hybridization on the array can then be transformed into reliable gene absence and presence predictions for the strain of interest by simple manipulations in an Excel file. In addition, gene amplifications can be easily detected. However, certain sequences in the genome under investigation will not be identified. These undetected features include genes that are not represented on the array and small deletions below the level of resolution of the array probes.

The technology has been used extensively for many bacterial genera (reviewed in **refs. *1–3***). For *Salmonella*, whole-genome array platforms have been developed based on at least three different sequenced strains ***(4–6)***. We have generated a nonredundant PCR product array that currently covers 99 % of all genes in the genomes of the *Salmonella* serovars Typhimurium LT2, Typhimurium SL1344, Typhi CT18, Typhi Ty2, Paratyphi A SARB42, and Enteritidis PT4. This array represents each gene in a separate spot, deposited in 50 % DMSO onto the amino silane-modified surface of bar-coded Corning Ultra-GAPS glass slides (cat. no. 40015, Corning). Each glass slide contains triplicate identical arrays, facilitating robust downstream statistical analysis, if desired. Pathogenic properties and host range vary significantly between different strains within the genus *Salmonella*. Despite this, there is very high homology on the sequence level between all *Salmonella* genomes. This is essential for successful data analysis—the observed close genetic relatedness between all strains allows the generation of comparable hybridization signals from any *Salmonella* strain that contains a close homolog of a sequence on the array. A significant loss of signal is usually observed if the sequence is 3 % or more divergent for the best 100 bp match, or 90 % divergent over the entire open reading frame (ORF) sequence ***(5)***.

2. Materials

2.1. Bacterial Growth and DNA Isolation

1. GenElute Bacterial Genomic DNA Kit (cat. no. NA2110, Sigma-Aldrich), or similar.
2. Luria-Bertani (LB) broth (BD Diagnostic Systems, Fisher cat. no. DF0446-07-5), or similar growth medium.

2.2. Fluorescent Tagging of the Nucleic Acid

1. *2.5×* Random primer/reaction buffer mix: 125 mM Tris–HCl (pH 6.8), 12.5 mM $MgCl_2$, 25 mM β-mercaptoethanol, and 0.6 μg/μL random hexamers (Sigma); stock solution of the random hexamers can range between 1 and 2 μg/μL; the buffer has to be made fresh every time just prior to the labeling reaction.
2. *10×* dNTP mix for labeled dCTP: 1.2 mM each dATP, dGTP, and dTTP, 0.6 mM dCTP, 10 mM Tris–HCl (pH 8.0), and 1 mM EDTA; can be stored for several weeks at −20 °C.
3. *10×* dNTP mix for labeled dUTP: 1.2 mM each dATP, dCTP, and dGTP, 0.6 mM dTTP, 10 mM Tris–HCl (pH 8.0), and 1 mM EDTA; can be stored for several weeks at −20 °C.
4. Cy3-dCTP or Cy3-dUTP (cat. no. PA53031/53032, GE Healthcare).
5. Cy5-dCTP or Cy5-dUTP (cat. no. PA55031/55032, GE Healthcare).
6. Klenow enzyme 5 U/μL (cat. no.M0210L, New England Biolabs).
7. 0.5 M EDTA (pH 8.0).
8. PCR product purification kit (cat. no. 28106, Qiagen).
9. Speed Vac (Savant, SC110), table-top centrifuge, heating block.

2.3. Hybridization to the Microarray

1. Vacuum desiccator (cat. no. 5312, Nalge/Sybron Corp.).
2. Drierite (cat. no. 22891–040, VWR).
3. Grease.
4. Opaque slide storage boxes.
5. Forceps.

2.3.1. Array Preparation

1. Prehyb solution: 25 % formamide, 5× SSC, 0.1 % SDS, and 0.1 mg/mL BSA; can be stored for several weeks at 4 °C, and reused twice.
2. Rinse solution: 0.1 % SDS.
3. Pressurized air containers (Office Depot).
4. Glass dish (500 mL).
5. Glass slide holders (cat. no. 900200, Wheaton Alcon).
6. Optional: Opaque slide containers for centrifugation (cat. no. WLS58834-A, VWR).
7. Rotating hybridization oven (Stratagene Hybridizer 600 or similar) set to 42 °C.
8. Table-top centrifuge (Sorvall RT6000B or similar), capable of spinning slides (either in containers or in holders).
9. 50-mL Falcon tubes.

2.3.2. Hybridization

1. Corning Hybridization Chamber (cat. no. 2551).
2. LifterSlip cover slips with 0.75 mm bar width (cat. no. 25x601-2-4789, Erie Scientific).

3. 2× Hyb solution: 50 % formamide, 10× SSC, and 0.2 % SDS; can be stored for several weeks at 37 °C (precipitates at room temperature) (*see* **Note 1**).
4. Water bath set to 42 °C, horizontal shaker.

2.3.3. Posthybridization Treatment

1. Wash solution I: 2× SSC and 0.1 % SDS.
2. Wash solution II: 0.1× SSC and 0.1 % SDS.
3. Wash solution III: 0.1× SSC.
4. 96 % Ethanol.

2.4. Data Acquisition

1. Laser-powered slide scanner with appropriate resolution (down to at least 10 μm) and capture spectrum (for the most common fluorescent dyes, Cy3 and Cy5): for example, ScanArray (Perkin Elmer), GenePix (Axon Instruments, Inc.), Microarray Scanner from Agilent Technologies, Alpha Scan (Alpha Innotech), or similar.
2. Acquisition software (usually comes with purchase of a scanner): for example, ScanArray Express (Perkin Elmer), Feature Extraction (Agilent Technologies), GenePix Pro (Axon Instruments, Inc.), ArrayEase (Alpha Innotech), and VersArray Analyzer (Bio-Rad).

2.5. Data Analysis

1. Microsoft Excel suite.
2. Optional: statistical analysis packages: for example, the freely available packages of WebArray *(7)*, significance analysis of microarrays (SAM) *(8)*, and variance-modeled posterior inference with regional exponentials (VAMPIRE) *(9)*.

3. Methods

3.1. Preparation of Genomic DNA From Bacterial Cells

1. Grow cells in LB in a shaker overnight at 30 or 37 °C.
2. Pellet 1.5 mL of an overnight bacterial culture (grown in LB) by centrifugation for 2 min at $12,000-16,000 \times g$ (14,000 rpm) (*see* **Note 2**). Remove the culture medium completely by pipetting, and discard. If a higher yield of genomic DNA is desired, add another 1 mL of culture to the same microfuge tube, spin 2 min as before and remove the supernatant.
3. Prepare genomic DNA using the Sigma-Aldrich GenElute Bacterial Genomic DNA Kit. Follow the manufacturer's protocol closely.
4. Elute with 100 μL 1 mM Tris–HCl (pH 8.0). Repeat the elution step.

3.2. Labeling of Bacterial Genomic DNA (See Note 3)

Fluorophores are light-sensitive and susceptible to photobleaching. Therefore, after the addition of fluorophores to the reaction tubes, these tubes have to be shielded from light whenever possible. Wrap aluminum foil around the tubes.

1. Add 1.5 μg genomic DNA of the sample to be labeled to a microfuge tube (*see* **Note 4**). If the volume exceeds 21 μL, use a speed vac to evaporate the excessive water by rotating at medium drying rate for 10–30 min, depending on the starting volume.
2. Add sterile water to bring the total volume to 21 μL. Then, add 20 μL of freshly prepared 2.5× random primer/reaction buffer. Incubate at >96 °C for 5 min, then place on ice. Spin down briefly, and replace on ice.
3. On ice, add 5 μL of the appropriate 10× dNTP mix.
4. Add 2 μL of the appropriate labeled nucleotide (Cy5-dCTP or Cy5-dUTP, Cy3-dCTP or Cy3-dUTP, 1 mM stocks).
5. Add 2 μL Klenow enzyme (5 U/μL). High concentration Klenow enzyme (50 U/μL), available from New England Biolabs (cat. no. M0210M), may produce better labeling efficiency.
6. Incubate at 37 °C overnight, in darkness.
7. Stop the labeling reaction by adding 5 μL 0.5 M EDTA (pH 8.0).
8. Purify the probes using the Qiagen PCR product purification kit, following the standard procedure of the manufacturer. The PE washing step should be repeated to ensure high purity of the labeled eluate. Elution in 30 μL 1 mM Tris–HCl (pH 8.0), or sterile water, will give good results (*see* **Note 5**). The naked eye should see color (pink shade for Cy3 and blue shade for Cy5).
9. Minimize volumes of probes to 20 μL in the speed vac by incubation for 5 min (or more, as desired) at medium drying rate. If the probe volume reduces to less than 20 μL, or dries out completely, add sterile water up to 20 μL.
10. Probes can now be used immediately, or stored at −20 °C protected from light for at least several months.
11. Combine the labeled reference probe (genomic DNA from the reference strain), with the labeled probe generated from the unknown strain (labeled with a different dye than the reference probe).

3.3. Hybridization to a Salmonella*-Specific Microarray*

1. Arrays have to be stored under vacuum, desiccated and kept in the dark at all times, until use. Storage can routinely be done in a well-greased desiccator that had been subjected to vacuum for at least 5 min and contains Drierite to ensure low humidity. Replace the Drierite when its color has changed from blue (dry state) to purple/pink (wet state) (*see* **Note 6**).
2. The presented hybridization protocol is adapted from http://www.corning.com/Lifesciences/technical_information/techDocs/gaps_ii_manual_protocol_5_02_cls_gaps_005.pdf.

3. Unless indicated otherwise, all washing and incubation steps are performed at room temperature in 500-mL glass dishes on a horizontal shaker (*see* **Note 7**). Make sure the solution always completely covers the slides when shaking.
4. Minimize carry-over from one wash container to the other: transfer only the slides, and not the slide holder, using forceps on the slide edges.
5. Never touch the surface of the arrays with hands. Use forceps whenever possible, on the slide edges. If needed, use gloves and transfer slides by hand, touching only the slide edges.

3.3.1. Prehybridization

1. Warm up the prehyb solution to 42 °C.
2. Recover an array slide from the vacuum-sealed desiccator storage, and wash the slide in rinse solution for 2 min.
3. Wash array slide in deionized water (resistance at least 18 MΩ-cm) for 2 min.
4. Repeat step 3.
5. Incubate arrays in prehyb solution for 45 min at 42 °C (rotate or shake). This step can be routinely performed in a 50-mL Falcon tube inside a rotating hybridization oven. Two array slides can be processed simultaneously in one Falcon tube, by placing them back-to-back into the tube.
6. Wash arrays by immersing, and shaking, in water for 1–2 min.
7. Repeat step 6. It is important that the SDS is completely removed from the arrays.
8. Place arrays in slide holders.
9. Dry arrays by centrifugation in a Sorvall RT6000B (or similar) table-top centrifuge for 5 min at low speed ($100 \times g$) at room temperature.
10. If dust speckles adhere to the slide surface after the cleaning procedure, carefully subject the slide surface to pressurized air. Do not use excessive air because a strong airflow may damage the slide surface structure.

3.3.2. Hybridization

Labeled probes should be kept in the dark throughout the entire procedure to minimize photobleaching of the fluorescent dye: use aluminum foil around the respective tubes, and do the washes in an unlit room. Once probes have been applied to the arrays, the arrays have to be light-protected at all times also. Because our array versions have labeled control spots, we protect the arrays from light at all times, even before the hybridization experiment.

1. Add 40 μL of labeled probe (mixture of labeled genomic DNAs from both reference and experimental strains, prepared in **Subheading 3.2.**) to 40 μL of 2× hyb buffer.
2. Incubate the probe solution at >96 °C for 5 min.
3. Centrifuge the probe for 30 s to collect condensation and let sample cool to room temperature.

4. Place array right side up into the Corning Hybridization Chamber. Pipette the probe onto the surface of the printed side of the slide. Carefully place the LifterSlip on top of the array. Lower the LifterSlip slowly to avoid the formation of air bubbles under the coverslip. Small air bubbles that may form usually dissipate during the overnight hybridization. It is important to use LifterSlips (i.e., cover slips that have thicker edges to allow some space between the slide and the slip surface) as opposed to the usual cover slips, to enable more uniform distribution of the hybridization solution across the surface of the array.
5. Pipette 20 μL water into both reservoir holes in the hybridization chamber (to prevent the probe solution from drying during hybridization). Assemble the chamber as described by the manufacturer.
6. Submerge the chamber in a 42 °C water bath or place in a hybridization oven set to 42 °C overnight. Make sure the chamber is shielded from light.

3.3.3. Posthybridization Washing

- Do not wash slides that have been hybridized with different probes in the same wash containers.
- Never allow slides to dry out between the washes.
- Perform all washing steps in the dark.

1. Warm up the wash solution I to 42 °C.
2. Disassemble the hybridization chamber right side up.
3. Remove the cover slip by immersing the array upside down in wash solution I (at 42 °C) until the slip moves freely away from the slide.
4. Place array in wash solution I for 5 min at 42 °C. This step can be performed in a 50-mL Falcon tube, inside a rotating hybridization oven.
5. Place array in wash solution II for 10 min at room temperature.
6. Place array in wash solution III for 1 min at room temperature.
7. Repeat step 6 four times, using fresh solution each time.
8. Place array in deionized water (18 MΩ-cm) for up to 10 s or less.
9. Place array in 96 % ethanol for 5 s.
10. Put arrays into a slide holder.
11. Dry arrays by centrifugation for 5 min at low speed (100×g).

3.4. Scanning and Data Acquisition

1. Cautiously use pressurized air to clean the slide from any particles that may have adhered after the final wash.
2. Insert the slide into the scanner.
3. Scan the slide surface at low resolution (50 μm) at wavelengths appropriate to the fluorescent dye used in the labeling reactions. Most scanners have preset wavelengths for the most common dyes, Cy3 and Cy5. If these two dyes are used, scan the more photosensitive dye Cy5 prior to the more stable Cy3.

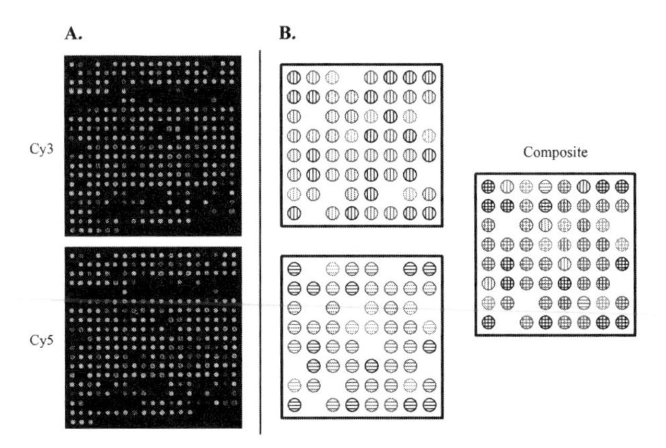

Fig. 1. Principle of arrays hybridized with Cy3-labeled and Cy5-labeled genomic DNA preparations. (**A**) Actual ScanArray image of the first of the 16 subarrays that make up the array. The depicted experiment is a comparative genomic hybridization of *Salmonella* enterica isolates Pullorum 99113 versus Enteritidis PT4. (**B**) Schematic representation of an image overlay. ⦶, Spot containing DNA present in strain 1 (labeled with Cy3); ⊜, Spot containing DNA present in strain 2 (labeled with Cy5); ⊕, Spot present in both strains (in composite image only). Signal strength is represented on a gray scale, with black being the strongest.

4. Evaluate the low-resolution scan. Adjust the laser power and (if present) photomultiplier (PMT) settings to the desired intensity: use settings that detect a maximum number of spots without excessive amounts of saturated signals. We usually leave the laser power constant, and change the PMT values to obtain the needed signal intensities.
5. Optional: after changing the settings, rescan at 50 μm resolution and re-evaluate. This process can be done several times until intensities are satisfactory. However, the Cy5 dye generally bleaches out during scans. Therefore, try to minimize the amount of low-resolution scans.
6. Once the settings are satisfactory, scan at 10 μm or higher resolution for the final images (*see* **Fig. 1**).

3.5. Data Analysis

3.5.1. Determination of Signal Intensities and Ratios

1. Use software of choice to calculate signal intensities for each spot in each channel, according to the manufacturer's instructions.
2. Export the measurements into an Excel spreadsheet.
3. Perform the following Excel calculations, using simple formulas:

a. If data acquisition software does not compute background corrected signals, calculate spot signal minus background signal around each spot.
b. Calculate sum of all (background corrected) spot signals in each channel.
c. Calculate percent contributions of each spot to total signal in each channel (i.e., background corrected spot signal × 100/background corrected total signal).
d. Calculate ratios of spot contributions of the experimental channel over the control channel for each spot (i.e., percent contribution of spot in $channel_{exp}$/percent contribution of spot in $channel_{ctrl}$).
e. Calculate median of the ratios for identical probes (median of three data points, because arrays are spotted in triplicates onto the glass slide) (*see* **Note 8**).

3.5.2. Graphical Representation

1. Sort spots according to their location on the reference genome (if this information is available).
2. Plot calculated median of ratios on a logarithmic scale graph, using the Excel chart function (*see* **Fig. 2**). Amplifications of genomic areas can be detected by eye as elevated ratios over several adjacent probes. Deletions can be spotted by sharp drops in signal ratios. The absence and presence of genes can be calculated in Excel as described in **Subheading 3.5.3**.

3.5.3. Calculation of Absence/Presence of Genes in the Strain of Interest

3.5.3.1. Spots Representing Elements Present on the Genome of the Control Strain

1. Discard spots with low signals in control sample: assign the lowest 5 % of median signals, and mark as "uncertain." This can be achieved by sorting all spots that are present in the control strain by their signal intensity and then replace the measured signal of the lowest 5 % of spots with a generic comment, like "uncertain."
2. Discard spots with high background values: identify the spots that exhibit background values higher than 20 % of the median (background corrected) spot signal in either of the two channels, and mark as "uncertain." This can be calculated by a series of formulas in Excel:

   a. Calculate the median of all spot signals in each channel (M_{ch1} and M_{ch2}).
   b. Calculate 20 % of these values ($M_{20,ch1}$ and $M_{20,ch2}$).
   c. Mark all spots that display ch1 background values higher than $M_{20,ch1}$ and/or ch2 background values higher than $M_{20,ch2}$.

3. Sort the data points according to their location on the genome of the control strain.
4. Set threshold parameters for absence/presence predictions: spots that display median ratios of >0.67 are called "present," spots that exhibit ratios <0.33 are called

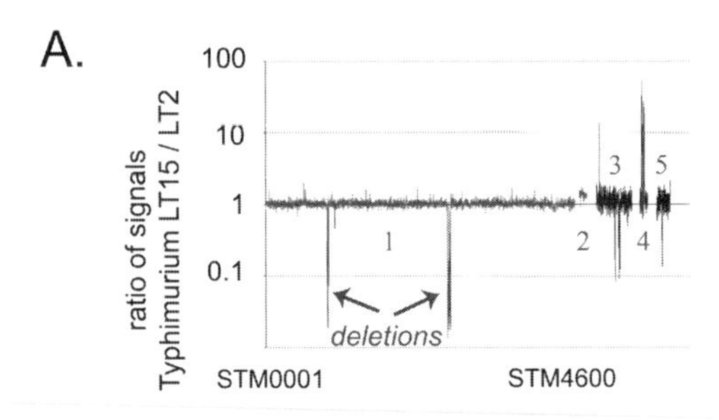

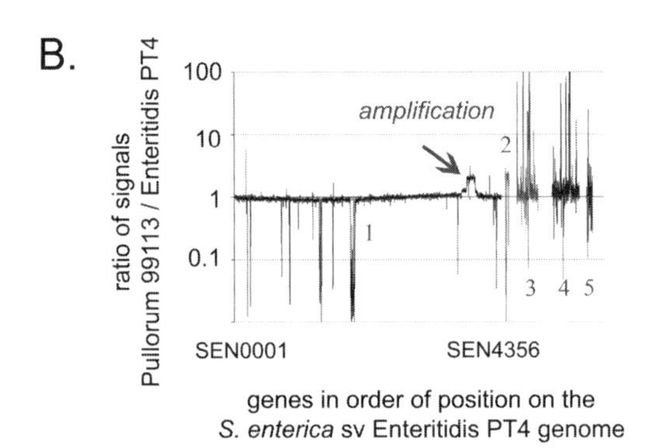

Fig. 2. Graphical representation of normalized signal ratios between experimental and control isolates for all probes present on the nonredundant Salmonella array. (**A**) Typhimurium LT15 versus Typhimurium LT2. 1, LT2 chromosomal genes, probes are sorted according to their locus on the LT2 genome, starting with gene STM0001; 2, plasmid pSLT genes; 3, genes that are present in Typhi CT18, but not in LT2; 4, genes present in Paratyphi A SARB42, but not in LT2 or CT18; 5, genes present in Enteritidis PT4, but not in LT2, CT18, or SARB42. Drops in the ratio indicate LT2 regions that are deleted in LT15. Ratio spikes in probes representing genes in isolates other than LT2 indicate possible presence in the experimental strain. (**B**) Pullorum 99113 versus Enteritidis PT4. 1, PT4 chromosomal genes, probes are in order of position on the PT4 genome, starting with gene SEN0001; 2, plasmid pSLT genes; 3, genes present in LT2, but not in PT4; 4, genes present in CT18, but not in PT4 or LT2; 5, genes present in SARB42, but not in PT4, LT2, or CT18. An area of amplification in the experimental isolate 99113 is marked by elevated ratios.

"absent," and spots that fall between these thresholds, or have been flagged in step 1 or 2, are called "uncertain" (*see* **Note 9**).

5. The different categories (present, absent, and uncertain) can be color-coded in Excel and imported into a graphical program such as Adobe Photoshop to generate outputs as in **Fig. 3**.

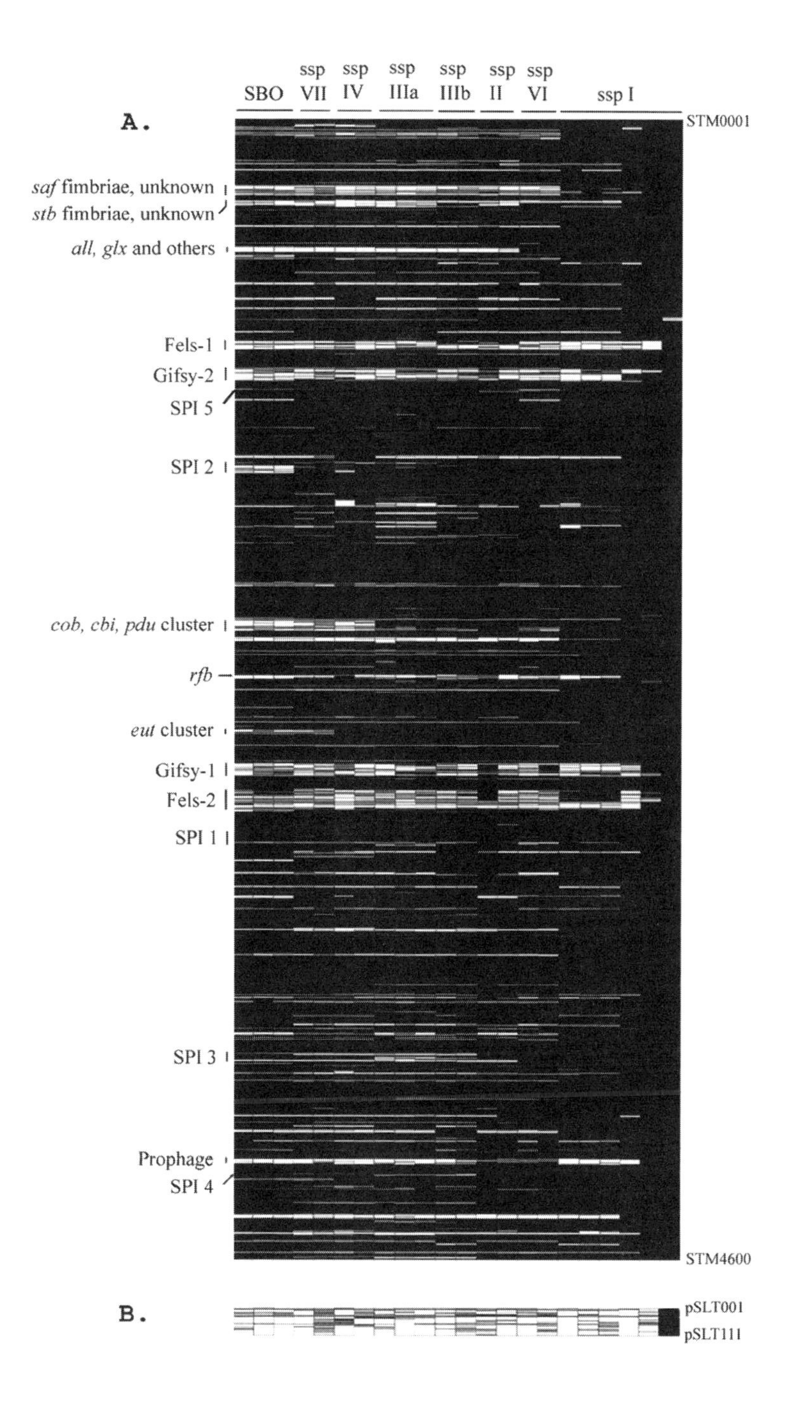

ssp ssp ssp ssp ssp ssp
SBO VII IV IIIa IIIb II VI ssp I
A.
STM0001
saf fimbriae, unknown
stb fimbriae, unknown
all, glx and others
Fels-1
Gifsy-2
SPI 5
SPI 2
cob, cbi, pdu cluster
rfb
eut cluster
Gifsy-1
Fels-2
SPI 1
SPI 3
Prophage
SPI 4
STM4600
B.
pSLT001
pSLT111

3.5.3.2. Spots Representing Plasmid Elements, and/or Genes Absent from the Control Strain (***See* Note 10**)

Predictions are purely based on the signal strength of the spot and therefore less reliable. In genomic hybridizations where the control strain is expected to hybridize to x% of all spots, elements that exhibit a signal that is among the top $(x - 15)$% of all spots on the array can generally be assumed to be present. Those that are among the lowest $(100 - x - 5)$% of all spots may be assumed to be absent.

3.5.4. Data Deposition

Routinely, microarray data should be deposited into a MIAME-compliant database ***(11,12)***. We deposit our raw data (outputs from the ScanArray Express software) into the Gene Expression Omnibus repository of the National Center for Biotechnology Information (http://www.ncbi.nlm.nih.gov/geo/), in the requested SOFT format specified on the website.

4. Notes

1. Addition of 30–50 μg human Cot-1 DNA (Gibco/BRL; 1 mg/mL stock; blocks hybridization to repetitive DNAs if present on array), 100 μg yeast tRNA (Gibco/BRL; make a 5 mg/mL stock; blocks nonspecific DNA hybridization), and 20 μg poly(dA)-poly(dT) (cat. no. P9764, Sigma; make a 5 mg/ml stock; blocks hybridization to poly A tails of cDNA array elements) has been recommended for human arrays (http://cmgm.stanford.edu/pbrown/protocols/4_genomic.html). However, in our routine *Salmonella* applications, these additions are unnecessary.
2. Stationary-phase bacterial culture is needed to ensure equal representation of the entire genome. Cultures in logarithmic growth phase will have an overrepresentation of the genomic region around the replication origin compared with the area around the replication terminus ***(5)***.
3. The DNA labeling technique presented uses Klenow enzyme, random hexamers, and two Cy-labeled nucleotides in a one-step labeling procedure. However, other labeling techniques have been used successfully, including a two-step method

Fig. 3. Presence and absence of LT2 protein coding sequence homologs in 22 strains representative of all Salmonella enterica subspecies and S. bongori (SBO). Strains are sorted from left to right with ascending relatedness to LT2. The gene status is gray scale-coded: black, present; gray, uncertain; white, absent. Some prominent regions are indicated. (**A**) The genes on the LT2 chromosome in order of position from STM0001 to STM4600. (**B**) The genes of the LT2 virulence plasmid pSLT. Figure modified from **ref. 10.**

where amine-modified nucleotides (aminoallyl-dUTP) are incorporated by DNA polymerase I in step 1, which are subsequently tagged by a chemically reactive fluorophore. (This strategy is employed, for example, in the ARES DNA Labeling Kit from Invitrogen.) Dyes that are being used in that protocol include Alexa Fluor 594, which has been shown to be a viable candidate to facilitate three-color experiments in conjunction with Cy3 and Cy5 ***(13)***. We have had promising results using a one-step labeling procedure with Chromatide Alexa Fluor 488-dUTP (cat. no. C11397, Invitrogen) as a third dye. However, low Alexa488 signal intensities sometimes occur that currently prevent this dye from being included in a standard protocol.

4. For high-complexity DNAs (e.g., human genomic DNA), the labeling reaction works more efficiently if the fragment size of the DNA is first reduced, which can be routinely accomplished by restriction enzyme digestion (usually *Dpn*II, though other four-cutters work as well). After digestion, the DNA should be cleaned up by phenol/chloroform extraction/ethanol precipitation, or by the Qiagen PCR purification kit. Alternatively, DNA may be sheared through a 27-ga needle several times before subjecting it to the labeling reaction. However, DNA fragment size reduction is not necessary for standard bacterial genomes.
5. Incorporation rates of the fluorophores can be estimated by UV spectrophotometry. Measurements of the absorbance at 260, 280, 550, and 650 nm can be converted into incorporation rates (FOI per 1000 nt) by the following calculations:
   Cy3: $OD_{550} \times volume(\mu L)/0.15 = M_{cy3}$ (pmol)
   Cy5: $OD_{650} \times volume(\mu L)/0.25 = M_{cy5}$ (pmol)
   Both: $M \times 324.5/mDNA(ng) = FOI$
   Generally, 1.5 μg of genomic DNA with an FOI > 6 in either dye will give sufficient results in subsequent hybridizations. Spectrophotometric measurements have to be performed in cuvettes that have been thoroughly cleaned with water and ethanol to avoid contamination of the probe.
6. The used wet Drierite can be dried and reused at least twice by incubating in an oven set to 70 °C overnight or longer. If Drierite becomes gray during the drying process, discard.
7. Coplin jars may be used instead of glass dishes. However, we recommend big jars that allow considerable volume and motion for the solutions to maximize washing efficiency.
8. Because the array is spotted in triplicate on each slide, we get three data points per probe in each hybridization experiment. With this relatively low number of data points, using the median is a better parameter than the average values because there may be defects on some spots (outliers), which may skew the average considerably, but have lower impact on medians. Other methods that measure the scatter of the data will simply remove the data from consideration when one spot is unacceptable, but this is almost always not necessary. We generally find one hybridization experiment adequate, but the user may wish to perform a second hybridization

with swapped dyes compared with the first hybridization, thereby recording six data points per probe. This would allow more robust statistics downstream as well, including more precise false discovery rate calculations and more accurate Bayesian calculations—methods that are often implemented in microarray statistical analysis tools.

9. A more accurate, but more complicated series of calculations can be applied as follows: For calculation of the presence baseline P, calculate the median of the ratios of all genes that display a ratio of >0.67 and which in addition are neighbored on the control strain genome by elements that also displayed ratios of >0.67. Calculate the standard deviation (SD_P) of these ratios. Similarly, medians and SDs for genes with ratios of <0.5 which were neighbored by elements with ratios of <0.5 should also be determined (absence baseline A and SD_A, respectively). Genes exhibiting ratios higher than the presence threshold ($P - 2SD_P$) should be scored as "present," whereas genes with ratios lower than the absence threshold ($A + 2SD_A$), should be scored as "absent." Genes that are outside of these thresholds and those that display ratios between 0.5 and 0.67 should be scored as "uncertain." If the hybridization was of high quality, that is, scans revealed crisp and clear signals in both channels with low background, the easy absence/presence predictions described in the general protocol will work accurately for >95 % of probes.
10. Absence predictions of elements not present on the genome of the control strain are much less reliable, because the spot quality is not verified by the control isolate, and a low signal may be the result of insufficient spot quality rather than the absence of a homologous gene sequence in the strain of interest.

Acknowledgments

This work was supported in part by NIH grant AI34829 (M.M.) and the generosity of Sidney Kimmel.

References

1. Ochman, H. and Santos, S. R. (2005) Exploring microbial microevolution with microarrays. *Infect. Genet. Evol.* **5,** 103–108.
2. Dorrell, N., Hinchliffe, S. J., and Wren, B. W. (2005) Comparative phylogenomics of pathogenic bacteria by microarray analysis. *Curr. Opin. Microbiol.* **8,** 620–626.
3. Schoolnik, G. K. (2002) Functional and comparative genomics of pathogenic bacteria. *Curr. Opin. Microbiol.* **5,** 20–26.
4. Chan, K., Baker, S., Kim, C. C., Detweiler, C. S., Dougan, G., and Falkow, S. (2003) Genomic comparison of *Salmonella enterica* serovars and *Salmonella bongori* by use of an *S. enterica* serovar Typhimurium DNA microarray. *J. Bacteriol.* **185,** 553–563.

5. Porwollik, S., Frye, J., Florea, L. D., Blackmer, F., and McClelland, M. (2003) A non-redundant microarray of genes for two related bacteria. *Nucleic Acids Res.* **31,** 1869–1876.
6. Thomson, N., Baker, S., Pickard, D., et al. (2004) The role of prophage-like elements in the diversity of *Salmonella enterica* serovars. *J. Mol. Biol.* **339,** 279–300.
7. Xia, X., McClelland, M., and Wang, Y. (2005) WebArray: an online platform for microarray data analysis. *BMC Bioinformatics* **6,** 306.
8. Tusher, V. G., Tibshirani, R., and Chu, G. (2001) Significance analysis of microarrays applied to the ionizing radiation response. *Proc. Natl. Acad.Sci. U. S. A.* **98,** 5116–5121.
9. Hsiao, A., Ideker, T., Olefsky, J. M., and Subramaniam, S. (2005) VAMPIRE microarray suite: a web-based platform for the interpretation of gene expression data. *Nucleic Acids Res.* **33,** W627–W632.
10. Porwollik, S., Wong, R. M., and McClelland, M. (2002) Evolutionary genomics of Salmonella: gene acquisitions revealed by microarray analysis. *Proc. Natl. Acad. Sci. U. S. A.* **99,** 8956–8961.
11. Ball, C. A., Brazma, A., Causton, H., et al. (2004) Submission of microarray data to public repositories. *PLoS Biol.* **2,** e317.
12. Brazma, A., Hingamp, P., Quackenbush, J., et al. (2001) Minimum information about a microarray experiment (MIAME)—toward standards for microarray data. *Nat. Genet.* **29,** 365–371.
13. Forster, T., Costa, Y., Roy, D., Cooke, H., and Maratou, K. (2004) Triple-target microarray experiments: a novel experimental strategy. *BMC Genomics* **5,** 13.

7

In Vivo Excision, Cloning, and Broad-Host-Range Transfer of Large Bacterial DNA Segments Using VEX-Capture

James W. Wilson and Cheryl A. Nickerson

Summary

The performance of many bacterial genetic experiments would benefit from a convenient method to clone large sets of genes (20–100+ kb) and transfer these genes to a wide range of other bacterial recipients. The VEX-Capture technique allows such large genomic segments to be cloned in vivo onto a broad-host-range IncP plasmid that is able to self-transfer to a wide variety of Gram-negative bacteria. The advantages of VEX-Capture are its efficiency, specificity, and use of common molecular biological techniques that do not require non-standard equipment and are easily applicable to many types of bacterial species. Here, we describe the VEX-Capture experimental protocol using *Salmonella typhimurium* as the source of the target DNA segment.

Key Words: *Salmonella typhimurium*; VEX-Capture; broad-host-range; IncP; R995; horizontal gene transfer; conjugation; bacterial genome; Cre-*lox*.

1. Introduction

The precise cloning of large (20–100+ kb) bacterial genomic segments onto a plasmid vector that can easily self-transfer to a broad range of host genera allows many molecular biological experiments to be performed more conveniently. An advantage to such a construct is that large numbers of genes that function together for a specific purpose can be cloned on a single DNA fragment. In addition, the cloned DNA fragment can be transferred to a wide array of different bacterial backgrounds, thus offering greater experimental

From: *Methods in Molecular Biology, vol. 394: Salmonella: Methods and Protocols*
Edited by: H. Schatten and A. Eisenstark © Humana Press Inc., Totowa, NJ

flexibility. The VEX-Capture system allows the researcher to perform such DNA manipulations (*see* **Figs. 1** and **2**) *(1)*.The VEX-Capture system has been used to successfully clone several large chromosomal sections from the *Salmonella typhimurium* genome, and this species is used as the target genome in the protocol described here *(1–3)*. However, the system should be easily applicable to any Gram-negative species that is able to express and utilize the R995 IncP transfer system and the IncQ plasmid replication system.

The VEX-Capture procedure takes place in three steps (*see* **Fig. 1**). First, differentially marked [spectinomycin- and chloramphenicol-resistant (Sp-r and Cm-r)] pVEX suicide vectors containing regions of homology to the target genomic sequence are integrated through homologous recombination at each end of the targeted *S. typhimurium* DNA section. This creates a structure termed the "double cointegrate" and serves to insert DNA sequences called *loxP* sites on both sides of the targeted section. A site-specific recombinase called Cre is able to excise any section of DNA between two *loxP* sites as a circular DNA piece. Second, a plasmid-encoded Cre recombinase is then expressed in the double cointegrate host, and the targeted DNA section is excised from the host genome as a non-replicating circle that carries the Sp-r marker of the pVEX-up plasmid. The excision is performed in the presence of another plasmid termed R995 VC, a self-transmissible, broad-host range IncP plasmid that contains homology to the excised, targeted DNA section. This allows the "capture" or cloning of the excised circle onto the R995 VC plasmid through homologous recombination. Third, in order to recover the newly formed R995 VC + excised circle construct, the plasmid is transferred to an *Escherichia coli* rifampicin-resistant (Rif-r) recipient host, which results in a cloned and isolated (or "captured") *S. typhimurium* DNA section. This transconjugant is selected using spectinomycin (excised circle), kanamycin (R995 VC), and rifampicin (*E. coli* recipient). The captured region can then be manipulated

Fig. 1. The VEX-Capture system. Excision and capture of a section of the *Salmonella typhimurium* genome is depicted to illustrate the functioning of the VEX-Capture system. In step 1, differentially-marked pVEX vectors containing DNA fragments homologous to the ends of the targeted genomic region are integrated at the desired locations to form a double cointegrate. In this structure, single *lox*P sites are located on either side of the targeted region. In step 2, the targeted region is excised from the genome by the Cre recombinase, and the excised circle is "captured" through homologous recombination with the R995 VC plasmid. Note that the capture fragment on R995 VC is shown as targeted to one end of the excised genomic region, but it can be targeted to any desired location on the excised region. In step 3, the R995 VC::excised

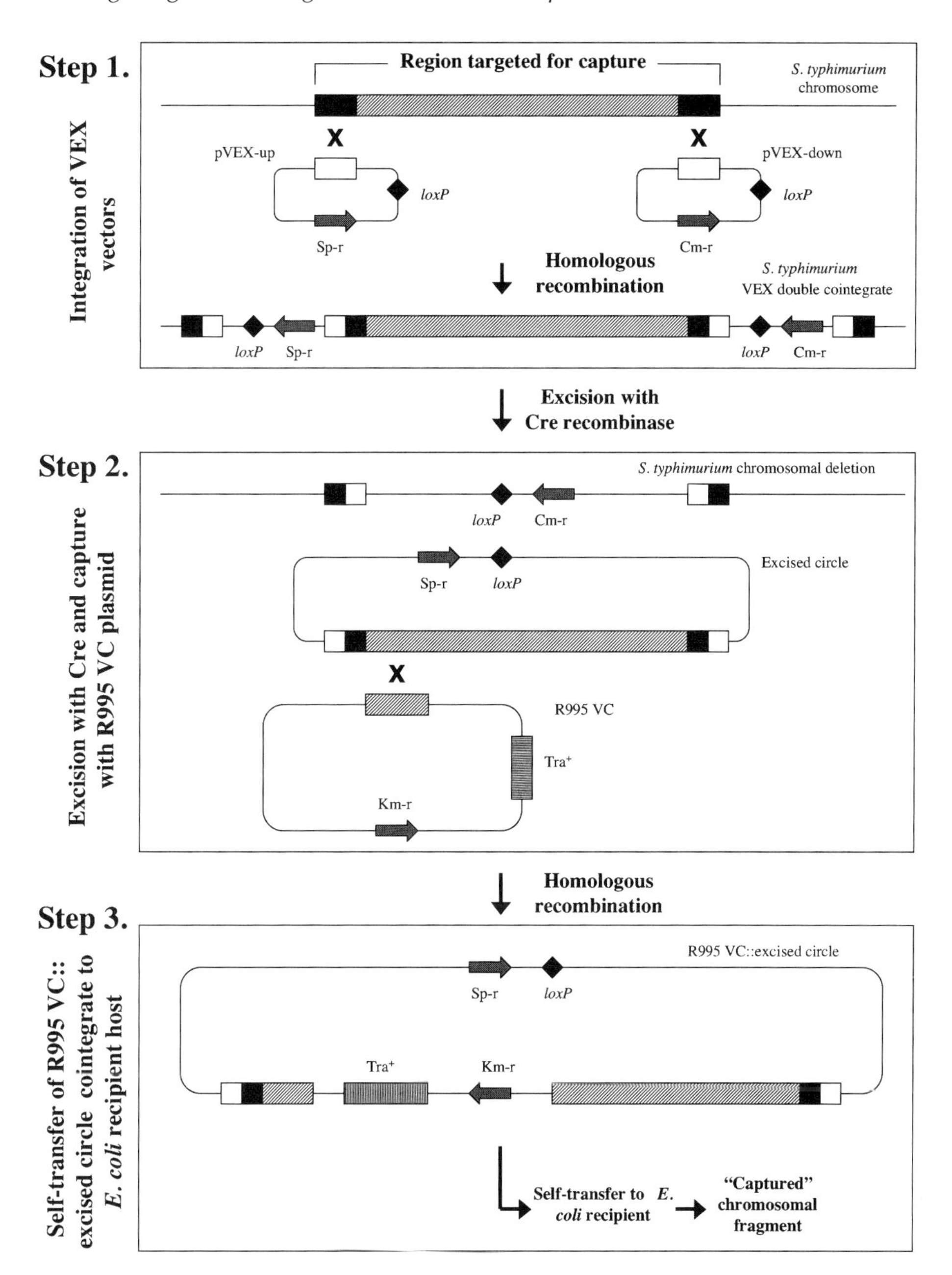

Fig. 1. circle plasmid is transferred to an *Escherichia coli* recipient to create a strain containing the captured genomic fragment. Please see the text in **Subheadings 3.2 and 3.3.** for further explanations of the individual steps. Diagram not drawn to scale.

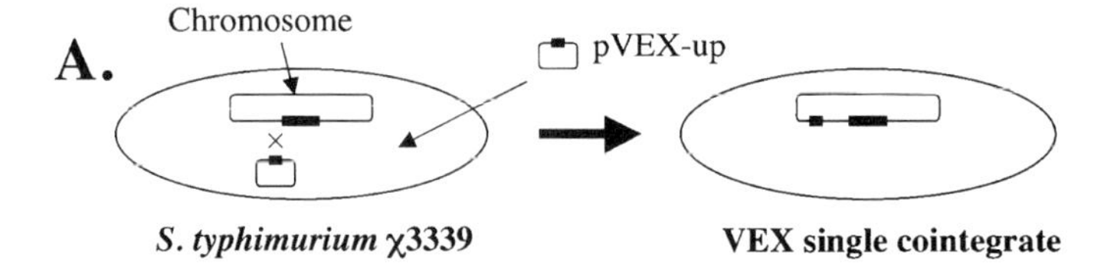

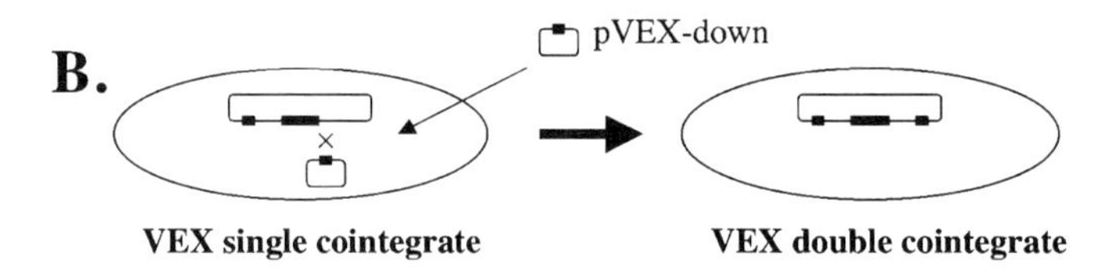

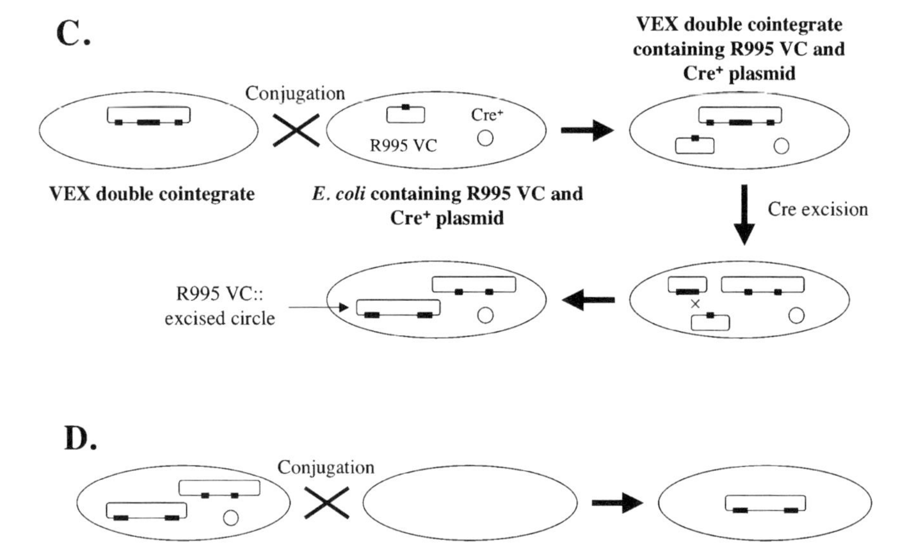

Fig. 2. Sequence of DNA transfer events between strains used in the VEX-Capture procedure. This figure describes the DNA transfer events of the VEX-Capture procedure in specific technical steps. Please see the text in **Subheadings 3.2. and 3.3.** for additional details. The large ovals represent bacterial cells and the rectangular and circular structures represent the DNA elements involved

and studied in *E. coli* or transferred to another host for subsequent studies. The resulting *S. typhimurium* strain that contains a genomic deletion of the targeted section can also be studied for the role of this section in cellular processes and responses.

Figure 2 describes the VEX-Capture procedure in specific technical steps. First, the single cointegrate structure in *S. typhimurium* is made with one of the pVEX plasmids (here, we show pVEX-up as the single cointegrate plasmid). Second, the double cointegrate is made by integration of the second pVEX plasmid. Third, the R995 VC plasmid and a mobilizable, broad-host-range IncQ plasmid constitutively expressing the Cre recombinase are transferred at the same time from an *E. coli* donor containing both plasmids to the *S. typhimurium* double cointegrate through conjugation. When this occurs, Cre is expressed in the double cointegrate, excision of the targeted region occurs in the presence of R995 VC, and recombination between the excised circle and the R995 VC plasmid takes place. Fourth, this conjugation mixture is then mixed with the *E. coli* Rif-r recipient, and transfer of the R995 VC + excised circle plasmid to this recipient can be selected using the antibiotics described in **Subheading 3.3**. The *E. coli* Rif-r strain containing the R995 VC + excised circle plasmid can then be used as a donor for transfer of this plasmid to other bacterial species.

All the bacterial strains and plasmids used to perform VEX-Capture, as well as answers to any additional questions involving this technique, can be obtained by contacting the authors using the provided contact information.

2. Materials

2.1. Construction of Vectors Used to Insert loxP Sites at Desired Genomic Locations

1. For PCR amplification: 10× PCR buffer (Invitrogen, Carlsbad, CA), 50 mM $MgCl_2$ (Invitrogen), 10 mM dNTPs (Invitrogen), DNA oligonucleotide primers (Integrated DNA Technologies), Taq polymerase (Invitrogen), and template DNA.

Fig. 2. in the procedure. Panels **A** and **B** depict the sequential electroporation of the pVEX-up and pVEX-down plasmids, respectively, into *Salmonella typhimurium* to form the VEX double cointegrate. Panel **C** depicts the use of conjugation to transfer R995 VC and the Cre-expressing plasmid to the *S. typhimurium* VEX double cointegrate strain and the subsequent excision and capture of the targeted genomic region. In panel **D**, the conjugation mixture in panel **C** is mixed with an *E. coli* Rif-r strain, and the R995 VC::excised circle plasmid is isolated through conjugation to this recipient strain.

2. For cloning of PCR fragment: agarose gel, Qiagen gel extraction kit (Qiagen, Valencia, CA), TOPO T/A cloning kit (Invitrogen), bacterial strain TOP10 (included with TOPO T/A cloning kit), and QIAprep Spin Mini Prep kit (Qiagen).
3. For subcloning of PCR fragment into pVEX vectors: specific restriction enzymes (Invitrogen), QIAquick gel extraction kit (Qiagen), 5× ligase buffer (Invitrogen), ligase (Invitrogen), spectinomycin (Sigma, St. Louis, MO), chloramphenicol (Sigma), kanamycin (Sigma), Luria–Bertani (LB) media (broth and solid) (Fisher Scientific, http://www.fishersci.com), agar (Difco), bacterial strains EKA260 and AS11 ***(1,4)***, and plasmids pVEX1212 (for pVEX-up, Sp-r) and pVEX2212 (for pVEX-down, Cm-r) ***(1,4)***.

2.2. Chromosomal Integration of loxP Sites

1. Qiagen Maxi DNA prep kit (Qiagen), 0.025-μm dialyzing filters (Millipore, Billerica, MA), TE buffer.
2. Electrocompetent target host cells (*S. typhimurium* in this example), glycerol (Fisher).
3. LB media as in **Subheading 2.2.**, step 3 with spectinomycin and chloramphenicol.

2.3. Capture of Excised Target DNA

1. Bacterial plasmids R995 *Xba*I and IncQ-Cre (pEKA30).
2. The following bacterial strains are required: target host strain containing double cointegrate (*S. typhimurium* double cointegrate, in this case), *E. coli* TOP10 containing plasmids R995 VC and IncQ-Cre, *E. coli* TOP10 containing plasmids R995 XbaI and IncQ-Cre (this will be a negative control strain), and *E. coli* TOP10 Rif.
3. LB media with indicated antibiotics.
4. Qiagen Mini DNA prep kit (Qiagen).

3. Methods

3.1. Construction of Vectors Used to Insert loxP Sites at Desired Genomic Locations

1. The procedure described here uses homologous recombination to insert *loxP* sites at locations flanking the targeted DNA region. Fragments of DNA homology are PCR amplified and cloned into the pVEX-up (pVEX1212) and pVEX-down (pVEX2212) *loxP*-containing suicide vectors. We have found that relatively large regions of homology (3–5 kb) work best for this objective. Design oligonucleotides to amplify the "up" and "down" regions of homology from a chromosomal DNA template. Include restriction sites in the primers for subsequent cloning of the PCR product into the polylinkers of the pVEX vectors. It is best to use different enzymes for each side of the PCR product, and it is important to make sure the enzymes you choose do

not cut the pVEX vectors. The maps of the pVEX plasmids and the sequence of the polylinker used in these plasmids are previously published ***(4,5)***.

2. Set up the PCR with the following materials: 3 μL chromosomal DNA template (about 50–250 ng), 10 μL 10× PCR buffer, 3 μL 50 mM $MgCl_2$, 2 μL 10 mM dNTPs, 1 μL primer 5′, 1 μL primer 3′, 75 μL dH_2O, and 5 μL Taq polymerase (2–5 units) (add Taq last after hot start; mix 1 μL Taq and 4 μL dH_2O).
   It is a good idea to set up three to four tubes of this reaction to maximize your yield of PCR product. After all reagents have been mixed except Taq, place the tubes in the thermocycler and heat to 94 °C for 5 min.
3. Add Taq polymerase and mix by pipetting up and down. It is helpful to dilute 1 μL Taq in 4 μL dH_2O so that a 5 μL volume is added to each PCR.
4. Let the PCR run with the following parameters: 1 min at 94 °C, 1 min at 65 °C (or appropriate annealing temperature), 1 min at 72 °C, cycle 31 times, 2 min at 72 °C, and hold at 4 °C.
5. Run the entire amounts of all the reactions in multiple lanes of a 1.5 % agarose gel in 1× Tris-Acetate-EDTA(TAE) buffer. Cut out the desired bands with a razor blade and distribute into microfuge tubes so that about 300 mg gel is in each tube.
6. Using a Qiagen gel extraction kit, extract the DNA from the agarose gel slices. Sequentially combine all the samples after the gel slices have solubilized to a single spin column. Elute the DNA in 35–50 μL elution buffer.
7. Check the yield of this sample using A260 reading or gel electrophoresis.
8. Ligate the PCR product to the pTOPO vector: 4 μL PCR product, 1 μL salt solution, and 1 μL pCR-TOPO vector. Let reaction incubate for 10–30 min at room temperature.
9. Dialyze the ligation reaction in Petri dish against 20mm Tris buffer, pH 7.5. Spot the ligation on a 0.025-μm filter that has been laid on the surface of the Tris buffer. Let this incubate for 15–30 min (or longer if desired). This step serves to remove salt ions that are present in the reaction that would cause the electroporation to "pop" or "arc."
10. Add 3 μL of the dialyzed ligation reaction to electrocompetent *E. coli* TOP10 (part of the TOPO vector kit). Electroporate the *E. coli*/ligation sample at 2.5 kV, 200 Ohms, and 25 μF.
11. Outgrow the transformed cells for 1 h at 37 °C with shaking. Plate the cells on LB agar plates containing 50 μg/mL kanamycin.
12. After 18–24 h of growth, pick isolated colonies to screen for the presence of PCR product insert. Inoculate 3–4 mL LB culture containing kanamycin selection, grow overnight, and harvest these cells for plasmid DNA preparation using a mini-prep plasmid kit (we have consistently used the QIAprep Spin Kit). Although there are a number of ways to screen for the insert, we recommend making DNA preparations for two reasons: (1) it allows visualization of the plasmid construct through gel electrophoresis to confirm its proper structure, and (2) it provides a source of

DNA for subsequent cloning steps. Confirm the presence of the PCR product using restriction enzyme digestion, PCR analysis, and/or DNA sequencing.

13. To subclone the PCR fragment into the corresponding pVEX vector, digest both the appropriate pVEX vector and the pCR-TOPO + PCR fragment plasmid with the appropriate restriction enzymes for ligation of the PCR product into the pVEX vector. Gel isolate the PCR fragment band from an agarose gel and check the yield as described in step 7. It is also a good idea to gel isolate the digested pVEX vector as well to help increase the likelihood of obtaining the desired construct in the ligation.
14. Ligate the pVEX vector and PCR product in a reaction as suggested below with a 5:1 to 10:1 (PRC product:pVEX vector) molar ratio: 5 μL PCR product, 5 μL pVEX vector, 4 μL 5× ligation buffer, 5 μL dH_2O, and 1 μL ligase (usually 5 units per μL).
15. Dialyze and electroporate ligation reaction as described in steps 9–11. Select for transformants on LB media containing antibiotics for corresponding pVEX vector (either spectinomycin or chloramphenicol).
16. Pick isolated colonies and screen for insertion of PCR insert as described in step 12. Be sure to confirm the proper structure of this plasmid as it will be the construct that will be used for chromosomal integration.

3.2. Chromosomal Integration of loxP Sites

1. Make a large-scale DNA preparation of each pVEX + PCR product plasmid for electroporation into the corresponding host for integration into the chromosome. This can be performed with the Qiagen Maxi prep kit using 500–1000 mL culture. This will typically yield 100–200 μg total DNA. If resuspended in 100 μL TE buffer, this will correspond to about 1–2 μg/μL. You will want to electroporate 1–3 μg DNA in the subsequent steps below.
2. Prepare the bacteria containing the genes targeted for VEX-Capture cloning for electroporation. One of the pVEX + PCR fragment constructs will be electroporated into this strain to obtain integration of this plasmid into the chromosome. To prepare electrocompetent cells, grow 40 mL cells to mid-late log phase in LB media, harvest the cells through centrifugation, pour off the supernatant, and resuspend the cells in a 1× volume of 10% glycerol. Recover the cells through centrifugation, and repeat this step twice using 1/2× and 1/10× volume washes. After pouring off the supernatant of the 1/10 volume glycerol wash, resuspend the cells in a 1/500 volume of 10% glycerol (80 μL). Distribute the cells in 20-μL aliquots in microfuge tubes. This will yield at least four tubes (and probably a few more because the mass of the cells will add volume to this sample as well).
3. After dialyzing the pVEX + PCR fragment construct DNA sample, add 2–3 μL of this plasmid to a tube of electrocompetent cells prepared in **step 2**. Electroporate this sample as described in **Subheading 3.1., steps 9–11**. Add 4 mL LB to the electroporated cells and outgrow this sample 12–16 h. This extended outgrowth will

allow ample time for the electroporated DNA to integrate into the chromosome through homologous recombination. Plate the entire sample on LB agar media containing the appropriate antibiotic (either spectinomycin or chloramphenicol), distributing the sample evenly on four different plates. Incubate the plates at the appropriate growth temperature and check the plates for desired colonies. If you do not see colonies after the first 16–18 h of growth, be patient and continue to check the plates over the course of the next week.

4. Screen any isolated clones from this electroporation to confirm proper insertion of the pVEX construct. This can most easily be performed through PCR, but Southern blot analysis can also be performed. Once confirmed, this isolate is termed the "single cointegrate."
5. The next step will be to integrate the other pVEX construct into its proper location in the chromosome of the single cointegrate. Repeat **steps 2–4** using the single cointegrate as the target strain. Once integration of the other pVEX construct is confirmed, this strain is termed the "double cointegrate." One convenient (and essential) way to confirm the double cointegrate is to electroporate the Cre-expressing plasmid pEKA30 into this strain, select for ampicillin-resistant colonies, and screen these colonies for loss of the spectinomycin marker and retention of the chloramphenicol marker. This strain can be saved as a mutant containing a deletion of the targeted region and replacement of this region with a chloramphenicol marker.
6. There are possible alternatives to the procedure described here for integration of *loxP* sites into a target genome, and these are described in **Notes 1** and **2**.

3.3. Capture of Excised Target DNA

1. The first step of this section will be to construct the R995 VC plasmid, which is the plasmid R995 XbaI containing a region of homology to the excised targeted DNA region. This region of homology will be obtained through PCR and cloned into the pCR-TOPO vector as described in **Subheading 3.2**. Design this region of homology (termed the "VC fragment") so that it will integrate into one of the outer sides of the targeted DNA and will not disrupt any critical genes that you want to analyze subsequently. To clone the VC fragment into R995, it is convenient to perform a "two-step" cloning, which is described in **Note 3** as an alternative to clone the "up" and "down" fragments into the pVEX vectors. To use the two-step method for cloning into the *Xba*I site of R995 *Xba*I, design PCR primers such that the VC fragment will contain *Xba*I and *Spe*I sites as shown in **Fig. 3**. Digest the pCR-TOPO + VC fragment plasmid with *Xba*I, ligate to *Xba*I-digested R995 plasmid, and electroporate into *E. coli* TOP10 cells as described in **Subheading 3.1**. Select for transformants on LB agar media containing tetracycline and ampicillin. This will select for constructs in which the two plasmids have ligated together (because co-transformation of the two separate plasmids is not a high-frequency event). Obtain plasmid DNA from the transformants using 15 mL culture with a Qiagen Mini DNA prep kit, and screen to confirm proper structure of the plasmid clones.

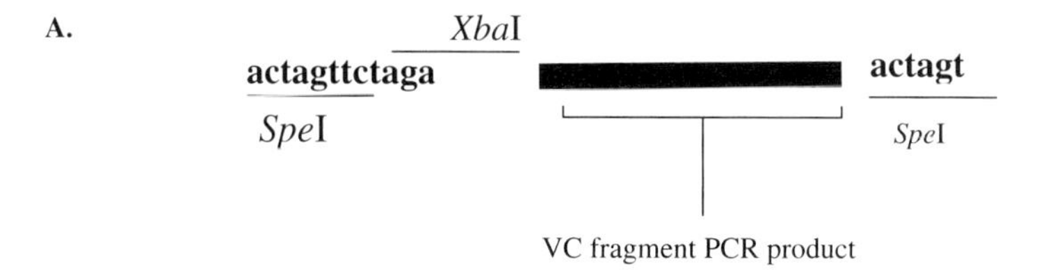

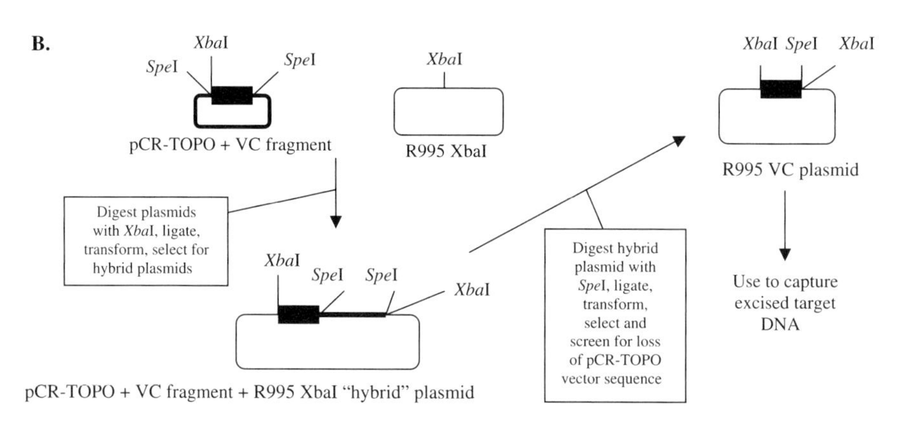

Fig. 3. Design of VC PCR fragment and diagram of "two-step" cloning procedure. **(A)** The arrangement of *Spe*I and *Xba*I restriction sites that should be engineered into oligonucleotides that will serve as primers to amplify the VC fragment for capture of the target DNA segment. The sequences of the *Spe*I and *Xba*I sites as they would occur in the primers are shown. The VC PCR fragment is represented by the solid black line. Note that the full primers would include homology to the VC PCR fragment that is not indicated by the restriction sequences shown. Also note that the drawing is not to scale. **(B)** The steps involved in the "two-step" cloning procedure for inserting the VC fragment into plasmid R995 *Xba*I. The plasmids and restriction sites involved in this procedure are indicated in the diagram. This procedure more easily facilitates subcloning of PCR products into large, low-copy plasmids such as R995 *Xba*I. This procedure can also be used as an alternative for subcloning PCR products into the pVEX vectors.

Resolve this "hybrid" plasmid by digesting with *Spe*I, ligating the products of this digestion together, and electroporating this ligation into *E. coli* TOP10 as described above. Select for transformants using LB agar media containing tetracycline only, and screen the transformants for absence of ampicillin resistance. Obtain plasmid DNA from the transformants and confirm loss of the pCR-TOPO vector and proper structure of the resulting R995 VC plasmid. This plasmid will be used to capture the excised target DNA as described below in steps 2–6.

2. Make overnight cultures of the four strains listed in **Subheading 2.3** with selection in LB media. The next day, spin down 1 mL of the first three listed strains and 4 mL TOP10 Rif and remove the supernatants. Resuspend each in 1 mL fresh LB to wash, spin again, and remove supernatant. Resuspend each in 900 μL fresh LB. Mix 100 μL donors with 200 μL *Salmonella* double cointegrate in separate tubes, but make two tubes of each mating. Thus, there will be four mating tubes total: two with R995 VC donors and two with R995 *Xba*I donors. Spin down the cells and take off supernatant, but leave behind about 100 μL LB. Resuspend the matings in the 100 μL LB and spot them on separate dry LB plates. Also, spot about 100 μL strains alone on LB plates as well. Incubate these plates at 37 °C for about 4–6 h.
3. Next, scrape up the matings with a sterile toothpick and transfer them to a microfuge tube. Resuspend them in 800 μL washed TOP10 Rif cells. Spin this mixture down and pull off supernatant, but leave behind about 100 μL LB. Resuspend the cells in the 100 μL LB and spot the mixtures on dry LB plates. Incubate these plates overnight at 37 °C.
4. The next day, scrape up the matings and transfer them to a microfuge tube. Resuspend them in 800 μL fresh LB. Distribute and plate the matings on four selection plates (LB agar containing Rif Km50 Sp100), using 200 μL for each plate. From the plates with the strains alone, streak some of the cells onto the selection plates. This will serve as a negative control with no resulting growth for these samples. Wait for colonies to appear on the plates. You will most likely get a mixture of colonies, both *Salmonella* and *E. coli*. The *Salmonella* colonies are likely spontaneous Rif-resistant mutants that received the R995 VC plasmid through conjugation and will be not be pursued. Be patient, as you may need to wait for smaller *E. coli* colonies that will grow up on the R995 VC matings, but not on the R995 *Xba*I matings. These sometimes take several days to grow up (3–5 days). Just keep the plates on the benchtop and check them each day (after initially having them at 37 °C for the first overnight).
5. When the colonies appear, pick them to fresh plates and streak them for single colony isolates. Also, streak some of the background *Salmonella* colonies that grew for comparison. We need to make sure the colonies of interest are *E. coli* and a good way to do this is to compare with a *Salmonella* streak. From four separate isolates, inoculate overnight cultures with 20 mL LB plus kanamycin and spectinomycin selection (no need to add Rif). Also, set up a culture of the strain TOP10 (R995 VC) for comparison. Obtain plasmid DNA with 15 mL culture using the Qiagen Mini DNA prep kit columns. Resuspend the DNA in 50 μL TE.
6. Run 12 μL non-digested plasmid DNA samples. Comparison of the capture isolates with the R995 VC plasmid on the agarose gel will yield a migration difference on the gel. Subsequently, perform a PCR using the isolated plasmid DNA for template and using primers that will amplify a region within the captured targeted DNA. It is also helpful to perform a PCR with primers that will amplify a fusion junction that can only be formed in the R995 + excised target DNA plasmid.

7. Once confirmed, the resulting strain will be *E. coli* TOP10 Rif containing the R995 + excised target DNA plasmid. This plasmid can be analyzed in this strain or conjugated to other Gram-negative bacteria for subsequent studies. It is best to include a selection for spectinomycin or streptomycin (the resistance determinant used here encodes resistance to both) when establishing any R995 + excised circle construct in another bacterial recipient because the captured target DNA could possibly "loop-out" in the absence of this selection.

4. Notes

1. An alternative to using electroporation for chromosomal integration of the pVEX constructs is conjugation. We have successfully used the IncP plasmid origin of transfer (*oriT*) to mobilize pVEX constructs into bacterial recipients to recover desired single and double cointegrates. The functional IncP *oriT* sequence can be obtained on a 300-bp PCR fragment that can be conveniently cloned onto the pVEX plasmids. The IncP *oriT* sequence can be obtained from GenBank Accession NC001621 and by using **ref.** ***6***. The transfer system contained on the *oriT*-deficient IncP plasmid pUZ8002 is used to mobilize *oriT*-containing constructs to bacterial recipients (J. W. Wilson and D. H. Figurski, unpublished results). This plasmid can be obtained from the authors.
2. A possible alternative strategy for integration of the *loxP* sites into a target chromosome is to utilize the lambda red recombination system to insert engineered PCR products into the desired genomic locations *(7,8)*. This efficient recombination system has been used to create genomic insertions in *E. coli* and *Salmonella*, but should be applicable to other bacterial hosts as well. A proper PCR product for such an approach should contain: (1) a single *loxP* site, (2) an antibiotic resistance marker, and (3) the required length of homology on either end to target the product for integration at the desired site (usually 36–50 bp at either end of the PCR fragment). We are currently developing this method for chromosomal integration of *loxP* sites.
3. There is an alternate way to clone the "up" and "down" PCR products into the pVEX vectors that has worked successfully in our laboratory called the "two-step" method. This method is also described in **Subheading 3.3** and **Fig. 3** for cloning the VC fragment into the R995 *Xba*I vector. This method involves cloning the PCR fragment into the pCR-TOPO vector as described above, but the subsequent cloning of this product into the pVEX vectors will be performed in two steps by making "hybrid" plasmids between the pCR-TOPO + PCR fragment plasmid and the appropriate pVEX vector. This is powerful because you can directly select for the desired hybrid plasmid using the different antibiotic selections on the plasmids. Then, you will resolve the hybrid by digesting with the other enzyme that is present on your PCR product to "drop-out" the pCR-TOPO vector. When you re-ligate this digestion, you will be able to recover the desired pVEX + PCR

fragment construct. First, separately digest both the pVEX vector and the pCR-TOPO + PCR fragment plasmid with one of the restriction enzymes you have engineered into the PCR primers. Ligate both of these digested products together, transform the ligation into *E. coli* TOP10, and select for the appropriate pVEX antibiotic resistance (either spectinomycin or chloramphenicol) and for kanamycin resistance encoded by the pCR-TOPO + PCR fragment plasmid. You will recover ligated "hybrid" plasmids from this transformation because co-transformation of the two separate plasmids is a very low-frequency event. Screen the DNA of these transformants for the correct orientation of the hybrid such that when you next cut with the other enzyme, the pCR-TOPO vector (and not the PCR product) will "drop-out." Digest the correct hybrid plasmid with the enzyme to be used for the resolution, ligate the products of this digestion, and transform the ligation into *E. coli* EKA260 or AS11 (depending on which pVEX plasmid you are using). Select for the antibiotic resistance for the corresponding pVEX plasmid, and then screen these colonies for sensitivity to the kanamycin. Obtain plasmid DNA from the corresponding colonies and check for the proper structure through restriction digest, PCR, or sequencing. This will be the pVEX + PCR product plasmid to be used for chromosomal integration. When performing this method, remember to meet the following requirements: (1) choose enzymes that will not cut the PCR product, (2) choose enzymes that will not digest the pVEX vector, and (3) at least one of the enzymes should not cut the pCR-TOPO plasmid.

References

1. Wilson, J. W., Figurski, D. H., and Nickerson, C. A. (2004) VEX-capture: a new technique that allows in vivo excision, cloning, and broad-host-range transfer of large bacterial genomic DNA segments. *J. Microbiol. Methods* **57**(3), 297–308.
2. Wilson, J. W. and Nickerson, C. A. (2005) Cloning of a functional Salmonella SPI-1 type III secretion system and development of a method to create mutations and epitope fusions in the cloned genes. *J. Biotechnol.* **122**(2), 147–160.
3. Wilson, J. W. and Nickerson, C. A. (2006) A new experimental approach for studying bacterial genomic island evolution identifies island genes with bacterial host-specific expression patterns. *BMC Evol. Biol.* **6**(1), 2.
4. Ayres, E. K., Thomson, V. J., Merino, G., Balderes, D., and Figurski, D. H. (1993) Precise deletions in large bacterial genomes by vector-mediated excision (VEX). The trfA gene of promiscuous plasmid RK2 is essential for replication in several gram-negative hosts. *J. Mol. Biol.* **230**(1), 174–185.
5. Brosius, J. (1989) Superpolylinkers in cloning and expression vectors. *DNA* **8**(10), 759–777.
6. Pansegrau, W., Lanka, E., Barth, P. T., et al. (1994) Complete nucleotide sequence of Birmingham IncP alpha plasmids. Compilation and comparative analysis. *J. Mol. Biol.* **239**(5), 623–663.

7. Datsenko, K. A. and Wanner, B. L. (2000) One-step inactivation of chromosomal genes in Escherichia coli K-12 using PCR products. *Proc. Natl. Acad. Sci. USA* **97**(12), 6640–6645.
8. Uzzau, S., Figueroa-Bossi, N., Rubino, S., and Bossi, L. (2001) Epitope tagging of chromosomal genes in Salmonella. *Proc. Natl. Acad. Sci. USA* **98**(26), 15264–15269.

8

Amplified Fragment Length Polymorphism Analysis of *Salmonella enterica*

Ruiting Lan and Peter R. Reeves

Summary

Amplified fragment length polymorphism (AFLP) is a powerful PCR-based fingerprinting method and has the capacity to reveal variation around the whole genome by selectively amplifying a subset of restriction fragments for comparison. The restriction fragments analyzed are small, and even mutation of 1 bp can be detected. The use of different sets of restriction enzymes or different primer combinations can generate large numbers of different AFLP fingerprints. AFLP is of particular value for studies of closely related strains, such as analysis of variation within a serovar of *Salmonella enterica*. We present here protocols for both radioactively labeled and fluorescent dye-labeled AFLP analyses that are also applicable to other bacterial species. Fluorescent AFLP has proved to be reproducible and capable of standardization.

Key Words: AFLP; *Salmonella enterica*; DNA fingerprinting.

1. Introduction

Amplified fragment length polymorphism (AFLP) is a DNA fingerprinting method developed to find molecular markers for plant genome mapping ***(1)***. It is based on the selective amplification of restriction fragments from digested genomic DNA by PCR, using restriction site/adaptor-specific primers under stringent conditions. Two restriction enzymes are used, and the nature of the adaptors ensures that only hetero-fragments (different enzyme cut at each end) are amplified, and the fragments amplified can be further restricted to those with a specific base or doublet or triplet of bases adjacent to either cut site. This is achieved by addition of what are called selective bases in one or both

From: *Methods in Molecular Biology, vol. 394: Salmonella: Methods and Protocols*
Edited by: H. Schatten and A. Eisenstark © Humana Press Inc., Totowa, NJ

primers. The number of fragments amplified is reduced on average fourfold for each selective base in the primer pair. Thus, using primers with appropriate selective bases, the number of fragments amplified can be tuned to achieve optimal separation. Furthermore, a combination of a six-base cutter and a four-base cutter will give two amplifiable fragments for most sites for the six-base cutter, but they will be in the size range for the four-base cutter, suitable for resolution at single-base level.

Many studies have shown the value of AFLP in typing of microorganisms (for review, *see* **ref. *2***). AFLP has been shown to be specific and reproducible. It enables the whole bacterial genome to be explored to find restriction site changes caused by mutations or absence/presence of genes. There are a large number of enzymes to choose from so there is almost unlimited potential to explore genomes for polymorphisms. AFLP is very useful for finding polymorphisms for typing, which has been its major application. However, it is also useful because of the relative ease of determining the molecular basis of the polymorphism. In this, it resembles multilocus sequence typing ***(3)***, but with the advantage that a much large proportion of the genome can be explored.

The procedure (*see* **Fig. 1**) involves digestion of DNA with *Eco*RI and *Mse*I or another combination of enzymes with six-base and four-base recognition sites. Double digestion with *Eco*RI and *Mse*I will produce around 2000 *Eco*RI–*Mse*I fragments, 18,000 *Mse*I–*Mse*I fragments, and very few *Eco*RI–*Eco*RI fragments for a genome of 4600 kb such as *Salmonella enterica*. This is followed by ligation of adaptors to both ends of the fragments, and amplification using primers based on the linkers is done so as to amplify *Eco*RI–*Mse*I fragments. This pre-AFLP amplification is followed by dilution and a second "selective" amplification using primers that include one or two bases in addition to the segment based on the linker/restriction site. Each additional base reduces the number of effective substrates fourfold. With one selective base on each primer, the number of bands is reduced 16-fold to about 125 in this case, a workable number of bands in a size range of 50–500 bp and resolvable to one base using polyacrylamide sequencing gels or equivalent separation media. A software tool is available on the web for choosing enzymes and selective bases for AFLP by in silico analysis of the sequenced genomes ***(4)***.

There are a large number of *S. enterica* studies using AFLP which can be categorized into three levels: species, serovar, and phage type. In our opinion, AFLP is not a good choice for species level typing. We have used it for studying variation within serovar Typhimurium ***(5,6)*** and detected variation between and within phage types. We found that AFLP is quite reliable after

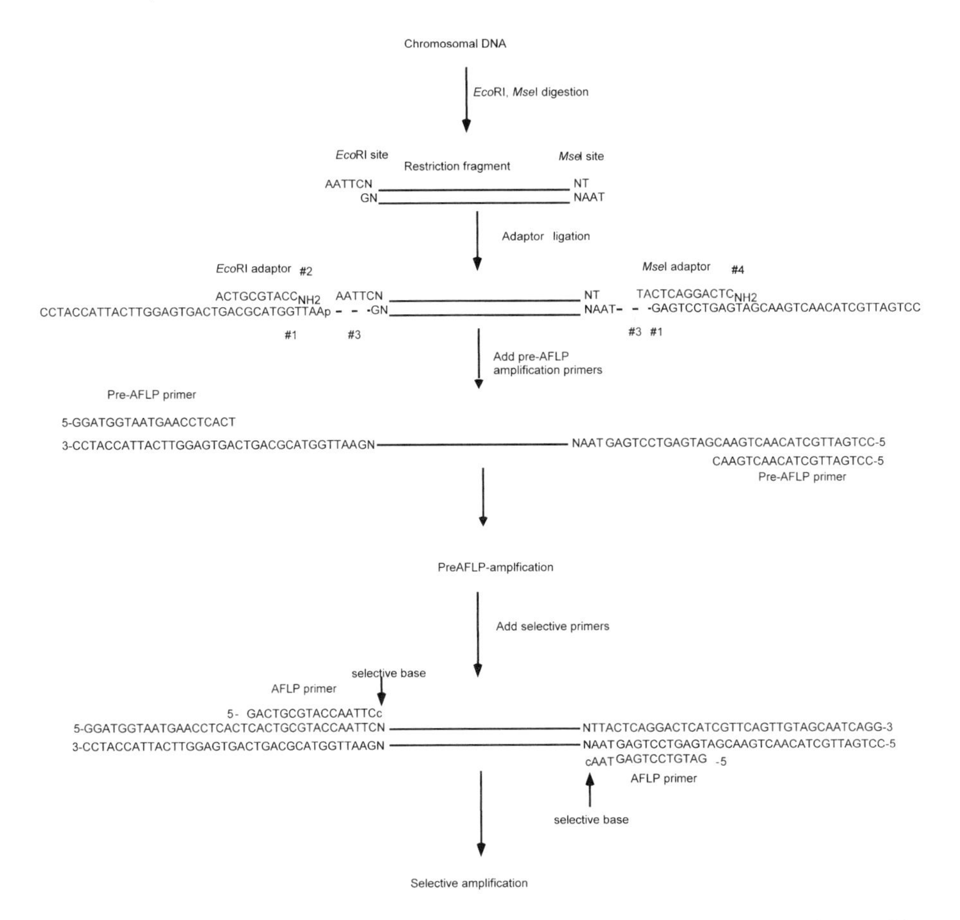

Fig. 1. Schematic representation of the amplified fragment length polymorphism (AFLP) technique. Chromosomal DNA is digested with *Eco*RI and *Mse*I, ligated with adaptors. A pre-AFLP amplification step eliminates *Eco*RI–*Eco*RI and *Mse*I–*Mse*I fragments. A sixteenth of the *Eco*RI–*Mse*I fragments is amplified using selective primers (+1) (shown are *Eco*RI +C and *Mse*I +C primers). See Lan and Reeves *(8)* for details of the adaptors and primers. The salient features of the adaptors are indicated and noted as follows: **1**, adaptors do not restore the *Eco*RI (GAATTC) or *Mse*I (TTAA) sites; **2**, the NH_2 group blocks the ligation of that strand, so that it falls off during the denaturation step in the first cycle of amplification; **3**, site of ligation, so that only one strand is ligated with an adaptor sequence at both ends and only *Eco*RI–*Mse*I hetero-fragments are amplified exponentially during pre-AFLP amplification; **4**, the NH_2 group blocks extension of that strand so blocks amplification of *Mse*I homo-fragments.

initial setup. An important feature of AFLP is that bands can be matched to genome-level differences relatively easily in comparison with other typing methods such as pulsed field gel electrophoresis. Therefore, any difference can be further analyzed *(7)*. The AFLP technique was performed essentially according to Vos et al. *(1)* except that adaptors and primers were designed according to Lan and Reeves *(8)*, and for fluorescent AFLP, one of the primers was labeled with a fluorescent dye. The protocol presented here is generically applicable to other bacterial species. We have used it for determining relationships of closely related isolates of *Escherichia coli* (unpublished) and *Vibrio cholerae (9)*.

2. Materials

2.1. Solutions

1. 10× One-Phor-All buffer (Amersham Pharmacia Biotech).
2. Bovine serum albumin (BSA), 20 mg/mL from Roche Applied Science.
3. 10× PCR buffer II and $MgCl_2$ solution (supplied with the AmpliTaq Gold [*see* **subheading 2.2.**].
4. Nucleotides for PCR (MBI Fermentas) as individual nucleotide solutions at 100 mM and prepared as a master mix of all four with each nucleotide at 10 mM concentration.
5. Ethanol, phenol, chloroform, and isoamyl alcohol (analytical grade, we used Sigma products but any supplier is fine).
6. Phase-dividing gel (15 mL, heavy, used to be sold by Sigma, now called phase-lock gel sold by Eppendorf).
7. 50 mM Tris–HCl, pH 8.0.
8. 0.4 M EDTA.
9. 10% Sodium dodecyl sulfate (SDS) (Sigma).
10. Phenol:chloroform:isoamyl alcohol (25:24:1).
11. TE: 10 mM Tris–HCl, 1 mM EDTA, pH.
12. Stabilized γ-^{33}P ATP (Stable Label Gold™) at a specific activity of ~2000 Ci/mmol (74 TBq/mL) and at a concentration of 10 mCi/mL (370 MBq/mL). (We purchased this from GeneWorks, a local supplier but not an international supplier.)
13. 19:1 Acrylamide/bisacrylamide at 40% (w/v) (Amresco, supplied by Astral Scientific).
14. *N, N, N, N'*-Tetramethylethylenediamine (TEMED) (Promega).
15. 10% (w/v) Ammonium persulfate (Promega).
16. Polyacrylamide gel mix: each 80 mL contains 40 g urea, 1 g resin, 12 mL 40% 19:1 acrylamide/bisacrylamide stock, 8 mL 10× TBE, and 20 mL Milli-Q water.

2.2. Enzymes

1. AmpliTaq Gold™ (ABI Applied Biosystems).
2. Restriction enzymes *Eco*RI and *Mse*I (New England Biolabs).
3. T4 DNA ligase at 5 U/μL (MBI Fermentas).
4. T4 Polynucleotide Kinase (Promega).
5. Proteinase K (20 mg/mL) (Roche Applied Science).
6. RNase (10 mg/mL) (Roche Applied Science).
7. Lysozyme (Amersham Pharmacia Biotech).

2.3. Oligonucleotides

Oligonucleotides are prepared commercially with reverse cartridge purification (*see* **Table 1**) and provided as lyophilized samples. These are resuspended at a concentration of 300 pmol/μL in 10 mM Tris-HCl (pH 8.0).

Table 1
Adaptors and Primers

Oligo name	Sequence (5′ → 3′)	Selective base	Enzyme ends
*Adaptors**			
924	actgcgtacc$_{NH2}$	N/A	*Eco*RI
925	P-aattggtacgcagtcagtga ggttcattaccatcc	N/A	*Eco*RI
930	tactcaggactc$_{NH2}$	N/A	*Mse*I
931	cctgattgctacaactgaacgat gagtcctgag	N/A	*Mse*I
pre-AFLP amplification primers			
926	ggatggtaatgaacctcact	None	*Eco*RI
932	cctgattgctacaactgaac	None	*Mse*I
Selective primers			
927	gactgcgtaccaattca	A	*Eco*RI
1269	gactgcgtaccaattct	T	*Eco*RI
1270	gactgcgtaccaattcg	G	*Eco*RI
1271	gactgcgtaccaattcc	C	*Eco*RI
933	gatgagtcctgagtaac	C	*Mse*I
1272	gatgagtcctgagtaaa	A	*Mse*I
1273	gatgagtcctgagtaat	T	*Mse*I
1274	gatgagtcctgagtaag	G	*Mse*I

* NH2: amino group; P, phosphate.

3. Methods

3.1. Isolation of Chromosome DNA

The quality of chromosomal DNA is critical in generating a good AFLP product. The protocol provided here is based on phenol–chloroform extraction, which gives good-quality DNA. We found that some commercial DNA purification kits were not well suited, as the quality of purified DNA deteriorated quickly during 4 °C storage. However, the situation would be different if DNA were needed only for immediate use and long storage was not a consideration. There are many kits in the market. Test before you commit to using a kit.

1. Bacteria are collected from 2 mL of a 10-mL overnight nutrient broth culture. Transfer 1 mL of the culture to an Eppendorf and spin for 30 s, then remove supernatant. Add another 1 mL culture and repeat the steps.
2. Resuspend bacterial pellet in 1 mL of 50 mM Tris-HCl (pH 8.0), spin for 30 s, remove supernatant and resuspend in 1 mL 50 mM Tris-HCl (pH 8.0). Add 40 μL 0.4 M EDTA and incubate at 37 °C for 20 min. Then, add 40 μL fresh lysozyme (20 mg/mL, prepare a stock and store in aliquots at −20 °C) and incubate for a further 20 min at 37 °C.
3. Add 5 μL proteinase K (20 mg/mL) and mix gently, then add 60 μL 10 % SDS and incubate at 50 °C for at least 1 h (preferably 2–5 h and can be left for overnight digestion).
4. Add 2 μL RNase (10 mg/mL) and incubate for 15 min at 65 °C.
5. Spin down phase-dividing gel tube (*see* **Note 1**). Transfer solution from Eppendorf tube to gel tube using a cut tip to avoid shearing DNA.
6. Add 1 mL of phenol:chloroform:isoamyl alcohol (25:24:1) and mix gently for 2 min. Spin for 20 min (*see* **Note 2**).

   After spinning, the organic and aqueous phases should be separated by the phase-dividing gel (*see* **Note 1**). Step 6 should be repeated one to two times. Phenol–chloroform–isoamyl alcohol is added to the aqueous phase directly without transferring to a new tube. The organic phase under the gel will remain separated.
7. Add 1 mL chloroform:isoamyl (24:1). Spin at 4000 rpm in a bechtop centrifuge for 10 min.
8. Remove top aqueous phase and transfer to a McCartney bottle.
9. Add 2× volume (~2.2 mL) of cold 100 % ethanol and invert bottle slowly to precipitate DNA, then spool DNA using a glass rod (easily made using a Pasteur pipette) and rinse in 70 % ethanol.
10. Dissolve DNA in 400 μL TE buffer and heat tube with lid open for 10 min at 65 °C to evaporate ethanol. Store at 4 °C. Let DNA to dissolve for a couple of days before determining DNA concentration.
11. DNA quantification is done by measuring the optical density (OD) at 260 and 280 nm. We used a Beckman DU 640 spectrophotometer. One OD_{260} equals

50 μg/mL double-stranded DNA *(10)*. Purity of DNA is indicated by ratio of OD_{260} to OD_{280} with an acceptable range being between 1.7 and 2.0.

3.2. Preparation of Adaptors

1. The adaptor for the *Eco*RI enzyme restriction site is prepared as follows:
   a. Mix in a 1.5-mL Eppendorf tube of 20 μL (300 pmol/μL) each of oligo 924 and 925 (*see* **Table 1**), 6 μL 10× One-Phor-All buffer and 74 μL Milli-Q water to make up 120 μL total.
   b. Place the mixture in a beaker of boiling water for 2 min, and then leave to cool down gradually to room temperature in the beaker to allow the two oligos to anneal. The mixture is then diluted 1:10 to give a final concentration of 5 pmol/μL.
2. The adaptor for the *Mse*I enzyme is prepared similarly, by mixing of 20 μL each of oligo 930 and 931 (*see* **Table 1**), 6 μL 10× One-Phor-All buffer and 74 μL Milli-Q water, except that this stock is not diluted, but used at a final concentration of 50 pmol/μL.
3. Adaptors are stored at −20 °C.

3.3. Digestion of Chromosomal DNA and Ligation of Adaptors

1. Genomic DNA is digested by *Eco*RI and *Mse*I and ligated to *Eco*RI and *Mse*I adaptors simultaneously (note that ligation to adaptors does not regenerate a full restriction site).
2. The reaction mixture contains 0.1 μg DNA, 5 U *Eco*RI, 5 U *Mse*I, 10 μL 10× One-Phor-All buffer, 1 U T4 ligase, 1 μL 10 mM ATP, 5 μg BSA, 1 μL (5 pmol) *Eco*RI adaptor, and 1 μL (50 pmol) *Mse*I adaptor.
3. The digestion/ligation step is performed at 37 °C for at least 5 h or overnight.
4. The ligation product is denatured at 94 °C for 2 min and snap-cooled.
5. Store product at −20 °C, and 1 μL is used for pre-AFLP amplification.

3.4. pre-AFLP amplification

1. pre-AFLP amplification is done using primers without selective nucleotides (Primer Pair 926-927, *see* **Table 1**).
2. The 20 μL of PCR mixture contains 2 μL 10× PCR buffer II, 1.6 μL 25 mM $MgCl_2$, 1 μL 4 mM dNTP, 0.4 μL 10 mg/mL BSA, 1 μL each of 6 μM *Eco*RI primer and 6 μM *Mse*I primer, 0.1 μL AmpliTaq Gold (5 U/μl), 11.9 μL Milli-Q water, and 1 μL of the above digestion/ligation mix as template (*see* **Note 3**).
3. The reaction mixture is overlaid with sterile pure liquid paraffin. We still prefer overlaying with paraffin oil despite using heated lid PCR machines. Without it, sometimes there is a loss of reaction volume, although it seems not to affect the AFLP-PCR result.

4. pre-AFLP amplification is run for 20 cycles in a thermal cycler (PC-960G, Corbett Research, Australia) with denaturation at 94 °C for 15 s except for 10 min for the first cycle to activate the AmpliTaq Gold (*see* **Note 4**), annealing at 56 °C for 30 s, and extension at 72 °C for 2 min except for 5 min for the last cycle.
5. Following PCR, 10 μL of each pre-AFLP product is run in a 1.5 % agarose gel to check that amplification has occurred. If amplification is successful, a visible smear in the 100–1500 bp range is seen (*see* **Note 5**).
6. Each pre-AFLP product is diluted 10-fold with TE buffer (pH 8.0) as template for selective amplification (*see* **Note 6**).

3.5. Selective Amplification

3.5.1. Primer Labeling

1. *Mse*I primer is either fluorescent dye labeled, or radioactively labeled by $\gamma{-}^{33}P$ ATP.
2. Fluorescent dye-labeled primer is synthesized commercially. We used 6-carboxy fluorescein (FAM; blue) that was synthesized by ABI Applied Biosystems (*see* **Note 7**).

The following steps (3–5) are only relevant to radioactive labeling:

3. To make 1 pmol/μL radioactively labeled *Mse*I primer, the 10 μL reaction mixture contains 3.3 μL 6 pmol/μL nonlabeled *Mse*I primer, 1 μL 10× One-Phor-All buffer, 2 μL $\gamma{-}^{33}P$ ATP (10 mCi/mL), 0.2 μL T4 polynucleotide kinase, and 3.5 μL Milli-Q water.
4. Incubate at 37 °C for 2 h. Then, an equal amount (10 μL) of Milli-Q water is added.
5. The labeled primer is stored at −20 °C until use.

3.5.2. Selective Amplification

1. One microliter of diluted pre-AFLP amplification product is used for the secondary selective amplification.
2. This is done using either 6-carboxy fluorescein (FAM; blue) fluorescent dye-labeled or radioactively labeled *Mse*I primer with one base selection (*Mse*I +1, *see* **Table 1** for the four different primers) and unlabeled *Eco*RI primer with one base selection (*Eco*RI +1, *see* **Table 1**). The other components of the mixture are the same as in pre-AFLP amplification.
3. The amplification is run for 30 cycles in PC-960G Gradient Thermal Cycler with denaturation at 94 °C for 15 s except for 10 min for the first cycle to activate AmpliTaq Gold, annealing at temperatures described in step 4 for 30 s, and extension at 72 °C for 1 min except for 15 min for the last cycle. Ramping rate is set at 1 °C/2 s.
4. Annealing temperatures are "touch down" for the first 10 cycles from 66 to 57 °C decreasing by 1 °C per cycle, to ensure specific primer matches and to increase the differences in product amount between correct and incorrect annealing, followed by 20 cycles at 56 °C. This is easily programmed in most PCR machines.

3.6. Visualization of AFLP Fingerprinting Patterns

3.6.1. Visualization of Fluorescent AFLP

1. One microliter of each fluorescent AFLP product is electrophoresed on an ABI DNA Sequencer equipped with the GeneScan version 3.5 Software (ABI) (we used our local service at the Sydney University Prince Alfred Macromolecular Analysis Center).
2. GeneScan TAMRA-500 internal size standard (ABI) is also loaded with each AFLP sample to enable precise size determination of amplified fragments.
3. The data for each lane are saved as an individual GeneScan file and displayed as an electropherogram (*see* **Fig. 2A**).

3.6.2. Visualization of Radioactive AFLP

1. The *MseI* primer was radioactively labeled by γ—^{33}P ATP as described in **Subheading 2.1**. Radioactively labeled AFLP products were denatured and separated by polyacrylamide gel electrophoresis as described in **Subheading 3.7.** and visualized through autoradiography (*see* **Fig. 2B**).

3.7. Polyacrylamide Gel Electrophoresis

Hazards: acrylamide is neurotoxic.

The procedure presented here is essentially the same as for preparing a manual sequencing gel but does not include basics in setting up a sequencing gel. One may need to consult Sambrook et al. ***(10)***. The gel prepared is approximately 0.4 mm × 32 cm × 40 cm. The comb we used can accommodate 48 samples at one run. We tried 96-well combs. However, the AFLP bands look like a dot rather than a band. For this reason, we do not recommend 96-well combs for initial AFLP runs.

1. Polyacrylamide gel electrophoresis is conducted using 6% polyacrylamide gel with 1× TBE as running buffer.
2. Prepare fresh 80 mL polyacrylamide gel mix. The mix is filtered through a 0.2-μm filter cup, which degases the solution at the same time.
3. To the gel mix, 400 μL 10% (w/v) ammonium persulfate and 45 μL TEMED are added. The solution was mixed by gentle rocking to start the polymerization of the gel and poured immediately into the gel mold.
4. Allow 45 min for polymerization. The gel can be used on the same or next day. We normally prepare the gel the day before. To store the gel overnight, wrap it with wet paper towel and leave at room temperature.
5. The gel is run in 1× TBE in an electrophoresis apparatus (Gibco BRL) at 600 V for 1 h to warm the gel before loading DNA samples.

A

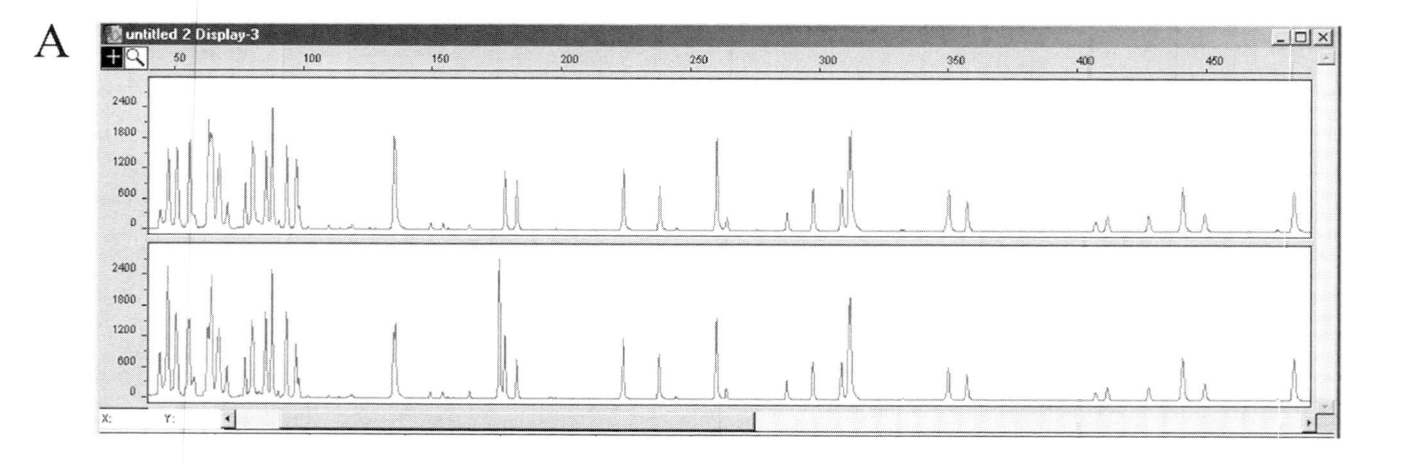

B

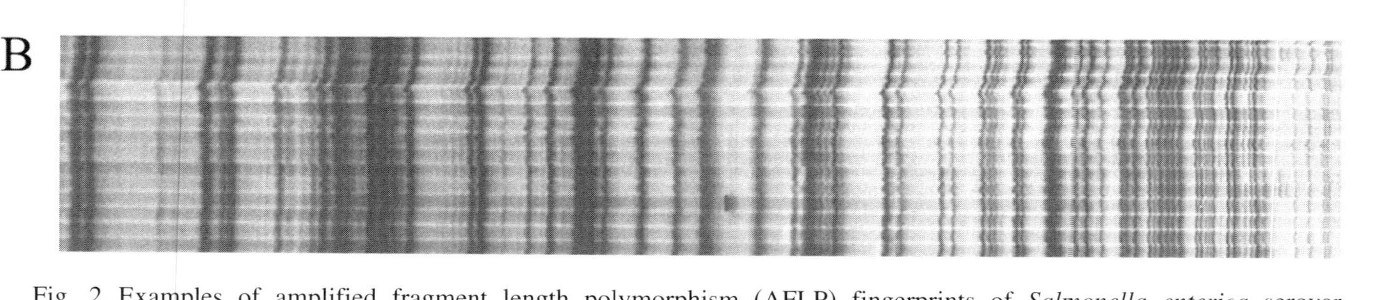

Fig. 2. Examples of amplified fragment length polymorphism (AFLP) fingerprints of *Salmonella enterica* serovar Typhimurium. (**A**) Fluorescent AFLP of two isolates as displayed by GeneScan (shown in gray scale but original in color). (**B**) A portion of an autoradiograph of radioactive-labeled AFLP.

6. DNA samples are denatured at 94 °C for 5 min and snap-cooled in ice before loading. Do not spin down as the evaporation actually serves to concentrate the samples.
7. SequaMark DNA ladder (Research Genetics Inc.) for isotopic sequencing method can also be loaded as external size standard. Follow the manufacturer's instructions for labeling the marker. However, we normally do not use a molecular marker.
8. After loading samples, the gel is run at 1200 V for 3 h.
9. Then, the gel is transferred from the glass plate onto Whatman 3MM paper, wrapped with Cling Wrap, and dried for 1 h under vacuum at 80 °C using a Bio-Rad Model 483 Slab Dryer. It is important to avoid wrinkles with the Cling Wrap, which may cause a high background.
10. Place in a 36.6 cm × 43.2 cm cassette and expose to a 35 cm × 43 cm sheet of BioMax film (Kodak) at −70 °C for autoradiography.

3.8. Computer Analysis

1. The approaches to the analysis of AFLP results differ widely in published studies. We performed the analysis manually. We needed to be confident of scores for each marker. The absence or presence of a band is scored as 0 or 1 and tabulated in a spreadsheet.
2. For radioactive AFLP, we manually score each gel (X-ray film) by comparison among the lanes. Bands on both ends of the gel are excluded, as they are more likely to be affected by variation of DNA concentration. We number each band including the nonvarying bands and the results are entered into Excel spreadsheet. For automated analysis, *see* **Note 8**.
3. For fluorescent AFLP, samples were displayed in GeneScan and peaks representing AFLP fragments from 40 to 600 bp were visually inspected and scored. Most peaks were easily scored for presence or absence, but some low-intensity peaks are judged by comparison with the same peak of other strains. Generally, peaks with intensity less than 1/10 of the peak with the highest intensity were not scored. For automated analysis, *see* **Note 9**.
4. Converting data to a dendrogram: We give here an outline of steps involved in converting data to a dendrogram using free software packages available. For commercial alternatives, *see* **Note 10**. To derive a dendrogram from the data, we first convert the raw data into dice coefficient, a most commonly used distance measure for AFLP data. We have written a program for the conversion, which is available upon request.

   a. The distance data are then used for constructing a dendrogram using the unweighted pair group method with arithmetic mean (UPGMA) method. The Neighbor program from the Phylip package (http://evolution.genetics.washington.edu/phylip.html) is used.

b. TreeView (http://taxonomy.zoology.gla.ac.uk/rod/treeview.html) is used for visualization and the output can be imported into Microsoft Powerpoint or other programs for presentation.

4. Notes

1. Use of phase-dividing gel is not essential. We found that the gel gives a good separation of the aqueous phase (DNA) and the organic phase enabling further extractions by adding phenol–chloroform–isoamyl alcohol mixture to the upper aqueous phase, mixing and again spinning which pushes the new phenol phase through the gel into the lower previous phenol phase. This avoids the transfer of the DNA to a new tube each time. The volume used in the steps 6 and 7 is designed to fit the 15-mL phase-dividing gel tube. The protocol can be scaled down to work with a 500-μL culture, and a 2-mL phase-dividing gel tube can then be used.
2. This step is critical to get pure DNA. A common fault is either not mixing long enough or not mixing well enough, leading to poor extraction of proteins by phenol, giving rise to dirty DNA.
3. It is best to prepare a master mix leaving out the template. A 5 % extra should be added to allow pipette volume variation. Add *Taq* polymerase last, immediately before distributing aliquots to tubes with template already added.
4. Hot start for pre-AFLP amplification is not suitable for adaptors designed based on the original configuration reported by Vos et al. ***(1)***. The Vos et al. adaptors are not phosphorylated, and one of the adaptor strands is not ligated to the target DNA that depends on *Taq* polymerase to remove/extend the unligated strand, otherwise no amplification will occur. The modified adaptors by Lan and Reeves ***(8)*** circumvent the problem and allow hot start to be used.
5. Half of the pre-AFLP product is left after the gel run but is enough for 100 AFLP reactions. We found that even if a good smear is not shown on gel, one may still get a good AFLP run. In other words, an observation of a perfect smear for pre-AFLP is not necessarily useful for troubleshooting purpose. If no AFLP bands at all, the sample must be repeated right from the beginning of restriction digestion. Gel checking may be omitted once the process becomes routine.
6. We dilute 1:10 uniformly for all samples, but if necessary, template concentration can be adjusted empirically at this stage based on the intensity of the smear on the pre-AFLP gel.
7. There are a number of dyes to choose from, but one must avoid the color dye used for the molecular size marker. Consult the genotyping/sequencing service before deciding which color dyes to use. We have only used one dye (FAM) per sample run. However, it is possible to use up to four dyes (FAM, NED, PET, and VIC) together, and hence four different samples can be mixed and run together to save cost. It should be noted that we have had experience using ABI machines only but other machines can also be used for AFLP analysis.

8. GelCompar from Applied Math is a program widely used for analysis of gel-based DNA fingerprints including AFLP. We initially used GelCompar to automatically score the bands, but we found it unsatisfactory for our purpose, as it did not recognize all occurrences of a given band. There were cases of bands not detected but clearly present by visual inspection. In other cases, multiple fragments that run close together were identified wrongly in some lanes but correctly in other lanes. The automated analysis is adequate for grouping similar isolates if a certain percentage of error is tolerated. Also, measures other than dice coefficient are available for measuring similarity between isolates, which may be more tolerant of such variation, but we did not explore these alternative measures. However, be warned of potential for errors if blindly using the output from an automatic analysis without looking at the raw data.
9. The latest software from ABI applied systems supports automated AFLP analysis, with GeneMapper replacing GeneScan and Genotyper. We used Genotyper for automatic scoring but found that the results were not satisfactory. GeneMapper 4.0 has a specific method for dealing with AFLP. It is beyond the scope here to detail AFLP analysis using GeneMapper, but output (calling of presence or absence of band) should still be checked after automatic processing.
10. Automated analysis is built into GelCompar, a commercial software alternative, for performing this and other AFLP analysis, but a substantial investment is required.

Acknowledgments

We thank Dr. Helen Hu for the raw fluorescent AFLP files presented in **Fig. 2A** and for reading of the manuscript. This research was supported by the National Health and Medical Research Council of Australia.

References

1. Vos, P., Hogers, R., Bleeker, M., et al. (1995) AFLP: a new technique for DNA fingerprinting. *Nucleic Acids Res.* **23,** 4407–4414.
2. Savelkoul, P. H., Aarts, H. J., de Haas, J., et al. (1999) Amplified-fragment length polymorphism analysis: the state of an art. *J. Clin. Microbiol.* **37,** 3083–3091.
3. Maiden, M. C., Bygraves, J. A., Feil, E., et al. (1998) Multilocus sequence typing: a portable approach to the identification of clones within populations of pathogenic microorganisms. *Proc. Natl. Acad. Sci. U. S. A.* **95,** 3140–3145.
4. Bikandi, J., San Millán, R., Rementeria, A., and Garaizar, J. (2004) *In silico* analysis of complete bacterial genomes: PCR, AFLP-PCR, and endonuclease restriction. *Bioinformatics* **20,** 798–799.
5. Hu, H., Lan, R., and Reeves, P. R. (2002) Fluorescent amplified fragment length polymorphism analysis of *Salmonella enterica* serovar Typhimurium reveals phage-type- specific markers and potential for microarray typing. *J. Clin. Microbiol.* **40,** 3406–3415.

6. Lan, R., Davison, A. M., Reeves, P. R., and Ward, L. R. (2003) AFLP analysis of *Salmonella enterica* serovar Typhimurium isolates of phage types DT 9 and DT 135: diversity within phage types and its epidemiological significance. *Microbes Infect.* **5,** 841–850.
7. Hu, H., Lan, R., and Reeves, P. R. (2006) Adaptation of multilocus sequencing to study variation within a major clone: evolutionary relationships of phage types in *Salmonella enterica* serovar Typhimurium. *Genetics* **172,** 743–750.
8. Lan, R. and Reeves, P. R. (2000) Unique adaptor design for AFLP fingerprinting. *Biotechniques* **29,** 745–746, 748, 750.
9. Lan, R. and Reeves, P. R. (2002) Pandemic spread of cholera: genetic diversity and relationships within the seventh pandemic clone of *Vibrio cholerae* determined by amplified fragment length polymorphism. *J. Clin. Microbiol.* **40,** 172–181.
10. Sambrook, J., Fritsch, E. F., and Maniatis, T. (1989) *Molecular Cloning: A Laboratory Manual*. Cold Spring Harbor Laboratory Press, Cold Spring Harbor, NY.

9

Salmonella Phages and Prophages—Genomics and Practical Aspects

Andrew M. Kropinski Alexander Sulakvelidze, Paulina Konczy, and Cornelius Poppe

Summary

Numerous bacteriophages specific to *Salmonella* have been isolated or identified as part of host genome sequencing projects. Phylogenetic analysis of the sequenced phages, based on related protein content using CoreGenes, reveals that these viruses fall into five groupings (P27-like, P2-like, lambdoid, P22-like, and T7-like) and three outliers (ε15, KS7, and Felix O1). The P27 group is only represented by ST64B; the P2 group contains Fels-2, SopEϕ, and PSP3; the lambdoid *Salmonella* phages include Gifsy-1, Gifsy-2, and Fels-1. The P22-like viruses include ε34, ES18, P22, ST104, and ST64T. The only member of the T7-like group is SP6. The properties of each of these phages are discussed, along with their role as agents of genetic exchange and as therapeutic agents and their involvement in phage typing.

Key Words: Bacteriophage; temperate; lytic; prophage; genome analysis; genetic map; genome evolution; P22-like phages; T7-like phages; lambdoid phages.

1. Introduction

Bacteriophages are the most abundant "life form" on this planet *(1)* and because of their diversity represent an incredible gene pool that has contributed significantly to host bacterial evolution. Bacteriophages are basically bacterial parasites that can exist by themselves but cannot grow or replicate except in bacterial cells. These viruses may go through either of two life cycles: the lytic and the lysogenic cycle. A phage in the lytic cycle converts a bacterial cell to a phage factory and produces many phage progeny. Such a phage adsorbs to a

From: *Methods in Molecular Biology, vol. 394: Salmonella: Methods and Protocols*
Edited by: H. Schatten and A. Eisenstark © Humana Press Inc., Totowa, NJ

specific receptor of the bacterial surface, and in the case of tailed phages, injects its DNA through the bacterial cell wall. Once in the cell, it prevents bacterial replication and/or transcription and subverts the cell to produce phage nucleic acids and proteins. It produces catalytic and structural proteins to replicate its DNA, to assemble the phage head, to package the phage DNA into the phage head, and to form a phage tail. It then releases the new phage particles by means of a membrane pore-forming protein (holin) and a lysin (often lysozyme) to dissolve the bacterial cell wall. A phage capable of only lytic growth is called a virulent or lytic phage. The number of phage particles produced by an infected cell is called the burst size.

Temperate phages can develop either by the lytic route or via the lysogenic cycle. A phage in the lysogenic cycle synthesizes an integration enzyme (integrase), turns off further viral transcription (repressor), and usually inserts itself into the DNA of the bacterium, which continues to grow and multiply whilst the phage genes replicate as part of the bacterial chromosome. A bacterium that contains a complete set of phage genes is called a lysogen, whereas the integrated viral DNA is called a prophage. A lysogen cannot be reinfected (superinfected) with a phage of the kind that first lysogenized the cell; it is immune to "superinfection" by virtue of the repressor. A lysogen may replicate the phage continuously and stably. However, when its DNA is damaged by ultraviolet light or the cells are treated with other inducing agents such as mitomycin C, the phage becomes derepressed and initiates a lytic cycle. This is called prophage induction. Most temperate phages form lysogens by integration at a unique attachment site in the host chromosome. Lysogenization of bacteria may result in lysogenic conversion if the lysogen obtains properties not present in the original bacterium. It may play a major role in conferring virulence properties to the lysogenized host *(2–4)*. Some phages can package host DNA and are called transducing phages. Generalized transducing phages produce particles that contain only bacterial DNA. Specialized transducing phages occasionally produce particles containing both phage and bacterial DNA sequences. Both types of transducing particles can inject their DNA into a host and transfer DNA from one bacterium to another. Such transfer is called transduction. Phage specificity is governed not only by the ability of the phage to adsorb to a specific host bacterial species but also by host restriction and host modification effected by strain-specific restriction nucleases and site-specific methylases, respectively.

In addition to playing a significant role in the development of the field of "molecular biology" *(5)*, *Salmonella* phages have practical significance: strain construction through transduction, phage typing for epidemiological purposes *(6–8)*, and the application of phage as therapeutic agents *(9)*. Lastly, phage genes and their products have contributed significantly to vector development (cosmids, integrative vectors, promoters, etc.) and as sources of molecular biologicals (DNA and RNA polymerases, ligase, nucleases, recombinases, restriction endonucleases, etc.).

Based on a compellation of bacteriophage names assembled by Abedon and Ackermann *(10)* and supplemented by a scan of the scientific literature up to 2005 reveals some 170 named *Salmonella* phages. Among the large tailed viruses of the order Caudovirales *(11)*, the breakdown is as follows: Myoviridae (phages with contractile tails), 44; Siphoviridae (viruses with long noncontractile tails), 65; and Podoviridae (short noncontractile tails), 63. This review will concentrate on the properties of the members of this order, particularly those that have been completely sequenced. The latter include phages and prophages ε15 *(12)*, ε34 *(13)*, ES18 *(14)*, Felix O1 *(15,16)*, Fels-1, Fels-2, Gifsy-1, and Gifsy-2 *(17–19)*, SopEϕ *(20)*, KS7 *(21)*, P22 *(22,23)*, PSP3 *(24)*, SP6 *(25)*, ST64B *(26)*, ST64T *(27)*, and ST104 *(28)*. Recognition that *Salmonella enterica* serovar Typhimurium carried "symbiotic bacteriophages" was first recognized by Boyd *(29)*. It is noteworthy that most salmonellae carry the genomes of temperate phages—*S*. Typhimurium strains usually carry four to five prophages, whereas *S. enterica* serovar Typhi strains Ty2 and CT18 both possess seven prophages *(30)*.

The phages are grouped and discussed here firstly based on their phylogenic relatedness deduced using a phage proteomic tree approach analogous to that of Rohwer and Edwards *(31)*. In this chapter, the phage proteomes were compared using CoreGenes *(32)*—a BlastP algorithm employing a cutoff score of 100 to ensure identification of homologs. Please note that the phylogenetic position of members of the order Caudovirales *(11,33)* as defined by the NCBI Taxonomy Browser (http://ncbi.nlm.nih.gov/Taxonomy/Browser/wwwtax.cgi?mode=Root) changes frequently to reflect newer data, and in certain cases, this chapter will conflict with the established grouping. Phage taxonomy has largely been based on electron microscopy, and where there is a conflict between the phylogenic position and the taxonomic position, the phages sharing a common morphotype will be discussed together.

CoreGenes analysis revealed that the *Salmonella* phages fall into five groupings (P27-like, P2-like, lambdoid, P22-like, and T7-like) and three outliers (ε15, KS7, and Felix O1) (*see* **Fig. 1**). The P27 group is only represented by ST64B; the P2 group contains Fels-2, SopEϕ, and PSP3; the lambdoid *Salmonella* phages

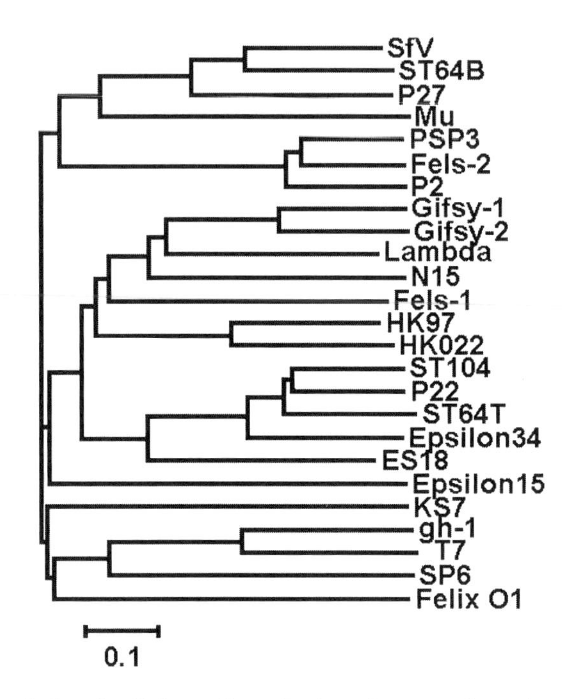

Fig. 1. GenBank data for *Salmonella* and other phages and prophages were compared using CoreGenes, and the number of homologous proteins (BlastP score $\geq$ 100) was recorded in an Excel spreadsheet. The SAB score was calculated as 2 (number of common proteins)/(sum of the number of proteins in the two phage genomes). These data were transformed into a dissimilarity score for tree drawing using the neighbor-joining algorithm in MEGA3 ***(34)***.

include Gifsy-1, Gifsy-2, and Fels-1. The P22-like viruses include ε34, ES18, P22, ST104, and ST64T. The only member of the T7-like group is SP6.

2. *Salmonella* Phage Groups

2.1. The P22-Like Phages

2.1.1. P22

In 1952, Zinder and Lederberg ***(35)*** demonstrated the transfer (generalized transduction) of genetic material between *S. enterica* serovar Typhimurium mutants involving a phage intermediary. The temperate phage vector, originally called PLT 22, is now commonly referred to as P22 and has continued to be the virus of choice for investigating the genetics of this bacterium. It is also one of the best-studied bacterial viruses. Many studies have suggested that

P22, despite its morphology, is a member of the lambdoid family. Indeed, the layout of its genes is very similar to that of other lambdoid phages (compare **Figs. 2** and **3**), viable λ-P22 hybrids exist, and 15 of its genes show close λ analogs. But, largely because of the morphological difference between λ and P22, the latter is considered the archetype of the P22-like phage genus by NCBI. This genus is also proposed to include *Salmonella* phages KS7, ε15, ST64T, and *Myxococcus xanthus* phage Mx8 (NC_003085). Although we support the similarity of P22 and ST64T, the other phages show little or no similarity at the protein level (*see* **Fig. 1**).

P22 adsorption is a multistep process ***(36)***, which is initiated by phage binding to its receptor (lipopolysaccharide [LPS] O side chains of *Salmonella* serotypes A, B, and D1) via the virion tailspike proteins (TSPs) ***(37)***. The latter possess endorhamnosidase activity, which digests the O antigen, permitting diffusion of the phage through the LPS barrier to the surface of the outer membrane, where tight binding occurs. The linear viral genome enters the host cell and circularizes, a reaction involving the terminally redundant ends of the molecules, the phage-encoded protein Erf, and host RecA and gyrase. This is the substrate for replication or integration into the host chromosome. The phage integration site, like that of many prophages, maps within a tRNA gene, *thrW*. Unlike coliphage λ integration, the integration of P22 does not require integration host factor (IHF) even though IHF recognition sites are present within *attP* ***(38)***.

In the lysogenic state, P22 expresses three different systems that may interfere with superinfection by homologous phages. These are (1) immunity

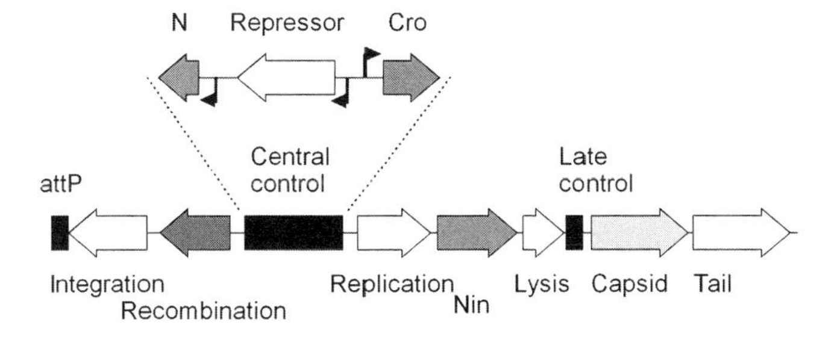

Fig. 2. Diagram of the gene layout of coliphage λ. For comparative purposes in all the genomic diagrams for the temperate phages, the integration features (*int* and *attP*) are presented on the left. The central control region specifies the repressor and its cognate operators, promoters, and those adjacent genes that control the lysis-lysogeny switch.

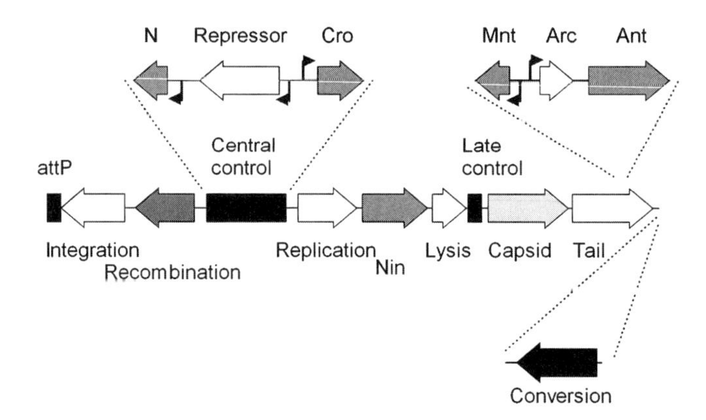

Fig. 3. Diagrammatic genetic map of *Salmonella* phage P22. Please note that the major differences between this and λ are the presence of the lysogenic conversion module (*gtrABC*) downstream of the integration cassette and the presence of the additional regulatory region, *immI*, inserted into the tail morphogenesis gene cluster.

conferred by the prophage repressor (*c2*; N.B. unfortunately the genetic designations for proteins of identical function in P22 and λ are frequently different. The lambda homolog in this case is *cI*), (2) superinfection exclusion mediated by the *sieA* and *sieB* genes, and (3) serotype conversion. The presence of the C2 protein represses the replication of homoimmune phage genomes, whereas the *sie* genes appear to function in preventing phage DNA injection ***(39,40)***. Lysogenization by P22 also results in a chemical change to the LPS O antigen, which results in a change in serotype from 4,[5],12 to 1,4,[5],12 and prevents the binding of P22 (*see* **Fig. 4**) ***(41,42)***. Sequence analysis ***(23)*** and cloning ***(22)*** have shown that three membrane-bound phage products are involved: GtrA, GtrB, and GrtC. GtrA is a 120-amino acid glucosyl undecanprenyl phosphate flippase, GtrB is a 310-amino acid bactoprenyl glucosyl transferase, and GtrC is a 486-amino acid conversion protein responsible for adding glucosyl residues to the side-chain repeat precursor.

Upon induction of lysogens, specialized transducing particles arise, carrying genes adjacent to the *attBP* sites, as well as a generalized transducing particle carrying only host DNA. This was most elegantly demonstrated experimentally by Ebel-Tsipis and coworkers ***(43)***.

Early events mimic those observed with coliphage λ in that transcription is initiated from two promoters, P_L and P_R, that flank the repressor (*c2*) gene. The early proteins are gp*24*, a λ N homolog that functions as a transcriptional antiterminator, and Cro, which functions to inhibit transcription from P_{RM} and generally downregulate transcription from P_L and P_R, thereby favoring

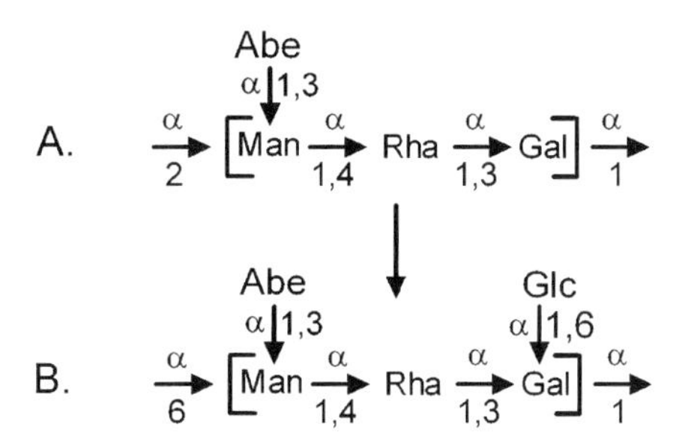

Fig. 4. The sequence of the O-antigenic repeat in *S.* Typhimurium is composed of a branched tetrasaccharide of Abe (abequose: 3,6-dideoxy-D-galactose), Man (mannose), Rha (rhamnose), and Gal (Galactose). The phage *gtrABC* cluster is responsible for the O1 antigen: a glucosyl side chain (Glc) on the terminal Gal residue.

lytic development. Another early transcript is initiated from P_{ant} in the unique *immI* region, giving rise to an antirepressor, Ant, which functions to inhibit *c2* repressor function. The synthesis of Ant is negatively regulated by Mnt binding to an operator (O_{mnt}) and preventing expression from P_{ant}. Late gene expression is regulated, as it is in λ in an antitermination-dependent mechanism involving gp*23*, a Q homolog ***(44)***. The late genes include a holin (gp*13*), a lysozyme homolog (gp*19*), and the genes involved in morphogenesis. The last have been extensively studied, revealing that, unlike the situation with λ phage morphogenesis, a unique scaffolding protein (gp*8*) is involved in the formation of a morphogenic core together with portal protein (gp*1*) and pilot proteins (gp*16*, gp*20*, and gp*7*) ***(45–49)***. The virus surface is composed almost exclusively of a single protein (gp*5*). The scaffold is reutilized in subsequent rounds of capsid assembly ***(50)***. Attention has been accorded the P22 tail structure and its assembly including very elegant cryo-electomicrographs ***(51,52)***. These have redefined our understanding of the tail machine showing the spatial relationship of tail-associated multimeric proteins gp*4* and gp*10*, TSP gp*9*, and the needle (gp*26*). A dodecamer of gp*4* subunits provides for binding of gp*10* (hexamer) to form the tail tube. To this, six tailspike trimers bind, and the tail is complete by the addition of a trimer of gp*26* which forms a 12.4 nm × 3.8 nm plug in the tail preventing leakage of DNA from the head and possibly, because it protrudes, playing a role in adsorption and/or injection ***(51)***.

DNA replication is initiated from an origin (Ori) located within the primase gene (gp*18*) ***(7)*** and involves both this protein and a helicase (gp*12*). Replication leads to the formation of concatemeric molecules as a result of rolling-circle replication. DNA packaging in P22 proceeds from a unique site (*pac*) located within gene *3* on the concatemeric substrate, resulting in the head-full packaging of a limited series of terminally redundant and circularly permuted genomes ***(53)***. P22 packages about 43.4 kb of DNA that is terminal redundant (0.9 kb, 2.2%).

2.1.2. ST104

Mitomycin C treatment of *S. enterica* serovar Typhimurium DT104 resulted in the induction and isolation of phage ST104, which, based on its close homology to phage P22 and the lack of a gene for a tail tape measure protein, is a member of the Podoviridae ***(28)***. The prophage is probably the same as the generalized transducer PDT17 identified by Schmieger and Schicklmaier in this bacterial strain ***(54)***. Its genome is 41.4 kb (47.3% G+C), and it encodes for at least 65 proteins. A high percentage of these are homologous to proteins of ε34, P22, and ST64T. These include conserved antirestriction (*abc*), conversion (*gtr*), superinfection exclusion (*sie*), *nin*, and *immC* loci. Furthermore, the putative *pac* site of this phage (GAAGACTTATCTGAGGTCGTTA) is identical to that of phage ε34, ST64T, PS119 (GenBank accession number BPS011581), PS3 (BPS011579), L ***(55)***, and LP7 ***(56)***. Antitermination protein gp*24* (N-anolog) exhibits 96% sequence identity with gp*24* of P22, suggesting that the early controls of both these phages are similarly controlled. The Q-analog (gp*23*) exhibits only 41% sequence identity with the highest degree of identity being associated with the N-terminal part of the protein.

2.1.3. ES18

Originally isolated in 1953, bacteriophage ES18 is a temperate, broad host range, generalized transducing virus of O antigen-containing (smooth) and O antigen-lacking (rough) strains of *S. enterica* ***(57,58)***. As such, it differs from other transducing phages for this genus which all require smooth LPS. The cellular receptor for ES18 is FhuA, the outer membrane protein involved in ferrichrome transport ***(59)***.

Morphologically, this phage possesses an isometric head 56 nm in diameter and a long tail (121 nm × 12 nm), making it a member of the Siphoviridae ***(14)***. One unusual property of the phage is the observation that it does not band in CsCl equilibrium gradients at the expected density of approximately 1.5 g/mL.

Using pulsed-field gel electrophoresis (PFGE) ***(169)*** demonstrated that the genome of ES18 formed a broad band at 51.5 kb, suggestive of variation in the lengths of the packaged genomes. Because the finished sequence of the DNA indicated a circular molecule of 46.9 kb (48.6% G+C), its genome must be circularly permuted and terminally redundant (ca. 10%). These features are a result of headful packaging. Detailed analysis of the process in ES18 reveals that the *pac* site is, as with other phages, located in the gene for the small subunit of terminase, but the cleavage reactions, which are a precursor to packaging from the concatemeric substrate, occur from 300 bp upstream to 700 bp downstream of this site. The consequence is that restriction digestion, coupled with agarose gel electrophoresis and ethidium bromide staining, fails to reveal a cluster of submolar fragments as are observed with phage P22. Hybridizations are required to reveal their presence.

As with all temperate phages, the ES18 genome is mosaic in nature. (It is noteworthy that ES18 has been shown to recombine with both Fels-1 and P22; *see* **ref. *60***.) The genome layout though follows the lambdoid model with its 79 genes arrayed in the following order: packaging, heads, tails, integration, recombination, central regulation, replication, and lysis. BlastX and CoreGenes analyses ***(32)*** reveal that 18 of the first 29 ES18 gene products, that is, those involved in packaging and capsid and tail formation, are related to prophage proteins in *Actinobacillus pleuropneumoniae* serovar 1 str. 4074 (GenBank accession number NZ_AACK01000018). The remainder of the genome is largely related to P22. Capsid morphogenesis has been used to subdivide the lambdoid phages into five types: HK97, Gifsy-2, 933W, λ, and P22. The analysis of its genome reveals that ES18 is the sole representative of a new group. The capsids of ES18, like those of coliphage λ, are composed of a major head protein plus a decorator protein. In the case of ES18, the major capsid protein (gp*9*) undergoes proteolytic removal of 51 amino acids, whereas the portal (gp*5*) is also shortened, in this case by 10 residues ***(14)***.

Transcription is regulated by repressor (gp*55*) and Cro (gp*56*) interactions with operators O_L (three sites), and O_R (three sites) affecting transcription from P_L, P_R, and the promoter for repressor maintenance (P_{RM}). All of these show sequence similarity to homologous sites and proteins in phage P22. Delayed early and late transcription are controlled by N-type and Q-type antiterminators (gp*54* and gp*73*, respectively). They display 63.3 and 37.8% identity to their P22 homologs, respectively. The excisionase (*xis*, gp*35*), integrase (*int*, gp*34*), and integration site (*attP*) show 96% sequence identity to the homologous region in phage P22, suggesting that ES18 and P22 share a common bacterial insertion site (*attB*) ***(14)***.

2.1.4. ST64T

This phage was mitomycin C induced, along with ST64B, from *S. enterica* serovar Typhimurium DT64 and, like P22, is a generalized transducer with a capsid 50 nm in diameter (M. W. Heuzenroeder, personal communication, 2005) ***(27)***. It is currently classified as a P22-like virus by NCBI. It is also a serotype-converting phage possessing homologs of the P22 *gtrABC* operon. Lysogenization by this phage is probably also responsible for converting *S.* Typhimurium phage type (DT) 9 to 64, 135 to 16, and 41 to 29 ***(61)***. Its genome is 40.7 kb with a G+C content of 47.5%. The 348 bp between *int* and *gtrA* is 99% identical to the similar region in ST104 and 93% to P22. Furthermore, the 128 bp proximal to the *int* gene of phage ES18 is 99% identical, suggesting that the integration sites (*attP*) of these four phages are identical.

Downstream of the late control protein (Q-homolog) we find as with P22 the lysis cassette—the lysin being 99% identical to gp*19* of *Salmonella* phage PS3 (CAA09701) and to a variety of other prophage and phage putative lysins, including those of the myoviruses RB49, RB43, and T5. The annotation of its homologs suggests that this protein may be an L-alanyl-D-glutamate peptidase. Downstream, ST64Tp47 possesses an N-terminal transmembrane domain and a high-level sequence identity to gp*15* of PS34 and Orf*66* from P22 and is most probably the Rz homolog. If one truncates the version of the annotated locus ST64Tp48 to represent a better initiation site and homologs it (*orf243*), then subsequent gene products (ST64Tp49 and ST64T50) exhibit 100% sequence identity to the products of contiguous ST104 genes (YP_006401, YP_006402, and YP_006403). Interestingly, the latter protein has been shown to be a capsid decoration protein by homology to *Salmonella* phage L (AAX21524) ***(55)***. Although its genomics are clearly related to P22, the *immI* region differs in that it lacks *arc* and *ant* genes.

2.1.5. ε34

This phage possesses an isometric head 62.5 nm in length, a neck 11 nm in width, and a short tail 4.4 nm in width by 5.5 nm in length ***(62)*** and is characterized by its ability to infect, lysogenize, and seroconvert *S. enterica* serovar Anatum ε15 lysogens.

The receptor and probable mechanisms of serotype conversion are shown in **Fig. 5**. A further common characteristic to its relative P22 is that the 60-kDa TSP (if unheated) migrates in SDS–polyacrylamide gels as a trimer and possesses LPS-depolymerizing activity ***(62)***. The first 113 amino acids of the TSP of this phage show a high level of sequence identity to analogous proteins of phages Sf6, HK620, P22, ST64T, and ST104. The conservation of the N-terminal region is a common feature among related phages because this region is associated with binding to the tail machine ***(51)***. It is noteworthy

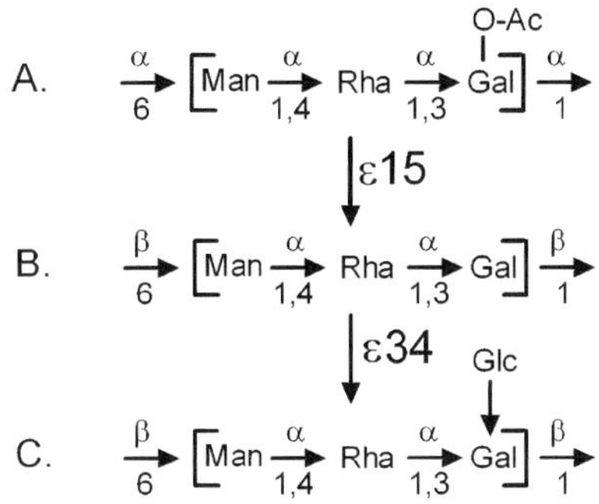

Fig. 5. Structure of the trimeric O-antigenic repeat in *Salmonella enterica* serovar Anatum by phages ε15 and ε34. The O-antigenic moiety is an *O*-acetyl group on the galactosyl residue.

that the $TSP_{\varepsilon 34}$ competes with TSP_{P22} for binding to P22 TSP-less particles, suggesting similar sites. But, noninfectious particles are formed.

The sequence of the genome of this bacteriophage was recently completed—43.0 kb and 47.3% G+C. As expected, it shows considerable overall sequence and spatial similarity to P22. The authors present the data that it integrates into the host *argU* tRNA gene ***(13)***.

2.2. Lambdoid Group

Three complete lambda-related prophages belonging to the family Siphoviridae (Fels-1, Gifsy-1, and Gifsy-2) have been identified in the sequenced *Salmonella* genomes (***19,63–66***).

2.2.1. Fels-1

Induction of *S.* Typhimurium LT2 led Yamamoto to discover two serologically and morphologically different phages (Fels-1 and Fels-2). Fels-1 at 41.7 kb is clearly a member of the lambdoid Siphoviridae (*see* **Figs. 1** and **2**). A preliminary analysis of its sequence suggests that as many as a dozen open reading frames (ORFs) may have been missed during the annotation of *S.* Typhimurium LT2 (Kropinski, unpublished results). It is integrated between host genes *ybjP* and STM0930 and carries two potential virulence genes: *nanH* (neuraminidase) and *sodC3* (superoxide dismutase) ***(19,64)***.

2.2.2. Gifsy-1

This mitomycin C and UV-inducible 47.8-kb prophage is integrated into the 5' end of the host *lepA* gene that encodes a ribosome-binding

GTPase *(67)*. The capsid measures approximately 60 nm in diameter and possesses a flexible tail approximately 133 nm in length (http://www.cgm.cnrs-gif.fr/salmonella/bossi_fr.html). Gifsy-1 derivatives exhibiting differing susceptibilities to superinfection have been identified in various *S*. Typhimurium isolates, suggesting that the immunity regions are different. Because phages commonly evolve through the horizontal exchange of gene modules or cassettes, this observation is not unexpected. The surface receptor for this phage and for Gifsy-2 is OmpC, and therefore, these phages propagate most easily on rough mutants of *Salmonella* in which this outer membrane protein is surface exposed rather than obscured by a coating of LPS ***(68)***. The annotated prophage contains a number of missed or potentially incorrectly annotated genes. STM2628 encodes a protein of 136 amino acids containing a helix–turn–helix (HTH) motif. This is separated from a 79-amino acid protein with an HTH motif with homology to phage Cro-like proteins. It is quite possible that STM2627 represents a CII-like protein. This requires experimental verification. Interestingly, the putative repressor STM2628 lacks the Ala-Gly or Cys-Gly sites associated with RecA-dependent UV induction ***(69,70)***, yet both Gifsy prophages are UV inducible ***(19)***. Both of these prophages also carry *dinI* homologs that negatively regulate induction. Gifsy-1 carries a number of potential virulence modulating genes including *gipA* in what is equivalent to the lambda *b2* region. The latter gene is involved in colonization of the small intestine, and its deletion results in reduced bacterial virulence.

2.2.3. Gifsy-2

This 45.5 kb prophage, integrated between *pncB* (nicotinate Phosphoribosyltransferase) and *pepN* (an aminopeptidase of peptidase family M1 [pfam01433]), is probably defective in strain LT2 but active in ATCC14028s ***(19)***. Induction results in the release of a siphovirus with a head diameter of approximately 55 nm. As with Gifsy-1, the 693 amino acid-containing major capsid gene (STM1033) contains an amino terminal Clp protease domain (pfam00574). The C-terminus shows homology to proteins from prophages Fels-1 (STM0912), CP-1639 (CAC83157), and CP-933K (NP_286515) and represents a novel head/scaffold/protease composite. Amino acid sequence analysis of the mature capsid protein suggests cleavage near residue 399 ***(19)***. In contrast, in Gifsy-1, STM2604 is probably a head decoration protein analogous to gp*D* in coliphage lambda, and STM2603 is the λ gp*E* homolog (pfam03864). Gifsy-2 also carries a range of potential virulence determinants, two of which—*gtgA* (identical to Gifsy-1 *gogA*) and *sodC1* (periplasmic superoxide dismutase)—have been implicated in the host pathogenesis. Deletion of *gtgA* results in a sevenfold reduction in virulence, whereas removal of *sodC1* attenuates virulence by fivefold *(71)*.

2.3. P27 Group

2.3.1. ST64B

Except for the lack of a tail structure, this phage is morphologically identical to ST64T and possesses a genome of 40.1 kb (51.3% G+C). Its genomic layout is again very similar to that of coliphage λ (*see* **Fig. 2**). The DNA packaging, capsid, and tail genes of ST64B are most closely related (and collinear) to those of ϕP27, a shiga toxin-carrying siphovirus *(72)*, and the serotype-converting *Shigella flexneri* phage V (aka SfV) that is a member of the Podoviridae *(73)* (*see* **Fig. 1**). This phage is defective in that it cannot propagate. This is most probably due to a frameshift resulting in two proteins Sb21 (NP_700394) and Sb22 (NP_700395) that are homologous to P27p52 (NP_543104), SfrVp19 (NP_599051), and Mup47 (NP_050651). The latter proteins are variably described as a hypothetical, uncharacterized homolog, tail protein, or baseplate J-like protein. An unusual feature of this phage is protein SfVp11 (NP_599043), which is defined as a "tail sheath protein." It is not only homologous to myovirus Mu tail sheath protein (gp*L*; pfam6274) but also to SfV and P27 proteins. It also encodes a 73-amino acid residue putative DNA inversion protein (Sb27; NP_700400) that possessed sequence similarity to but is considerably shorter than P1 Cin or Mu Gin, suggesting that it might not be active.

The *attP* site is located downstream of *int*, and evidence has been found that this phage integrates into *serU*. The *immC* region, encoding the *cI*-like, *cro*-like, and *cII*-like genes, has been cloned and shown to mediate a change in the phage type from DT 41 to DT 44 *(74)*.

2.4. T7 Group

2.4.1. SP6

Coliphage T7 is one of the best-studied virulent (lytic) bacteriophages and one of the first to have been completely sequenced *(75)* (**Fig. 6**). Its life cycle is described here briefly: after binding to its cell receptor, LPS, viral DNA is translocated into the cell through a virus-derived pore in a novel transcription-dependent process *(76)*. Host RNA polymerase (*Ec*RNP) transcribes the leftmost 20% of the T7 genome. The product of early gene *0.3* (Ocr) is a small protein that mimics B-form DNA and binds to, and inhibits, type I restriction endonucleases *(77,78)*. Early gene *0.7* produces functions in host gene shutoff *(79)* and as a protein kinase that phosphorylates

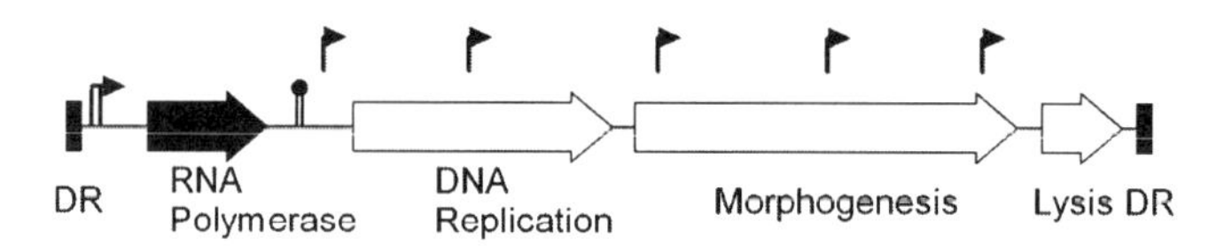

Fig. 6. Gene diagram of T7-like phages showing genome terminated by direct repeats (DRs), the presence of host-dependent (↱) and phage-dependent (↱) promoters and the rho-independent terminator at which the host RNP transcription terminates (⫯).

host elongation factors G and P and ribosomal protein S6 ***(80)***. The other, highly significant early gene is *1* that encodes the rifampicin-resistant phage-specific RNA polymerase (ϕRNP). This protein recognizes specific promoters and is responsible for middle (DNA replication) and late (morphogenesis and lysis) gene expression. At the molecular level, it is the possession of ϕRNP and their cognate promoters that have been used to define the "T7 group" of phages ***(81)***. One of the interesting features of this group of phages is the under-representation of certain common restriction sites—for example, both T7 and SP6 that are similar in mass and mol% GC content but show no DNA homology lack sites for *Bam*HI, *Pst*I, *Sac*I, *Sac*II, *Sal*I, *Sma*I, and *Sph*I.

SP6 has a 43.8-kb (47.2% G+C) genome with 174 bp direct terminal repeats ***(25,82)***. Although NCBI currently lumps it with "unclassified T7-like viruses" using Chen and Schneider's ***(81)*** criteria, it is a T7-like phage. Scholl et al. ***(82)*** describe it as belonging to "an estranged subgroup of the T7 super-group" showing the closest relationship to coliphage K1-5. CoreGenes analysis reveals that it only shares nine genes in common with T7 (*see* **Fig. 1**), whereas *Pseudomonas putida* phage gh-1 and coliphage T7 share 27 common genes. An interesting property of SP6 is that gp*49* is an acidic 59-kDa protein with homology to the C-termini of gp*9* of *Salmonella* phages ST104, ST64T, and P22 and to Orf49 of KS7. The P22 homolog has been demonstrated to be the TSP and possesses endorhamnosidase activity at its carboxy terminus, the portion of gp*49* that shares 48.3% sequence identity.

The biotechnological significance of this group of phages, and SP6 in particular, is that their promoters and polymerases have been used in an extensive array of general cloning (e.g., pGEM) and protein expression (pET and pALTER) vector systems as well as in the production of RNA hybridization probes and RNA interference and synthesis of mRNA for *in vitro* translation.

2.5. P2 Group

2.5.1. PSP3

The P2-like phages are temperate members of the Myoviridae and include coliphages P2 and 186, *Pseudomonas* phage ϕCTX, *Haemophilus* phages HP1 and HP2, and *Salmonella* phages PSP3 and SopEϕ. The genomic layout is shown in **Fig. 7**. The central control region of P2 contains the lysogenic repressor (C) and Cox. The latter protein, which is analogous to Cro of the lambdoid phages, is involved with the inhibition of the lysogenic promoter, but unlike Cro, it is not essential for lytic development. Integration into the *Escherichia coli* K12 genome occurs at $attB_{P2}$ located between genes *yegQ* and *b2083*, and it catalyzes, as with coliphage λ, by Int and IHF. Interestingly, in P2 the function of excisionase (Xis) during excision is provided by Cox. Lysogens also express three conversion genes (also known as morons): *fun(Z), old* (a predicted ATP-dependent endonuclease), and *tin* that result in the cell becoming resistant to coliphages T5, λ, and T4, respectively. Replication involves a complex of proteins A and B and results in single circular molecules (rather than concatemers), which are the substrate for packaging. Late gene expression is under positive regulation by Ogr from four promoters which display limited sequence homology to RpoD-dependent *E. coli* promoters and possess novel imperfect 17 bp inverted repeats at –55. Experimental evidence suggests that Ogr interacts with the RpoA subunit of the host RNA polymerase complex to enable recognition of these promoters ***(83)***. An interesting feature of P2 is that the usually collocate terminase genes (*terS* and *terL*) are in reverse order and interspersed with scaffold and capsid-encoding genes.

Phage PSP3 was originally isolated from *Salmonella potsdam* but can lysogenize *E. coli* ***(170)***. The 30.6 kb genome (52.8% G+C) has 19-bp 5'-extended

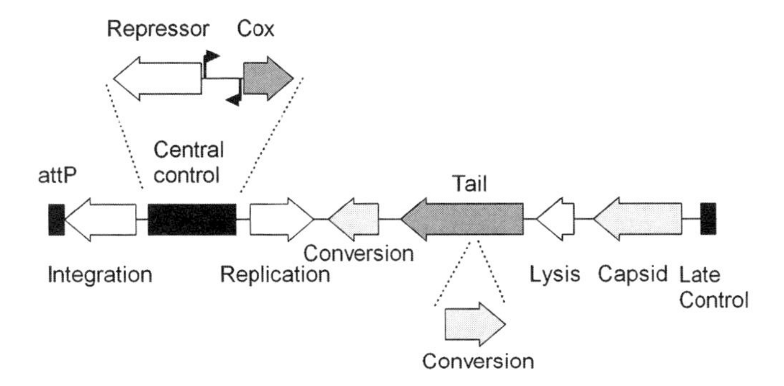

Fig. 7. Diagram of the gene layout of coliphage P2.

cohesive termini and encodes 42 proteins, 30 of which are P2 homologs. This phage is a closer relative of 186 and possesses no homologs to *old, tin*, or *fun*.

2.5.2. Fels-2

The Fels-2 phage has a head 55 nm in diameter and a contractile tail 110 nm × 20 nm ***(65)***. The prophage genome is 33.7 kb (52.5% G+C) and appears to be bounded by 47 bp direct repeats. As a prophage, it is in the genome of *Salmonella*. Typhimurium LT2 integrated between residues 2844427 and 2879233; and, shares 32 homologs with PSP3 and 29 with P2. The prophage contains several unannotated and missannotated genes including an ORF that is related to PSP3 TumB (NP_958095). The prophage integrates into the 3' end of the host *ssrA* genes. A unique feature of this phage is the presence of a gene encoding a DAM methylase (STM2730). The host is also Dam positive. Lastly, Yamamoto ***(65)*** demonstrated that the Fels-2 prophage could recombine with the morphologically unrelated P22 phage to give rise to F22 that is morphologically and serologically related to Fels-2 but carries the P22 *c* genes ***(84)***.

2.5.3. SopEϕ

SopEϕ also has the same *attB* site as Fels-2 ***(85)*** and shares many genes in common with the latter phage, and 25 with P2. Induction, with mitomycin C, from *S*. Typhimurium strain DT204 resulted in particles with heads 58 nm in diameter and contractile tails 133 nm × 19 nm ***(20)***. These particles infect sensitive *Salmonella* strains in an FhuA, TonB-independent manner and exhibit a burst size of approximately 8. The genome is approximately 34.7 kb (51.3% G+C) and carries the type III effector protein SopE that binds to, and transiently activates, eukaryotic RhoGTPases ***(86)***. *Salmonella* mutants lacking this protein show reduced invasiveness.

2.6. Orphans

2.6.1. KS7

The 40.8 kb genome of this phage has been recently sequenced by a group at Korea National Institute of Health, Laboratory of Enteric Infections, in Seoul (South Korea) (NC_006940). The data entry suggests that this is a member of the P22-like viruses. This is not correct because it is not supported by dotplot nucleotide similarity analysis nor by the presence of homologs revealed by BLASTP analysis (*see* **Fig. 1**). BLASTP analysis at NCBI indicates that the closest relative maybe phage MB78, but it should be noted that only limited sequence data are available on MB78. (Twenty-two proteins are found in GenBank.) An analysis of its sequence against other enterobacterial phage genomes using CoreGenes failed to indicate a significant level of sequence similarity. It displays limited sequence relatedness to a prophage in *Bordetella bronchiseptica* RB50.

Although classified as an "unclassified bacteriophage" at NCBI ***(21)***, the electron micrographs of Joshi et al. ***(87)*** clearly indicate that MB78 is a member of the Siphoviridae: head approximately 60 nm in diameter and tail 90–100 nm in length. The double-stranded circularly permuted and terminally redundant genome of this phage is 42 kb. The combined restriction maps of Murty et al. ***(88)*** and Khan et al. ***(89)*** include the cleavage sites for 12 restriction endonucleases. Although this phage exhibits no serological cross-reactivity with P22, and indeed, their genomes show no cross hybridization, genetic hybrids have been obtained ***(90,91)***. Additionally, MB78 is virulent (lytic), whereas P22 is temperate. An interesting feature of this phage is the putative presence of a restriction (GenBank accession number CAC81910) and modification (CAC81909) system. Unfortunately, both the former and latter proteins are related to phage (KS7, ST104, ST64T, and P22) TSPs, suggesting a frameshift sequencing error. This was confirmed by BlastX analysis of GenBank sequence AJ277754 that appears to contain frameshift errors as well as inframe-stop codon errors.

2.6.2. ε15

The receptor for phage ε15 is the O antigen of *S. enterica* serovar Anatum that belongs to serogroup E1 expressing serotype factors 3 and 10. The receptor structure is shown in **Fig. 5A**. Lysogenization of cells resulted in a serological change to serogroup E2 (*see* **Fig. 5B**), a phenomenon first explored by Cletake and Colleagues ***(92–94)***, and then Robbins and his coworkers ***(95–100)***. Electron micrographs of this phage reveal an isometric head 50 nm in diameter ***(101)***. The tail, like that of phage P22, possesses endorhamnosidase activity, capable of degrading Group E1 *S. enterica* O-polysaccharide polymers down to a D-*O*-acetyl-galactosyl-α1 → 4-D-mannosyl-β1 → 4-L-rhamnose trisaccharide end-product ***(102,103)***. The function of this enzymatic activity is probably to allow the virus particle closer access to the surface of the outer membrane prior to tight binding and DNA injection ***(104)***.

The unique genome sequence of ε15 is 39.7 kb and 50.8 mol%G+C ***(12)*** (**Fig. 8**). Because the original estimation of mass, based on restriction analysis, was 40.3 kb, the genome is circularly permuted with a terminal redundancy of approximately 0.6 kb ***(101)***. This phage contains few homologs to its 50 potential gene products amongst existing *Salmonella* phage proteins. Until

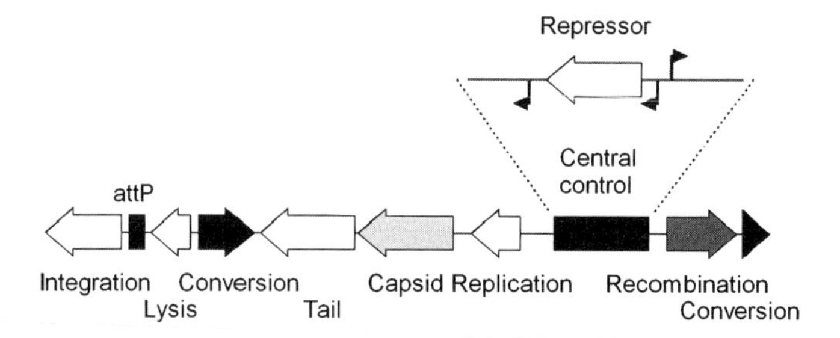

Fig. 8. Gene map of phage ε15. Please note that the conversion genes are not clustered and that *attP* is upstream of *int*.

recently, the nearest phage relatives of ε15 appeared to be *Photobacterium profundum* prophage PϕPpr1 (11 genes in common, *see* **ref. *105***), *Burkholderia cepacia* phage BcepC6B (10 genes in common, *see* **ref. *106***), and *Bordetella bronchiseptica*-specific phage, BPP-1 (9 genes in common, *see* **ref. *107***). In each case, the homologs are to other known or putative members of the Podoviridae. In the case of all three homologous phages, it is the morphogenic genes that are conserved. Recently, L. L. Perry and B. M. Applegate (Purdue University) submitted the sequence on an *E. coli* O157:H7-specific phage, ϕV10, to GenBank (NC_007804). This phage displays considerable sequence similarity to ε15 but lacks the conversion genes of this phage.

Although the protein product of gene *38* exhibits poor sequence similarity with other known phage repressors, it is similar in size (198 amino acids) and contains an HTH motif of the type that typically serves in operator recognition. A clear plaque mutation (ε15*vir*) maps to this gene. In typical manner, the repressor genes of Lambda, D3 ***(108)***, and phage r1t ***(109)***, ε15 gene *38* lacks an identifiable ribosome binding site. This phage is not UV inducible, despite containing a potential cleavage site (Ala120-Gly121) for the RecA protein-stimulated autodigestion of the major repressor proteins ***(69,70,110)***. Interestingly, we have identified ten 17 bp, hyphenated, inverted repeats, which are represented by the consensus sequence ATTACCWDWWNGGTAAT. Of particular interest was our observation that three of the putative operator sites lie to the left of gene *38* and two to its right. This suggests that, as with λ, divergent transcription is modulated by protein binding to sites on either side of the repressor gene. The significance of the other sites is, as yet, unknown. We have no biochemical, homology, or protein motif data to support the notion that either of the upstream genes (gp*39* or gp*40*) encodes a Cro homolog.

The two other noteworthy features of ε15 are putative presence of an upstream *attP* site and the genes involved in conversion. The phage integration site *attP* is almost always located downstream of the *int* gene. The only known exceptions are *M. xanthus* phage Mx8, where *attP* is located within *int* ***(111)***, and enterobacterial phages HK620 ***(112)*** and Sf6 ***(113)***, where *attP* is located upstream of *int*. Interestingly, there is no evidence for an Xis homolog.

Lysogenic conversion is brought about through the concerted activity of three intimate membrane proteins gp*21*, gp*22*, and gp*28*. The latter 60-amino acid protein acts to inhibit the transacetylase activity found in the wildtype cells, whereas the 66-amino acid-encoding gene *22* inhibits the α-polymerase activity responsible for adding the repeat units in an $\alpha 1 \rightarrow 6$ pattern. The 43-kDa product of gene *21* encodes a new $\beta 1 \rightarrow 6$ polymerase. Interestingly, only two of these genes are linked. The small size of the membrane inhibitors mimics that of the *Pseudomonas aeruginosa* serotype-converting phage D3 LPS polymerase inhibitor ***(3)***.

In a brilliant study by W. Jiang and colleagues, the structure of the entire ε15 virion has been studied by cryoelectron microscopy and proteomics ***(114)***. The latter analysis revealed that the virus particles contain capsid protein (gp*7*), which is partially proteolytically processed and may also be, in part, cross-linked, a 12-subunit portal complex (gp*4*), six bulbous tailspikes (gp*20*), and a tail hub, probably composed of gp*15*, gp*16*, and gp*17*.

2.6.3. Felix O1

Originally known as phage O1, this virus is a member of the Myoviridae with an icosahedral head 73 nm in diameter and a contractile tail (17 nm × 113 nm) terminating in six straight tail fibers ***(115)***. It was originally isolated by Felix and Callow ***(116)***. This phage lysed well over 99% of *Salmonella* strains it was tested on ***(117)***. Its surface receptor is the terminal *N*-acetylglucosamine residue of the LPS core. Therefore, deep rough mutants of this genus are Felix O1 resistant ***(118)***. Because this phage infects almost all *Salmonella* isolates, it has been used as a diagnostic reagent ***(119)***. In addition, a derivative of Felix O1 carrying the *luxAB* genes has been constructed to examine *Salmonella* in food samples ***(120)***. Furthermore, because of its broad host range, it has been investigated as a way of reducing contamination of foodstuffs by *Salmonella* ***(15)***.

The phage genome is 86.2 kb (39% G+C) and had been reported ***(121)*** to have a limited number of restriction sites. The sequence reveals that the high AT-content-recognizing restriction enzymes *Bcl*I, *Bgl*II, and *Eco*RI do not cleave. The annotated genome encodes 22 tRNA genes ***(16)*** and 130 protein-encoding genes (Whichard, Kropinski, and Sriranganathan, manuscript in preparation).

Most of these genes do not show homologs. Among genes that share homologs, one finds the myovirus ubiquitous *rIIAB* genes, which formed the basis of Seymour Benzer's pioneering work on the nature of genes, together with the subunits of aerobic and anaerobic ribonucleoside triphosphate reductases. Of the Felix O1 proteins with homologs, 12 are T4-like, 9 T5-like, and 4 T7-like.

3. *Salmonella* Phage—Practical Aspects

3.1. Genetic Manipulations

One of the prime uses of P22 is based on its ability to transfer host DNA (i.e., generalized transduction). These transducing particles are normally present at 1–3% of the plaque-forming particles, but high transducing derivatives (P22HT) have been isolated in which 50% of the particles can transduce ***(122–124)***. Classically, this was used for genetic mapping, but it is now more often used for strain construction. Other derivatives of P22 include the hybrid Mud-P22 phages that contain the termini of transposable coliphage Mu with the packaging features of P22 ***(125,126)***. These randomly insert into the host genome and upon induction specifically package DNA adjacent to the integration site.

The function of the *ImmI* region has been exploited by Stanley Maloy and coworkers to create the challenge phage system for studying protein–DNA interactions ***(127–130)***. A *mnt*::Km derivative engineered to carry a specific DNA-binding motif rather than O_{Mnt} is tested for its ability to lysogenize, that is, generate kanamycin-resistant colonies. Expression of the cognate DNA-binding protein will result in repression of Ant expression and enhanced lysogenization.

3.2. Phage Typing

3.2.1. Reasons for Characterization of Salmonella Isolates by Phage Typing

Phage typing entails a detailed subtyping of *Salmonella* isolates belonging to a particular *Salmonella* serovar. Further characterization of *Salmonella* isolates is of great benefit because it enables the investigator to trace cases and outbreaks of salmonellosis to its source and to recommend measures to eliminate the source and prevent reoccurrence of the disease. More than 2500 *Salmonella* serovars are now recognized ***(131,132)***. However, human and animal infections are primarily caused by a small subset of commonly occurring serovars that almost all belong to *S. enterica* subsp. *enterica*.

3.2.2. The Necessity to Serotype Salmonella *Isolates Before Phage Typing is Attempted*

Phage typing is a well-established, effective, and commonly used diagnostic technique producing reliable results and maximum differentiation of *Salmonella* isolates *(8)*. Before phage typing is attempted, some certainty must be obtained that the isolate indeed belongs to the *Salmonella* species. Therefore, isolates are first characterized as *Salmonella* by their ability to grow in selective enrichment broths such as tetrathionate brilliant green broth or Rappaport–Vassiliadis broth and by displaying specific biochemical reactions when grown on brilliant green sulfa agar or lysine decarboxylase agar, when streaked on urea slants and when stabbed into a triple sugar iron and Simmons citrate slants. Identity of *Salmonella* may be surmised based on a characteristic colony morphology and color when grown on media to distinguish *Salmonella* from other Enterobacteriaceae. This is often confirmed by picking a colony, preferably from a nonselective agar, and performing a slide agglutination reaction with polyvalent *Salmonella* antisera. When agglutinating, the isolate is likely to be a *Salmonella* strain and submitted to a *Salmonella* reference laboratory for serotyping. Serotyping is a method of subtyping of *Salmonella* isolates by examining the antigenic properties of the isolate. This is carried out by determining the O or somatic antigens and the H or flagellar antigens with the aid of group and serovar-specific O and H antisera. Serotyping the *Salmonella* isolate is a prerequisite and is generally carried out before phage typing is being attempted. The interaction of bacteriophages with their bacterial host is more or less specific for the *Salmonella* serovar. Serotyping often identifies surface structures of the *Salmonella* isolate that are unique for attachment by specific bacteriophages.

3.2.3. Phage Typing Procedures

Phage typing is carried out by infecting a *Salmonella* isolate of a serovar, for example, *S*. Typhi or *S*. Typhimurium, with a number of phages listed in the phage typing scheme for that serovar. The media employed in phage typing are phage broth consisting of 20 g Difco Nutrient Broth and 8.5 g sodium chloride that are mixed in 1 L distilled water and boiled to dissolve the ingredients, and then dispensed in 3.5 mL volumes in tubes and autoclaved. To prepare a solid medium, 13 g Difco Bacto Agar is added to the phage broth and after mixing, boiling, autoclaving and cooling to 50 °C, poured into scored square 15 cm × 15 cm Petri plates. The pH of both media should be about 6.8. The plates are dried before use. To type a *Salmonella* strain, the isolate is first grown in broth to a barely visible turbidity. The broth is flooded onto the agar surface with the aid of a sterile plastic pipette, and the broth is then sucked up with the same pipette and returned to its tube. The plate is dried again, and the typing phages are dropped with the aid of a set of loops dipped into each of the typing phages or from a set of syringes containing the phages onto the surface of the phage agar. The plates are incubated overnight at 37 °C. The phage typing plates are usually read after 24 h incubation. The readings are done with a ×10 hand lens through the bottom of the plates using transmitted oblique illumination. The method of recording degrees of lysis on the *Salmonella* phage typing plates, including identification criteria of the plaque sizes, plaque numbers, and the kind of lysis have been described ***(133,134)***. The spots where the phages have been dropped and where they caused lysis are described as confluent lysis (CL), semiconfluent lysis (SCL), intermediate degrees of lysis (<CL and <SCL), and opaque lysis (OL, a confluent lysis with a heavy central opacity because of secondary growth of lysogenized bacteria) (Callow, 1959) (*see* **Fig. 9**). The phages used for typing are in the routine test dilution (RTD). The RTD of a phage is the highest dilution that produces confluent or semiconfluent lysis on its homologous type strain, the strain on which the phage had been propagated. If the phages were to be used undiluted, many of the reactions might be nonspecific (Anderson and Williams, 1956). Some of the phages producing small or minute plaques may only have to be diluted to 10^{-3}, whereas others that produce large plaques can be diluted to 10^{-5} or 10^{-6} to obtain the RTD. Different sets of phages are being used for the phage typing of different *Salmonella* serovars.

Fig. 9. Examples of phage typing from the OIÉ Reference Laboratory for Salmonellosis, Laboratory for Foodborne Zoonoses (Public Health Agency of Canada; courtesy of Dr. A. Muckle): left (*Salmonella* Heidelberg) and right (*Salmonella* sp. I:4,12:i).

3.2.4. Specificity of Salmonella *Bacteriophages*

The specificity of the phages for a serovar or for isolates of the same serovar may differ dramatically. The specificity of a phage for the bacterium depends on a number of factors including the ability of the phage to adsorb to the surface structures of the host. Some phages display a high degree of host specificity, whereas others have a wide host range. A phage with high specificity is phage ϕX174, which grows well on *E. coli* strain C but fails to grow on most other *E. coli* laboratory strains. A phage with a very wide host range for *Salmonella* is the Felix O1 bacteriophage ***(116)***. It lyses more than 99% of all *Salmonella* isolates ***(117,135)*** and is used to distinguish *Salmonella* isolates of *S. enterica* subspecies *enterica, salamae, diarizonae*, and *indica* that are lysed, from those belonging to *S. enterica* subspecies *arizona* and *houtenae* that are not lysed ***(131)***. Many of the *Salmonella* serovar or serogroup specific phages attach to receptors in the O side chain the LPS of the outer membrane of Gram-negative bacteria, and a striking relationship has been observed in *Salmonella* between O antigens and phage susceptibility ***(136)***. Phage typing of *Salmonella* is mostly performed with sets of phages that are more or less specific for the serovar to which the isolates belong.

3.2.5. Phage Typing Schemes for Salmonella *Serovars*

Phage typing of *Salmonella* has been employed since the late 1930s. The phage typing method and the first phage typing scheme, the Vi-phage typing system, were developed by Craigie and Yen ***(137,138)***. The Vi antigen is a virulence capsular exopolysaccharide consisting of an acetylated polymer of galactosaminuronic acid. Other bacteria that are known to express this antigen are *S*. Paratyphi C, some strains of *S*. Dublin, and a few strains of *Citrobacter freundii* ***(139)***. The Vi antigen is encoded by the *viaB* operon. Nair et al. ***(140)*** showed that *S*. Typhi strains that possess the Vi antigen were lysed with adapted Vi II phages and with unadapted Vi I+IV phages, all of which use the Vi antigen for adsorption to the cell by Vi-specific phages, but that strains that lacked the Vi antigen were not lysed by these phages. They further noticed that 112 of 120 *S*. Typhi strains with the Vi antigen contained a 137-kb *Salmonella* Pathogenicity Island 7 (SPI7) encoding the *viaB* operon and possessing genes for type IVB pili, for putative conjugal transfer, and for SopE bacteriophage. The eight remaining strains had a complete or partial deletion of SPI7 and did neither possess the *viaB* locus for Vi exopolysaccharide nor the associated genes, and these were not lysed by the Vi-specific phages. The *S*. Typhi phage typing scheme was followed by the development of a typing scheme for *S*. Paratyphi B by Felix and Callow ***(116)***. These schemes were instrumental in the identification of chronic fecal carriers and contaminated food products ***(116)***. A phage typing scheme for *S*. Typhimurium was developed in the 1940s and 1950s, and an extended scheme recognized 80 phage types ***(134)***. *Salmonella* Typhimurium has been the predominant cause of salmonellosis in humans and is has also been the most frequently isolated *Salmonella* serovar from cattle and pigs, food products, animal feeds, and environmental sources ***(134,141)***. The *S*. Typhimurium typing scheme was again extended, and definitive numbers were given to the phage types; the present scheme consists of 34 phages and identifies 207 phage types ***(7)***. Large egg-associated outbreaks of *S*. Enteritidis infection in humans have occurred worldwide during the late 1980s and 1990s and prompted the development of a phage typing scheme for this serovar ***(142)***. Presently, it consists of 16 typing phages that distinguish 77 phage types. A few of the phage types may account for more than 80% of the isolates ***(143)***. The acquisition of plasmids may cause a phage type conversion. An example thereof is the conversion of *S*. Enteritidis PT4 to PT24 following acquisition of an incompatibility group N (incN) drug resistance encoding plasmid ***(171)***. During 1990–2005, *S*. Heidelberg has caused a large number of infections in chickens and humans in Canada and the United States. An *S*. Heidelberg phage typing scheme has recently been developed ***(144)***. It employs 11 typing phages and distinguishes 49 phage types. Many phage typing schemes including those for typing *S*. Typhimurium and *S*. Enteritidis and associated techniques have been developed at the Central Public Health Laboratory at Colindale, London, United Kingdom. This laboratory is the World Health Organization (WHO) Reference Laboratory for the phage typing of *Salmonella*.

3.3. Phagotherapy

Recent advances in understanding the genomics and life cycles of various *Salmonella* phages have raised intriguing and far-reaching possibilities for their practical applications, both as research tools and for "phage therapy." Nontherapeutic applications of *Salmonella* phages are described early in this chapter. This section includes a brief overview of the therapeutic use of *Salmonella* phages and their possible practical applications in various clinical or agricultural settings.

3.3.1. Initial Studies of Therapy With Salmonella *Phages*

Salmonella phages were the first phages examined for their ability to prevent and treat bacterial infections in various settings. The studies were conducted by Felix d'Herelle in 1919, a few years after the independent discovery of bacteriophages by Frederick Twort and Felix d'Herelle in 1915 and 1917, respectively ***(145)***. During the initial pilot experiment, d'Herelle isolated *Salmonella* bacteriophages from chickens, and he used them to treat birds experimentally infected with "*Salmonella gallinarum.*" It was a very small-scale study: only four birds were included in the phage-treated group, and two birds served as phage-untreated controls. Nonetheless, the results were promising: the phage-treated birds survived, but the untreated birds died from the experimental infection. d'Herelle extended his phage therapy studies almost immediately after the completion of his initial pilot study by conducting major field trials during 1919–1920. The phage preparations for the field trials were prepared by propagating *Salmonella* phages having the most potent lytic activity against *S. gallinarum*. The resulting phage lysates were packaged (0.5 mL volumes in sealed ampoules) and distributed to veterinarians in various regions of France, with instructions to administer them to chickens by subcutaneous injection. The treatment resulted in cessation of the epidemic and the recovery of most of the sick chickens ***(146)***. Encouraged by these results, d'Herelle expanded his therapy studies with other phages on other animals and humans. The success of d'Herelle's early phage therapy studies also prompted other investigators to begin examining the value of various bacteriophages in dealing with various infections of bacterial origin. Some of those studies in which *Salmonella*-specific bacteriophages were used are briefly reviewed in 3.3.2, 3.3.3 and 3.3.4. A more extensive review of various agricultural applications of *Salmonella* and non-*Salmonella* phages is available in the literature ***(147)***.

3.3.2. Salmonella *Phages for Preventing and Treating Salmonellosis in Laboratory Animals*

d'Herelle's early phage therapy studies triggered strong initial interest in the possibility of using *Salmonella* phages to prevent and treat salmonellosis in animals. To give just a few examples, Topley et al. ***(148,149)*** orally administered *Salmonella* phages in order to evaluate their efficacy in mice experimentally infected with *S.* Typhimurium. However, in contrast to d'Herelle's observations with *Salmonella*-infected chickens, phage administration did not reduce mortality of mice. One possible explanation for the discrepancy between these studies and d'Herelle's earlier observations is that *Salmonella* phages, with weak lytic activity, were used by Topley et al., whereas d'Herelle used phages with very strong lytic activity against the targeted bacteria. In this context, when Fisk ***(150)*** injected antityphoid phages possessing strong in vitro activity against the challenge *Salmonella* strain into mice before challenge with typhoid bacilli, phage administration strongly protected the mice. Also, in another study ***(151)***, when phages that specifically recognized the Vi antigen (a major virulence factor of *S.* Typhi) were injected into mice challenged with

S. Typhi, the mortality rate was reduced from 93% in the phage-untreated control group to 6% in the phage-treated group (452 mice were included in that part of the study). Interestingly, the study further emphasized the importance of using phages with good lytic activity against the challenge bacterium. For example, the mortality rates after treatment with boiled culture supernatant fluids and after treatment with a non-*S.* Typhi phages were not significantly different, and they ranged from 93 to 100% (200 mice were included in that part of the study). The protection afforded by the *S.* Typhi phage treatment was concentration dependent, and it was best when preparations containing more than 1×10^5 viable "phage particles" per mouse were employed.

The results of the aforementioned, briefly described studies were encouraging, and, together with the apparent discrepancy in results obtained by various investigators, they normally would have prompted additional research. However, the initially strong interest in phage therapy gradually decreased in the West after antibiotics became increasingly available, and research examining the ability of *Salmonella* phages to prevent or treat salmonellosis in animals was not actively pursued during the 1950s–1980s. This situation began to change in the 1990s, when a renewal of interest in phage therapy prompted several investigators to revisit the idea of using *Salmonella* phages to deal with *Salmonella* infections. Some of their studies utilized the traditional approach of administering phages to treat *Salmonella* infections. In addition, other studies explored the efficacy of an environmental approach for preventing or reducing the frequency of *Salmonella* infections; that is, applying *Salmonella* phages onto various food products, in order to reduce their contamination with salmonellae. A good example of the traditional therapeutic approach is the study by Berchieri et al. ***(152)***, who used *Salmonella* bacteriophages to treat chickens experimentally infected with *S.* Typhimurium. The treatment significantly reduced mortality, compared with the mortality of phage-untreated control birds. The efficacious phage treatment required using concentrated phage preparations (ca. 10^{10} PFU/mL), and phage administration shortly after bacterial challenge was significantly more effective than was delaying the treatment. Some of the studies examining an environmental approach for preventing or reducing the frequency of salmonellosis are reviewed briefly under **Subheading 3.3.4**.

3.3.3. Salmonella *Phages for Preventing and Treating Salmonellosis in Humans*

Although enteric infections arguably have been the most common targets for traditional phage therapy applications, using *Salmonella* phages to prevent or treat human salmonellosis has been relatively limited, even in countries (e.g., the former Soviet Union and some Eastern European countries) where phage therapy continued to be utilized during the antibiotic era ***(153)***. One such study ***(154)*** examined the efficacy of intravenously administered *Salmonella* phages (in an isotonic glucose solution) in treating 56 typhoid patients (infected with *Salmonella* serotype Typhi, previously known as *S. typhosa* and *Eberthella typhosa*) at Los Angeles County General Hospital. The authors used the same phages previously reported ***(151)*** to be effective in treating mice experimentally infected with *S*. Typhi, and efforts were made to select phages possessing strong in vitro lytic activity against the strain of *S*. Typhi isolated from each patient. The treatment reduced patient mortality from 20 to approximately 5%, and the authors concluded that treatment with bacteriophages offers a "promising and safe procedure against typhoid fever."

Shortly after the study by Knouf et al. ***(154)***, Desranleau ***(155)*** successfully used Vi antityphoid bacteriophages to treat 20 typhoid patients. He used a cocktail of at least four distinct *Salmonella* bacteriophages, and the phages were administered intravenously in an isotonic glucose solution. During a larger subsequent study ***(156)***, he used an expanded version of the previous phage cocktail (which now included two additional *Salmonella* phages, for a total of six distinct phages in the preparation) to treat *S*. Typhi infections in approximately 100 typhoid patients in the province of Quebec, Canada. The treatment continued to be effective, reducing the mortality from approximately 20% to approximately 2%. An interesting aspect of the latter study was that *Salmonella* antigens (released by phage-mediated lysis while preparing phage lysates in vitro and during the lysis of *Salmonella* in vivo) in the phage cocktail were proposed to be directly responsible for the observed clinical improvement, whereas direct antibacterial action of bacteriophages in vivo was proposed to be only an "indirect cause of the cure as regards symptoms." The idea of using phages to prepare phage lysates with strong immunostimulating activity has been advanced by several investigators since the 1920s, and it has been gaining increased attention lately, including the development of "ghost vaccines" for preventing bacterial infections (including *Salmonella* infections) in various agriculturally important animals. More information dealing with this subject is available in the literature ***(147,157)***.

In the former Soviet Union, *Salmonella* phages have been effectively used to prevent salmonellosis in children ***(158)***. Investigators at the Hirszfeld Institute of Immunology and Experimental Therapy in Poland also have reported clinical applications for *Salmonella* phages. Since its founding in 1952, the institute has used its large collection of bacteriophages to treat various bacterial infections in several hospitals in Poland. The most commonly targeted bacterial pathogens included *S. aureus, P. aeruginosa*, and *E. coli*, but phages lytic for *Salmonella* also were employed successfully to treat human infections ***(159)***. At the present time, therapeutic *Salmonella* bacteriophages are commercially produced by at least one company in Russia. ImBio currently manufactures several phage-based therapeutics, including a *Salmonella* phage cocktail ("Bacteriophagum salmonellae gr.ABCDE liquidum") targeting approximately 10 different *Salmonella* serotypes (http://home.sinn.ru/~imbio/Bakteriofag.htm).

3.3.4. *Salmonella* Phages for Improving Food Safety

A possible and novel application of bacteriophages, which recently has been generating increased interest, is to apply them directly onto food products or onto environmental surfaces in food processing facilities, in order to reduce the levels of foodborne bacterial pathogens in foods. The first report in which this approach was studied with *Salmonella* was recently published by Leverentz et al. ***(160)***. The authors examined the ability of phages to reduce experimental *Salmonella* contamination of fresh-cut melons and apples stored at various temperatures mimicking real-life settings. Treatment of the experimentally contaminated fruit with aliquots (25 μL per fruit slice, applied with a pipette) of a *Salmonella*-specific phage preparation reduced *Salmonella* populations by approximately 3.5 logs on honeydew melon slices stored at 5 and 10 °C, and by approximately 2.5 logs on slices stored at 20 °C (*see* **Fig. 10**), which was superior to the reduction usually achieved with commonly used chemical sanitizers. However, the phage preparation was significantly less effective on

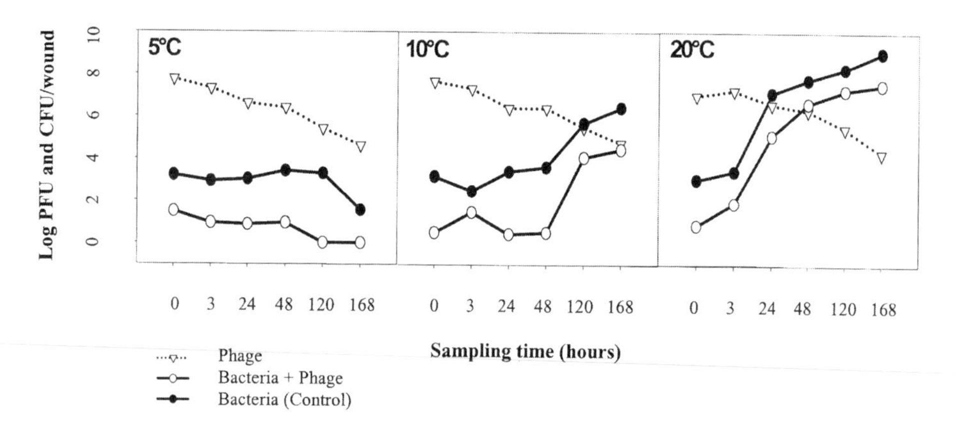

Fig. 10. Reduction of *Salmonella* serotype Enteritidis levels on honeydew melon slices treated with Salmonella-specific bacteriophages (from **ref.** ***14***). Reprinted with permission from the *J Food Prot* (Copyright held by A. Sulakvelidze).

fresh-cut Red Delicious apples, possibly because of the rapid inactivation of phages by the apple slices' more acidic pH (pH 4.2, as opposed to pH 5.8 for the melon slices). A description of one of the *Salmonella* phages (designated SPT-1) used in the earlier study was recently published ***(25)***. SPT-1 is an O1 species phage possessing broad-spectrum lytic activity against several *Salmonella* serotypes. It is a member of the family Myoviridae and contains abnormal tails that do not appear to result from in vitro assembly of dissociated phages. The phage genome has been fully sequenced, is 87,069 bp in size, and contains 129 ORFs, 95 (ca. 74%) of which encode genes of unknown function (A. Sulakvelidze, unpublished data).

A similar phage-based approach also may be of value in the poultry industry—for example, to reduce contamination of raw chicken carcasses with salmonellae prior to packaging. The results of a recent study ***(161)*** supporting that idea indicated that applying *Salmonella* phages onto chicken skin experimentally contaminated with *Salmonella* significantly reduced (by ca. 99%, $P \leq 0.01$) the number of salmonellae on the skin compared with that on phage-untreated control skin. In addition, when the level of initial *Salmonella* contamination was low, phage treatment yielded *Salmonella*-free skin. Another recent study ***(15)*** found that applying Felix O1 bacteriophage, or its mutant possessing increased in vitro lytic activity against *S.* Typhimurium strain DT104, onto chicken frankfurters experimentally contaminated with the bacterium reduced its concentration by approximately 1.8 and 2.1 logs, respectively, compared with that on phage-untreated control frankfurters ($P = 0.0001$).

The results of the studies briefly described in this subheading suggest that using phages to reduce the levels of salmonellae on various foods has merit. However, the practical applicability of that approach may be complicated by the narrow host range of phages—which may be of particular concern with *Salmonella*, a highly heterogeneous genus comprising more than 2400 serotypes ***(132)***. Thus, future research in this area may focus on developing phage preparations specifically targeting the *Salmonella* strains or serotypes known to be most often responsible for human salmonellosis or to be most virulent, such as *Salmonella* serotypes Enteritidis and Typhimurium (including *S.* Typhimurium definitive phage type 104/DT104 strains).

3.3.5. Safety Considerations

Phage-based intervention strategies may be some of the most environmentally friendly approaches for controlling *Salmonella* contamination in various settings. Phages are very specific to their bacterial hosts, and they do not infect eukaryotic cells and strains of unrelated bacteria. The total number of phages on

Earth is estimated to be 10^{30} to 10^{32} ***(162)***, and, although it is difficult to estimate what percentage of this phage population consists of *Salmonella* phages, it is likely that they are some of the most common phages in the environment. For example, the amount of F-specific phages for *S.* Typhimurium strain WG49 in U.S. sewage has been roughly estimated to be 1×10^{18} PFU per day ***(147)***.

Although the history of phage therapy strongly suggests that phages are very safe for humans, various strategies may be employed to further ensure the safety and consistency of modern therapeutic phage preparations. One concern with potentially therapeutic phages is the possible presence of undesirable, potentially harmful genes (e.g., bacterial toxin-encoding genes) in their genomes. Although such genes are very unlikely to endanger individual patients treated with phages, it is sensible to exclude phages containing them from commercial phage preparations, in order to limit their release into the environment. Thus, it is prudent to ensure that *Salmonella* phage preparations do not contain bacterial toxin genes in their genomes before they are used for traditional phage therapy or for improving food safety. From the practical standpoint, this should be feasible. For example, the increasing availability of high-throughput sequencing techniques should permit full-genome sequencing of all phages considered for therapeutic applications and their rapid screening for the presence of undesirable genes.

3.3.6. Concluding Remarks

Salmonellae are one of the leading causes of foodborne disease worldwide, causing 1.3–1.4 million cases annually in the United States alone ***(163)***, with the associated costs estimated to be as high as $12.8 billion per year (in 1998 dollars) ***(164)***. Despite the importance of salmonellae as human pathogens, studies evaluating the efficacy of *Salmonella*-specific phages in preventing or treating human salmonellosis are relatively rare. Several reasons may account for this phenomenon. One likely reason is that *Salmonella* infections have low fatality rates, and, when necessary, they can be relatively easily controlled by commonly available antibiotics. Thus, the lack of a pressing need for an alternative treatment modality may explain the lack of strong interest in developing therapeutic *Salmonella* phages for human salmonellosis, even in countries where phage therapy has continued to be utilized during the antibiotic era. Another possible explanation is that because salmonellae are intracellular pathogens, they must be targeted by therapeutic bacteriophages in the gut before they have been internalized and protected from the lytic effect of phages. Therefore, initiating phage treatment after the onset of clinical symptoms may be impractical, and such approaches have not been actively pursued. On the contrary, *Salmonella* phages may provide an intriguing tool for preventing salmonellosis in humans; for example, as a food safety intervention tool. Indeed, as briefly discussed in this chapter, applying *Salmonella*-specific bacteriophages to various foods and food preparation surfaces may provide a natural means for significantly reducing *Salmonella* contamination of various foods and thereby significantly improving public health.

The recent increase in the incidence of human salmonellosis, which is due to the ingestion of contaminated poultry and other foods indicates that developing and implementing novel approaches for reducing contamination of foods with *Salmonella* are of clear public health importance. A *Salmonella* phage-based approach may be an important part of an overall program for *Salmonella* control. Therefore, additional research in this area is indicated, and it is likely to generate critical data needed for the design and implementation of phage-based intervention strategies that optimally reduce the occurrence of human salmonellosis. In addition, improving our understanding of the genetic make-up and lytic cycles of various *Salmonella* phages (whose major groups were discussed earlier in this chapter) is likely to provide us with the tools and knowledge required to formulate lytic phage preparations optimal for the prevention and treatment of salmonellosis. A good example of the potential findings resulting from such research is the extensive characterization of the broad-range *S.* Typhimurium phage IRA by investigators at the Eliava Institute of Bacteriophage in Georgia ***(165)***. They cloned selected genes of the IRA phage into a plasmid vector, and they found that the recombinant plasmid pKI71 expression elicited lethal structural changes in the *Salmonella* cell wall. Similar studies (albeit not for *Salmonella* phages) have been recently gaining increased popularity in the United States ***(166–168)***, and further characterizing this and similar mechanisms is likely to yield new information about phage–bacterial host cell interactions. On a more long-term basis, they are also likely to aid in developing effective phage-based preparations and intervention strategies for *Salmonella* control and in identifying novel phage-encoded gene products of potential diagnostic or therapeutic value.

Addendum: Since writing this chapter ε15 is non longer a genomic orphan. *Escherichia coil* phage φV10 is closely related to it. (Perry, L. L. and Applegate, B. M. Sr. 2005. Complete nucleotide sequence of *Escherichia coil* O157:H7 bacteriophage φV10. GenBank accession number NC_007804).

Acknowledgments

A.M.K. would like to thank Don Seto for help with the CoreGenes analysis, Robert Villafane and Sherwood Casjens for an early look at the ε34 data, Michael McConnell for the ε15 data, and Nammalwar Sriranganathan for data on Felix O1. He also acknowledges the funding provided by the Natural Sciences and Engineering Research Council of Canada.

References

1. Rohwer, F. (2003) Global phage diversity. *Cell* **113,** 141.
2. Broudy, T. B. and Fischetti, V. A. (2003) *In vivo* lysogenic conversion of Tox(−) Streptococcus pyogenes to Tox(+) with lysogenic streptococci or free phage. *Infect. Immun.* **71,** 3782–3786.
3. Newton, G. J., Daniels, C., Burrows, L. L., Kropinski, A. M., Clarke, A. J., and Lam, J. S. (2001) Three-component-mediated serotype conversion in *Pseudomonas aeruginosa* by bacteriophage D3. *Mol. Microbiol.* **39,** 1237–1247.
4. Zhou, Y., Sugiyama, H., and Johnson, E. A. (1993) Transfer of neurotoxigenicity from *Clostridium butyricum* to a nontoxigenic *Clostridium botulinum* type E-like strain. *Appl. Environ. Microbiol.* **59,** 3825–3831.
5. Cairns, J., Stent, G. S., and Watson, J. D. (1966) *Phage and the Origins of Molecular Biology*. Cold Spring Harbor Laboratory Press, Cold Spring Harbor, NY.
6. Nicolle, P., Vieu, J. F., and Diverneau, G. (1970) Supplementary lysotyping of Vi-positive strains of *Salmonella typhi*, insensitive to all the adapted preparations of Craigie's Vi II phage (group I+IV). *Arch. Roum. Pathol. Exp. Microbiol.* **29,** 609–617.
7. Anderson, E. S., Ward, L. R., De Saxe, M. J., and De Sa, J. D. H. (1977) Bacteriophage-typing designations of *Salmonella* Typhimurium. *J. Hyg.* **78,** 297–300.
8. Anderson, E. S. (1964) The phage typing of *Salmonella* other than *S*. Typhi, in *The World Problem of Salmonellosis* (Van Oye, E., ed), Dr.W.Junk Publishers, The Hague, pp. 89–109.
9. Fischetti, V. A. (2001) Phage antibacterials make a comeback. *Nat. Biotechnol.* **19,** 734–735.
10. Ackermann, H.-W. and Abedon, S. T. (2000) Bacteriophage Names 2000: A compilation of known bacteriophages. http://mansfield.osu.edu/∼sabedon/names.htm (may 16, 2007).
11. Ackermann, H. W. (1998) Tailed bacteriophages: the order *Caudovirales. Adv. Virus Res.* **51,** 135–201.
12. Kropinski, A. M., Kovalyova, I. V., Billington, S. J., et al. Virology (in press)
13. Villafane, R., Casjens, S. R., and Kropinski, A. M. (2005) Sequence of *Salmonella enterica* serovar Anatum-specific bacteriophage Epsilon34. Unpublished results.
14. Casjens, S. R., Gilcrease, E. B., Winn-Stapley, D. A., et al. (2005) The generalized transducing *Salmonella* bacteriophage ES18: complete genome sequence and DNA packaging strategy. *J. Bacteriol.* **187,** 1091–1104.
15. Kuhn, J., Suissa, M., Chiswell, D., et al. (2002) A bacteriophage reagent for *Salmonella*: molecular studies on Felix 01. *Int. J. Food Microbiol.* **74,** 217–227.
16. Sriranganathan, N., Whichard, J. M., Pierson, F. W., Kapur, V., and Weigt, L. A. (2004) Bacteriophage Felix O1: genetic characterization (GenBank accession number NC_005282). Unpublished results.

17. McClelland, M., Sanderson, K. E., Spieth, J., et al. (2001) Complete genome sequence of *Salmonella enterica* serovar Typhimurium LT2. *Nature* **413,** 852–856.
18. Reen, F. J., Boyd, E. F., Porwollik, S., et al. (2005) Genomic comparisons of *Salmonella enterica* serovar Dublin, Agona, and Typhimurium strains recently isolated from milk filters and bovine samples from Ireland, using a Salmonella microarray. *Appl. Environ. Microbiol.* **71,** 1616–1625.
19. Bossi, L. and Figueroa-Bossi, N. (2005) Prophage arsenal of *Salmonella enterica* serovar Typhimurium, in *Phages: Their Role in Bacterial Pathogenesis and Biotechnology* (Waldor, M. K., Friedman, D. I., and Adhya, S. L., eds.), ASM Press, Washington, DC, pp. 165–186.
20. Mirold, S., Rabsch, W., Rohde, M., et al. (1999) Isolation of a temperate bacteriophage encoding the type III effector protein SopE from an epidemic *Salmonella typhimurium* strain. *Proc. Natl. Acad. Sci. U. S. A.* **96,** 9845–9850.
21. Kim, S., Kim, S. H., Chun, S. G., Lee, S. W., Kang, Y. H., and Lee, B. K. (2005) Therapeutic effect of bacteriophage from *Salmonella* Typhimurium (GenBank Accession Number NC_006940). Unpublished results.
22. Vander Byl, C. and Kropinski, A. M. (2000) Sequence of the genome of *Salmonella* bacteriophage P22. *J. Bacteriol.* **182,** 6472–6481.
23. Pedulla, M. L., Ford, M. E., Karthikeyan, T., et al. (2003) Corrected sequence of the bacteriophage P22 genome. *J. Bacteriol.* **185,** 1475–1477.
24. Christie, G. E. and Xu, P. (2002) Bacteriophage PSP3, complete genome (GenBank accession number NC_005340). Unpublished results.
25. Dobbins, A. T., George, M., Jr., Basham, D. A., et al. (2004) Complete genomic sequence of the virulent *Salmonella* bacteriophage SP6. *J. Bacteriol.* **186,** 1933–1944.
26. Mmolawa, P. T., Schmieger, H., and Heuzenroeder, M. W. (2003) Bacteriophage ST64B, a genetic mosaic of genes from diverse sources isolated from *Salmonella enterica* serovar typhimurium DT 64. *J. Bacteriol.* **185,** 6481–6485.
27. Mmolawa, P. T., Schmieger, H., Tucker, C. P., and Heuzenroeder, M. W. (2003) Genomic structure of the *Salmonella enterica* serovar Typhimurium DT 64 bacteriophage ST64T: evidence for modular genetic architecture. *J. Bacteriol.* **185,** 3473–3475.
28. Tanaka, K., Nishimori, K., Makino, S., et al. (2004) Molecular characterization of a prophage of *Salmonella enterica* serotype Typhimurium DT104. *J. Clin. Microbiol.* **42,** 1807–1812.
29. Boyd, J. S. (1950) The symbiotic bacteriophages of *Salmonella typhimurium. J. Pathol. Bacteriol.* **62,** 501–517.
30. Thomson, N., Baker, S., Pickard, D., et al. (2004) The role of prophage-like elements in the diversity of *Salmonella enterica* serovars. *J. Mol. Biol.* **339,** 279–300.

31. Rohwer, F. and Edwards, R. (2002) The Phage Proteomic Tree: a genome-based taxonomy for phage. *J. Bacteriol.* **184,** 4529–4535.
32. Zafar, N., Mazumder, R., and Seto, D. (2002) CoreGenes: a computational tool for identifying and cataloging "core" genes in a set of small genomes. *BMC Bioinformatics* **3,** 12.
33. Fauquet, C. M., Mayo, M. A., Maniloff, J. Dresselberger, V., and Ball, L. A. eds. (2005) *VIIIth Report of the International Committee on Taxonomy of Viruses.* Academic Press, London, England.
34. Kumar, S., Tamura, K., and Nei, M. (2004) MEGA3: Integrated software for molecular evolutionary genetics analysis and sequence alignment. *Brief. Bioinform.* **5,** 150–163.
35. Zinder, N. D. and Lederberg, J. (1952) Genetic exchange in *Salmonella. J. Bacteriol.* **64,** 679.
36. Venza Colon, C. J., Vasquez Leon, A. Y., and Villafane, R. J. (2004) Initial interaction of the P22 phage with the *Salmonella typhimurium* surface. *P. R. Health Sci. J.* **23,** 95–101.
37. Steinbacher, S., Miller, S., Baxa, U., Weintraub, A., and Seckler, R. (1997) Interaction of *Salmonella* phage P22 with its O-antigen receptor studied by X-ray crystallography. *Biol. Chem.* **378,** 337–343.
38. Cho, E. H., Nam, C. E., Alcaraz, R., Jr., and Gardner, J. F. (1999) Site-specific recombination of bacteriophage P22 does not require integration host factor. *J. Bacteriol.* **181,** 4245–4249.
39. Hofer, B., Ruge, M., and Dreiseikelmann, B. (1995) The superinfection exclusion gene (*sieA*) of bacteriophage P22: identification and overexpression of the gene and localization of the gene product. *J. Bacteriol.* **177,** 3080–3086.
40. Ranade, K. and Poteete, A. R. (1993) Superinfection exclusion (*sieB*) genes of bacteriophages P22 and lambda. *J. Bacteriol.* **175,** 4712–4718.
41. Iseki, S. and Kashiwagi, K. (1955) Induction of somatic antigen 1 by bacteriophage in *Salmonella* group B. *Proc. Jpn. Acad.* **31,** 558–564.
42. Rundell, K. and Shuster, C. W. (1975) Membrane-associated nucleotide sugar reactions: influence of mutations affecting lipopolysaccharide on the first enzyme of O-antigen synthesis. *J. Bacteriol.* **123,** 928–936.
43. Ebel-Tsipis, J., Botstein, D., and Fox, M. S. (1972) Generalized transduction by phage P22 in *Salmonella typhimurium.* I. Molecular origin of transducing DNA. *J. Mol. Biol.* **71,** 433–448.
44. Poteete, A. R. (1988) Bacteriophage P22, in *The Bacteriophages* (Calendar, R., ed.), Plenum Press, New York, pp. 647–682.
45. Parent, K. N., Doyle, S. M., Anderson, E., and Teschke, C. M. (2005) Electrostatic interactions govern both nucleation and elongation during phage P22 procapsid assembly. *Virology* **340,** 33–45.

46. Weigele, P. R., Sampson, L., Winn-Stapley, D., and Casjens, S. R. (2005) Molecular genetics of bacteriophage P22 scaffolding protein's functional domains. *J. Mol. Biol.* **348,** 831–844.
47. Kang, S. and Prevelige, P. E., Jr. (2005) Domain study of bacteriophage p22 coat protein and characterization of the capsid lattice transformation by hydrogen/deuterium exchange. *J. Mol. Biol.* **347,** 935–948.
48. Anderson, E. and Teschke, C. M. (2003) Folding of phage P22 coat protein monomers: kinetic and thermodynamic properties. *Virology* **313,** 184–197.
49. Cingolani, G., Moore, S. D., Prevelige, P. E., Jr., and Johnson, J. E. (2002) Preliminary crystallographic analysis of the bacteriophage P22 portal protein. *J. Struct. Biol.* **139,** 46–54.
50. Casjens, S. and Weigele, P. (2005) DNA packaging by bacteriophage P22. In *Viral Genome Packaging Machines: Genetics, Structure, and Mechanisms.* (Catalano, C. E., ed), pp. 80–88, Landes Bioscience, Georgetown, TX.
51. Tang, L., Marion, W. R., Cingolani, G., Prevelige, P. E., and Johnson, J. E. (2005) Three-dimensional structure of the bacteriophage P22 tail machine. *EMBO J.* **24,** 2087–2095.
52. Andrews, D., Butler, J. S., Al-Bassam, J., et al. (2005) Bacteriophage P22 tail accessory factor GP26 is a long triple-stranded coiled-coil. *J. Biol. Chem.* **280,** 5929–5933.
53. Wu, H., Sampson, L., Parr, R., and Casjens, S. (2002) The DNA site utilized by bacteriophage P22 for initiation of DNA packaging. *Mol. Microbiol.* **45,** 1631–1646.
54. Schmieger, H. and Schicklmaier, P. (1999) Transduction of multiple drug resistance of Salmonella enterica serovar Typhimurium DT104. *FEMS Microbiol. Lett.* **170,** 251–256.
55. Gilcrease, E. B., Winn-Stapley, D. A., Hewitt, F. C., Joss, L., and Casjens, S. R. (2005) Nucleotide sequence of the head assembly gene cluster of bacteriophage L and decoration protein characterization. *J. Bacteriol.* **187,** 2050–2057.
56. Petri, J. B. and Schmieger, H. (1990) Isolation of fragments with pac function for phage P22 from phage LP7 DNA and comparison of packaging gene 3 sequences. *Gene* **88,** 47–55.
57. Kuo, T. T. and Stocker, B. A. (1970) ES18, a general transducing phage for smooth and nonsmooth *Salmonella typhimurium. Virology* **42,** 621–632.
58. Le, M. L. and Chalon, A. M. (1975) Sensitivity to bacteriophage ES18 of strains of "S. dublin", "S. enteritidis" and "S. blegdam" and related serotypes. *Ann. Microbiol.* **126,** 327–331.
59. Killmann, H., Braun, M., Herrmann, C., and Braun, V. (2001) FhuA barrel-cork hybrids are active transporters and receptors. *J. Bacteriol.* **183,** 3476–3487.
60. Yamamoto, N. (1978) A generalized transducing salmonella phage ES18 can recombine with a serologically unrelated phage Fels 1. *J. Gen. Virol.* **38,** 263–272.

61. Mmolawa, P. T., Willmore, R., Thomas, C. J., and Heuzenroeder, M. W. (2002) Temperate phages in *Salmonella enterica* serovar Typhimurium: implications for epidemiology. *Int. J. Med. Microbiol.* **291,** 633–644.
62. Greenberg, M., Dunlap, J., and Villafane, R. (1995) Identification of the tailspike protein from the *Salmonella newington* phage epsilon 34 and partial characterization of its phage-associated properties. *J. Struct. Biol.* **115,** 283–289.
63. Figueroa-Bossi, N., Coissac, E., Netter, P., and Bossi, L. (1997) Unsuspected prophage-like elements in *Salmonella typhimurium. Mol. Microbiol.* **25,** 161–173.
64. Figueroa-Bossi, N., Uzzau, S., Maloriol, D., and Bossi, L. (2001) Variable assortment of prophages provides a transferable repertoire of pathogenic determinants in *Salmonella. Mol. Microbiol.* **39,** 260–271.
65. Yamamoto, N. (1969) Genetic evolution of bacteriophage. I. Hybrids between unrelated bacteriophages P22 and Fels 2. *Proc. Natl. Acad. Sci. U. S. A.* **62,** 63–69.
66. Yamamoto, N. (1967) The origin of bacteriophage P221. *Virology* **33,** 545–547.
67. Caldon, C. E., Yoong, P., and March, P. E. (2001) Evolution of a molecular switch: universal bacterial GTPases regulate ribosome function. *Mol. Microbiol.* **41,** 289–297.
68. Ho, T. D. and Slauch, J. M. (2001) OmpC is the receptor for Gifsy-1 and Gifsy-2 bacteriophages of*Salmonella. J. Bacteriol.* **183,** 1495–1498.
69. Little, J. W. (1991) Mechanism of specific LexA cleavage: autodigestion and the role of RecA coprotease. *Biochimie* **73,** 411–421.
70. Roberts, J. W., Roberts, C. W., and Mount, D. W. (1977) Inactivation and proteolytic cleavage of phage lambda repressor in vitro in an ATP-dependent reaction. *Proc. Natl. Acad. Sci. U. S. A.* **74,** 2283–2287.
71. Ho, T. D., Figueroa-Bossi, N., Wang, M., Uzzau, S., Bossi, L., and Slauch, J. M. (2002) Identification of GtgE, a novel virulence factor encoded on the Gifsy-2 bacteriophage of *Salmonella enterica* serovar Typhimurium. *J. Bacteriol.* **184,** 5234–5239.
72. Recktenwald, J. and Schmidt, H. (2002) The nucleotide sequence of Shiga toxin (Stx) 2e-encoding phage ϕP27 is not related to other Stx phage genomes, but the modular genetic structure is conserved. [Erratum in *Infect. Immun.* 2002; **70,** 4755] *Infect. Immun.* **70,** 1896–1908.
73. Allison, G. E., Angeles, D., Tran-Dinh, N., and Verma, N. K. (2002) Complete genomic sequence of SfV, a serotype-converting temperate bacteriophage of *Shigella flexneri. J. Bacteriol.* **184,** 1974–1987.
74. Tucker, C. P. and Heuzenroeder, M. W. (2004) ST64B is a defective bacteriophage in *Salmonella enterica* serovar Typhimurium DT64 that encodes a functional immunity region capable of mediating phage-type conversion. *Int. J. Med. Microbiol.* **294,** 59–63.

75. Dunn, J. J. and Studier, F. W. (1983) Complete nucleotide sequence of bacteriophage T7 DNA and the locations of T7 genetic elements. *J. Mol. Biol.* **166,** 477–535.
76. Molineux, I. J. (2001) No syringes please, ejection of phage T7 DNA from the virion is enzyme driven. *Mol. Microbiol.* **40,** 1–8.
77. Walkinshaw, M. D., Taylor, P., Sturrock, S. S., et al. (2002) Structure of Ocr from bacteriophage T7, a protein that mimics B-form DNA. *Mol. Cell* **9,** 187–194.
78. Sturrock, S. S., Dryden, D. T., Atanasiu, C., et al. (2001) Crystallization and preliminary X-ray analysis of ocr, the product of gene 0.3 of bacteriophage T7. *Acta Crystallogr. D Biol. Crystallogr.* **57,** 1652–1654.
79. Marchand, I., Nicholson, A. W., and Dreyfus, M. (2001) High-level autoenhanced expression of a single-copy gene in Escherichia coli: overproduction of bacteriophage T7 protein kinase directed by T7 late genetic elements. *Gene* **262,** 231–238.
80. Robertson, E. S., Aggison, L. A., and Nicholson, A. W. (1994) Phosphorylation of elongation factor G and ribosomal protein S6 in bacteriophage T7-infected *Escherichia coli. Mol. Microbiol.* **11,** 1045–1057.
81. Chen, Z. and Schneider, T. D. (2005) Information theory based T7-like promoter models: classification of bacteriophages and differential evolution of promoters and their polymerases. *Nucleic Acids Res.* **33,** 6172–6187.
82. Scholl, D., Kieleczawa, J., Kemp, P., et al. (2004) Genomic analysis of bacteriophages SP6 and K1-5, an estranged subgroup of the T7 supergroup. *J. Mol. Biol.* **335,** 1151–1171.
83. Nilsson, A. S. and Haggard-Ljungquist, E. (2006) The P2-like bacteriophages, in *The Bacteriophages* (Calendar, R., ed.), Oxford University Press, New York, pp. 365–390.
84. Yamamoto, N. and McDonald, R. J. (1986) Genomic structure of phage F22, a hybrid between serologically and morphologically unrelated *Salmonella typhimurium* bacteriophages P22 and Fels 2. *Genet. Res.* **48,** 139–143.
85. Pelludat, C., Mirold, S., and Hardt, W. D. (2003) The SopEPhi phage integrates into the ssrA gene of *Salmonella enterica* serovar Typhimurium A36 and is closely related to the Fels-2 prophage. *J. Bacteriol.* **185,** 5182–5191.
86. Rudolph, M. G., Weise, C., Mirold, S., et al. (1999) Biochemical analysis of SopE from *Salmonella typhimurium,* a highly efficient guanosine nucleotide exchange factor for RhoGTPases. *J. Biol. Chem.* **274,** 30501–30509.
87. Joshi, A., Siddiqi, J. Z., Rao, G. R., and Chakravorty, M. (1982) MB78, a virulent bacteriophage of *Salmonella typhimurium. J. Virol.* **41,** 1038–1043.
88. Murty, S. S., Pandey, B., and Chakravorty, M. (1998) Mapping of additional restriction enzyme cleavage sites on bacteriophage MB78 genome. *J. Biosci.* **23,** 151–154.

89. Khan, S. A., Murty, S. S., Zargar, M. A., and Chakravorty, M. (1991) Replication, maturation and physical mapping of bacteriophage MB78 genome. *J. Biosci.* **16,** 161–174.
90. Verma, M. and Chakravorty, M. (1985) Hybrid between temperate phage P22 and virulent phage MB78. *Biochem. Biophys. Res. Commun.* **132,** 42–48.
91. Verma, M., Siddiqui, J. Z., and Chakravorty, M. (1985) Bacteriophage P22 helps bacteriophage MB78 to overcome the transcription inhibition in rifampicin resistant mutant of Salmonella typhimurium. *Biochem. Int.* **11,** 177–186.
92. Uetake, H. and Uchita, T. (1959) Mutants of *Salmonella* ε15 with abnormal conversion properties. *Virology* **9,** 495–505.
93. Uetake, H., Luria, S. E., and Burrous, J. W. (1958) Conversion of somatic antigens in *Salmonella* by phage infection leading to lysis or lysogeny. *Virology* **5,** 68–91.
94. Uetake, H., Nakagawa, T., and Akiba, T. (1955) The relationship of bacteriophage to antigenic changes in group E Salmonellas. *J. Bacteriol.* **69,** 571–579.
95. Bray, D. and Robbins, P. W. (1967) Mechanism of ε15 conversion studied with bacteriophage mutants. *J. Mol. Biol.* **30,** 457–475.
96. Losick, R. and Robbins, P. W. (1967) Mechanism of ε15 conversion studied with a bacterial mutant. *J. Mol. Biol.* **30,** 445–455.
97. Robbins, P. and Uchida, T. (1962) Studies on the chemical basis of the phage conversion of O-antigens in the E-group salmonellae. *Biochemistry* **1,** 325–335.
98. Robbins, P. and Uchida, T. (1965) Chemical and macromolecular structure of O-antigens from *Salmonella anatum* strains carrying mutants of bacteriophage Epsilon 15. *J. Biol. Chem.* **240,** 375–383.
99. Robbins, P., Keller, J. M., Wright, A., and Bernstein, R. L. (1965) Enzymatic and kinetics studies on the mechanism of O-antigen conversion by bacteriophage Epsilon 15. *J. Biol. Chem.* **240,** 384–390.
100. Uchida, T., Robbins, P. W., and Luria, S. E. (1963) Analysis of the serological determinant groups of the *Salmonella* E-group O-antigens. *Biochemistry* **2,** 663–668.
101. McConnell, M., Walker, B., Middleton, P., et al. (1992) Restriction endonuclease and genetic mapping studies indicate that the vegetative genome of the temperate, *Salmonella*-specific bacteriophage, epsilon 15, is circularly-permuted. *Arch. Virol.* **123,** 215–221.
102. Kanegasaki, S. and Wright, A. (1973) Studies on the mechanism of phage adsorption: interaction between Epsilon 15 and its cellular receptor. *Virology* **52,** 160–173.
103. Takeda, K. and Uetake, H. (1973) *In vitro* interaction between phage and receptor lipopolysaccharide: a novel glycosidase associated with phage Epsilon 15. *Virology* **52,** 148–159.

104. McConnell, M. R., Reznick, A., and Wright, A. (1979) Studies on the initial interactions of bacteriophage Epsilon 15 with its host cell, *Salmonella anatum. Virology* **94,** 10–23.
105. Vezzi, A., Campanaro, S., D'Angelo, M., et al. (2004) Genome analysis of *Photobacterium profundum* reveals the complexity of high pressure adaptations. (GenBank accession number NC_006370). Unpublished results.
106. Summer, E. J., Gonzalez, C. F., Bomer, M., et al. (2006) Divergence and mosaicism among virulent soil phages of the *Burkholderia cepacia* complex. *J. Bacteriol.* **188,** 255–268.
107. Liu, M., Gingery, M., Doulatov, S. R., et al. (2004) Genomic and genetic analysis of *Bordetella* bacteriophages encoding reverse transcriptase-mediated tropism-switching cassettes. *J. Bacteriol.* **186,** 1503–1517.
108. Kropinski, A. M. (2000) Sequence of the genome of the temperate, serotype-converting, *Pseudomonas aeruginosa* bacteriophage D3. *J. Bacteriol.* **182,** 6066–6074.
109. van Sinderen, D., Karsens, H., Kok, J., et al. (1996) Sequence analysis and molecular characterization of the temperate lactococcal bacteriophage r1t. *Mol. Microbiol.* **19,** 1343–1355.
110. Craig, N. L. and Roberts, J. W. (1980) *E. coli recA* protein-directed cleavage of phage lambda repressor requires polynucleotide. *Nature* **283,** 26–30.
111. Magrini, V., Storms, M. L., and Youderian, P. (1999) Site-specific recombination of temperate *Myxococcus xanthus* phage Mx8: regulation of integrase activity by reversible, covalent modification. *J. Bacteriol.* **181,** 4062–4070.
112. Clark, A. J., Inwood, W., Cloutier, T., and Dhillon, T. S. (2001) Nucleotide sequence of coliphage HK620 and the evolution of lambdoid phages. *J. Mol. Biol.* **311,** 657–679.
113. Casjens, S., Winn-Stapley, D. A., Gilcrease, E. B., et al. (2004) The chromosome of *Shigella flexneri* bacteriophage Sf6: complete nucleotide sequence, genetic mosaicism, and DNA packaging. *J. Mol. Biol.* **339,** 379–394.
114. Jiang, W., Chang, J., Jakana, J., et al. (2006) Structure of epsilon15 bacteriophage reveals genome organization and DNA packaging/injection apparatus. *Nature* **439,** 612–616.
115. Ackermann, H.-W. and DuBow, M. S. (1987) Natural groups of bacteriophages, in *Viruses of Prokaryotes, Volume II* (Ackermann, H.-W. and DuBow, M. S., eds.), CRC Press, Boca Raton, FL, pp. 85–100.
116. Felix, A. and Callow, B. R. (1943) Typing of paratyphoid B bacilli by means of Vi bacteriophage. *Br. Med. J.* **2,** 4308–4310.
117. Kallings, L. O. (1967) Sensitivity of various salmonella strains to felix 0-1 phage. *Acta Pathol. Microbiol. Scand.* **70,** 446–454.
118. Hudson, H. P., Lindberg, A. A., and Stocker, B. A. (1978) Lipopolysaccharide core defects in *Salmonella typhimurium* mutants which are resistant to Felix O phage but retain smooth character. *J. Gen. Microbiol.* **109,** 97–112.

119. Hirsh, D. C. and Martin, L. D. (1983) Rapid detection of *Salmonella* spp. by using Felix-O1 bacteriophage and high-performance liquid chromatography. *Appl. Environ. Microbiol.* **45,** 260–264.
120. Kuhn, J., Suissa, M., Wyse, J., et al. (2002) Detection of bacteria using foreign DNA: the development of a bacteriophage reagent for *Salmonella. Int. J. Food Microbiol.* **74,** 229–238.
121. Whichard, J. M., Sriranganathan, N., and Pierson, F. W. (2003) Suppression of Salmonella growth by wild-type and large-plaque variants of bacteriophage Felix O1 in liquid culture and on chicken frankfurters. *J. Food Protect.* **66,** 220–225.
122. Maloy, S. R., Stewart, V. P., and Taylor, R. K. (1996) *Genetic Analysis of Pathogenic Bacteria.* Cold Spring Harbor Laboratory Press, Cold Spring Harbor, NY.
123. Davis, R. W., Botstein, D., and Roth, J. R. (1980) *Advanced Bacterial Genetics: A Manual for Genetic Engineering.* Cold Spring Harbor Laboratory Press, Cold Spring Harbor, NY.
124. Schmieger, H. (1972) Phage P22-mutants with increased or decreased transduction abilities. *Mol. Gen. Genet.* **119,** 75–88.
125. Benson, N. R. and Goldman, B. S. (1992) Rapid mapping in *Salmonella typhimurium* with Mud-P22 prophages. *J. Bacteriol.* **174,** 1673–1681.
126. Youderian, P., Sugiono, P., Brewer, K. L., Higgins, N. P., and Elliott, T. (1988) Packaging specific segments of the *Salmonella* chromosome with locked-in Mud-P22 prophages. *Genetics* **118,** 581–592.
127. Chen, L. M., Goss, T. J., Bender, R. A., Swift, S., and Maloy, S. (1998) Genetic analysis, using P22 challenge phage, of the nitrogen activator protein DNA-binding site in the Klebsiella aerogenes put operon. *J. Bacteriol.* **180,** 571–577.
128. Szegedi, S. S. and Gumport, R. I. (2000) DNA binding properties in vivo and target recognition domain sequence alignment analyses of wild-type and mutant RsrI [N6-adenine] DNA methyltransferases. *Nucleic Acids Res.* **28,** 3972–3981.
129. Ashraf, S. I., Kelly, M. T., Wang, Y. K., and Hoover, T. R. (1997) Genetic analysis of the *Rhizobium meliloti nifH* promoter, using the P22 challenge phage system. *J. Bacteriol.* **179,** 2356–2362.
130. Pfau, J. D. and Taylor, R. K. (1996) Genetic footprint on the ToxR-binding site in the promoter for cholera toxin. *Mol. Microbiol.* **20,** 213–222.
131. Popoff, M. Y. (2001) *Antigenic Formulas of the Salmonella Serovars.* WHO Collaborating Centre for Reference and Research on *Salmonella*, Pasteur Institute, Paris, France.
132. Popoff, M. Y., Bockemuhl, J., and Gheesling, L. L. (2004) Supplement 2002 (no. 46) to the Kauffmann-White scheme. *Res. Microbiol.* **155,** 568–570.
133. Anderson, E. S. and Williams, R. E. (1956) Bacteriophage typing of enteric pathogens and staphylococci and its use in epidemiology. *J. Clin. Pathol.* **9,** 94–127.

134. Callow, B. R. (1959) A new phage-typing scheme for *Salmonella typhimurium. J. Hyg.* **57,** 346–359.
135. Poppe, C., McFadden, K. A., and Demczuk, W. H. (1996) Drug resistance, plasmids, biotypes and susceptibility to bacteriophages of *Salmonella* isolated from poultry in Canada. *Int. J. Food Microbiol.* **30,** 325–344.
136. Lindberg, A. A. (1973) Bacteriophage receptors. *Annu. Rev. Microbiol.* **27,** 205–241.
137. Craigie, J. and Yen, C. H. (1938) The demonstration of types of *B. typhosus* by means of preparations of type II Vi phage. I. Principles and technique. *Can. J. Public Health* **29,** 448–484.
138. Craigie, J. and Yen, C. H. (1938) The demonstration of types of *B. typhosus* by means of preparations of type II Vi phage. II. The stability and epidemiological significance of V form types of *B. typhosus. Can. J. Public Health* **29,** 484–496.
139. Selander, R. K., Smith, N. H., Li, J., et al. (1992) Molecular evolutionary genetics of the cattle-adapted serovar *Salmonella dublin. J. Bacteriol.* **174,** 3587–3592.
140. Nair, S., Alokam, S., Kothapalli, S., et al. (2004) *Salmonella enterica* serovar Typhi strains from which SPI7, a 134-kilobase island with genes for Vi exopolysaccharide and other functions, has been deleted. *J. Bacteriol.* **186,** 3214–3223.
141. Mitchell, E., O'Mahony, M., Lynch, D., et al. (1989) Large outbreak of food poisoning caused by *Salmonella typhimurium* definitive type 49 in mayonnaise. *BMJ* **298,** 99–101.
142. Ward, L. R., de Sa, J. D., and Rowe, B. (1987) A phage-typing scheme for *Salmonella enteritidis. Epidemiol. Infect.* **99,** 291–294.
143. Khakhria, R., Duck, D., and Lior, H. (1991) Distribution of *Salmonella enteritidis* phage types in Canada. *Epidemiol. Infect.* **106,** 25–32.
144. Demczuk, W., Soule, G., Clark, C., et al. (2003) Phage-based typing scheme for *Salmonella enterica* serovar Heidelberg, a causative agent of food poisonings in Canada. *J. Clin. Microbiol.* **41,** 4279–4284.
145. Duckworth, D. H. (1976) Who discovered bacteriophage? *Bacteriol. Rev.* **40,** 793–802.
146. Summers, W. C. (1999) The hope of phage therapy, in *Felix d'Herelle and the Origins of Molecular Biology*, Yale University Press, New Haven, CT, pp. 108–124.
147. Sulakvelidze, A. and Barrow, P. (2005) Phage therapy in animals and agribusiness, in *Bacteriophages: Biology and Application* (Kutter, E. and Sulakvelidze, A., eds.), CRC Press, Boca Raton, FL, pp. 335–380.
148. Topley, W. W. C. and Wilson, J. (1925) Further observations of the role of the Twort-d'Herelle phenomenon in the epidemic spread of murine typhoid. *J. Hyg.* **24,** 295–300.
149. Topley, W. W. C., Wilson, J., and Lewis, E. R. (1925) Role of Twort-d'Herelle phenomenon in epidemics of mouse typhoid. *J. Hyg.* **24,** 17–36.

150. Fisk, R. T. (1938) Protective action of typhoid phage on experimental typhoid infection in mice. *Proc. Soc. Exp. Biol. Med.* **38,** 659–660.
151. Ward, W. E. (1942) Protective action of VI bacteriophage in *Eberthella typhi* Infections in mice. *J. Infect. Dis.* **72,** 172–176.
152. Berchieri, A. J., Lovell, M. A., and Barrow, P. A. (1991) The activity in the chicken alimentary tract of bacteriophages lytic for *Salmonella typhimurium. Res. Microbiol.* **142,** 541–549.
153. Sulakvelidze, A. and Kutter, E. (2005) Bacteriophage therapy in humans, in *Bacteriophages: Biology and Application* (Kutter, E. and Sulakvelidze, A., eds.), CRC Press, Boca Raton, FL, pp. 381–436.
154. Knouf, E. G., Ward, W. E., Reichle, P. A., Bower, A. W., and Hamilton, P. M. (1946) Treatment of typhoid fever with type-specific bacteriophage. *J. Am. Med. Assoc.* **132,** 134–136.
155. Desranleau, J. M. (1948) The treatment of typhoid fever by the use of Vi antityphoid bacteriophages. *Can. J. Public Health* **39,** 317.
156. Desranleau, J. M. (1949) Progress in the treatment of typhoid fever with Vi phages. *Can. J. Public Health* **40,** 473–478.
157. Jalava, K., Hensel, A., Szostak, M., Resch, S., and Lubitz, W. (2002) Bacterial ghosts as vaccine candidates for veterinary applications. *J. Control. Release* **85,** 17–25.
158. Kiknadze, G. P., Gadua, M. M., Tsereteli, E. V., Mchedlidze, L. S., and Birkadze, T. V. (1986) Efficiency of preventive treatment by phage preparations of children's hospital salmonellosis, in *Intestinal Infections* (Kiknadze, G. P., ed.), Soviet Medicine, Tbilisi, GA, pp. 41–44.
159. Slopek, S., Weber-Dabrowska, B., Dabrowski, M., and Kucharewicz-Krukowska, A. (1987) Results of bacteriophage treatment of suppurative bacterial infections in the years 1981–1986. *Arch. Immunol. Ther. Exp.* **35,** 569–583.
160. Leverentz, B., Conway, W. S., Alavidze, Z., et al. (2001) Examination of bacteriophage as a biocontrol method for salmonella on fresh-cut fruit: a model study. *J. Food Protect.* **64,** 1116–1121.
161. Garcia, P., Mendez, E., Garcia, E., Ronda, C., and Lopez, R. (1984) Biochemical characterization of a murein hydrolase induced by bacteriophage Dp-1 in Streptococcus pneumoniae: comparative study between bacteriophage-associated lysin and the host amidase. *J. Bacteriol.* **159,** 793–796.
162. Brussow, H. and Hendrix, R. W. (2002) Phage genomics: small is beautiful. *Cell* **108,** 13–16.
163. Mead, P. S., Slutsker, L., Dietz, V., et al. (1999) Food-related illness and death in the United States. *Emerg. Infect. Dis.* **5,** 607–625.
164. Frenzen, P. D., Drake, A., Angulo, F. J., and Emerging Infections Program FoodNet Working Group (2005) Economic cost of illness due to *Escherichia coli* O157 infections in the United States. *J. Food Protect.* **68,** 2623–2630.

165. Adamia, R. S., Matitashvili, E. A., Kvachadze, L. I., et al. (1990) The virulent bacteriophage IRA of *Salmonella typhimurium*: cloning of phage genes which are potentially lethal for the host cell. *J. Basic Microbiol.* **30,** 707–716.
166. Garcia, P., Garcia, E., Ronda, C., Lopez, R., and Tomasz, A. (1983) A phage-associated murein hydrolase in *Streptococcus pneumoniae* infected with bacteriophage Dp-1. *J. Gen. Microbiol.* **129,** 489–497.
167. Nelson, D., Loomis, L., and Fischetti, V. A. (2001) Prevention and elimination of upper respiratory colonization of mice by group A streptococci by using a bacteriophage lytic enzyme. *Proc. Natl. Acad. Sci. U. S. A.* **98,** 4107–4112.
168. Schuch, R., Nelson, D., and Fischetti, V. A. (2002) A bacteriolytic agent that detects and kills *Bacillus anthracis. Nature* **418,** 884–889.
169. Casjens, S. R., E. B. Gilcrease, D. A. Winn-Stapley, P. Schicklmaier, H. Schmieger, M. L. Pedulla, M. E. Ford, J. M. Houtz, G. F. Hatfull, and R. W. Hendrix. 2005. The generalized transducing Salmonella bacteriophage ES18: complete genome sequence and DNA packaging strategy. Journal of Bacteriology 187:1091-1104.
170. Bullas LR, Mostaghimi AR, Arensdorf JJ, Rajadas PT, Zuccarelli AJ. 1991. Salmonella phage PSP3, another member of the P2-like phage group. Virology 185:918-921.
171. Frost JA, Ward LR, Rowe B. 1989. Acquisition of a drug resistance plasmid converts Salmonella enteritidis phage type 4 to phage type 24. Epidemiol Infect. 103:243-248.

10

Salmonella Typhimurium Phage Typing for Pathogens*

Wolfgang Rabsch

Summary

Phage typing provides a rapid, accurate, and cheap method of investigating *Salmonella* strains for epidemiological use. *Salmonella* strains within a particular serovar may be differentiated into a number of phage types by their pattern of susceptibility to lysis by a set of phages with different specificity. Characterization based on the pattern of phage lysis of wild strains isolated from different patients, carriers, or other sources is valuable in epidemiological study. The phages must have well-defined propagation strains that allow reproducible discrimination between different *Salmonella* Typhimurium strains. Different schemes have been developed for this serovar in different countries. The Felix/Callow (England) and Lilleengen typing systems (Sweden) used for laboratory-based epidemiological analysis were helpful for control of salmonellosis. More recently, the extended phage-typing system of Anderson (England) that distinguishes more than 300 definitive phage types (DTs) has been used worldwide in Europe, the United States, and Australia. The use of this method for decades show us that some phage types (DT204 in the 1970s and DT104 in the 1990s) have a broad host range and are distributed worldwide, other phage types such as DT2 or DT99 are frequently associated with disease in pigeons, indicative of a narrow host range.

Key Words: Bacteriophages; definitive phage type (DT); *Salmonella*; epidemiological tool; lysogenic conversion, phage adsorption, super infection exclusion.

---

*This chapter is dedicated to the 50th jubilee of the establishment of the Central Laboratory for Phage Typing in Wernigerode in 1955.

From: *Methods in Molecular Biology, vol. 394: Salmonella: Methods and Protocols*
Edited by: H. Schatten and A. Eisenstark © Humana Press Inc., Totowa, NJ

1. Introduction

1.1. Non-Typhoidal Salmonellosis

Annually, approximately 40,000 cases of non-typhoidal salmonellosis are reported to the Centers for Disease Control and Prevention (CDC), Atlanta, GA, United States *(1)*. These numbers underestimate the magnitude of the problem, as many cases of salmonellosis are not reported because of any of the following reasons: the ill person does not visit a physician, no specimen is obtained for laboratory tests, or the laboratory findings are not communicated to the CDC *(2)*. Taking into account the degree of underreporting, the CDC estimates the annual number of non-typhoidal salmonellosis cases in the United States to be approximately 1.4 million *(3)*. Including both culture-confirmed infections and those not confirmed by culture, they estimated that *Salmonella* infections resulted in 15,000 hospitalizations and 400 deaths annually. These estimates indicate that salmonellosis presents a major ongoing burden to public health *(3)*. The most common vehicles of transmission are meat, meat products, eggs, and egg products that contain *Salmonella* because animals are infected or because fecal contamination occurs during processing *(4,5)*. The majority of human cases are caused by only a few non-typhoidal serovars. For instance, approximately 60 % of human cases reported to the CDC in 1995 were caused by four serovars, *Salmonella* Enteritidis (24.7 %), *Salmonella* Typhimurium (23.5 %), *Salmonella* Newport (6.2 %), and *Salmonella* Heidelberg (5.1 %) *(6)*. These four serovars also represented 46.4 % of strains isolated from non-human sources that year. The dominance of only a few serovars is even more pronounced in Germany, where *S*. Enteritidis (64.3 %) and *S*. Typhimurium (19.9 %) accounted for more than 80 % of human isolates reported to the National Reference Centre for *Salmonella* and other bacterial enterics at the Robert Koch Institute in 2004 (survstat@rki.de).

1.2. Significance of Salmonella Phages

1.2.1. Phage Typing of Salmonella

Phage typing is a simple test that has been developed to aid in *Salmonella* characterization for epidemiological purposes. The first work with phages for typing was done by Sonnenschein in 1934 with different *Salmonella* Typhi strains *(7,8)*. Craig Yen and Felix *(9–11)* developed a typing scheme for *S*. Typhi by adaptation of the Vi II phage to different wild strains. Vieu in Paris used the French word lysotype for phage typing, whereas Rische in Wernigerode and Brandis in Bonn used the German word Lysotyp. Vi-phages were used worldwide until now and distributed by the International Federation for Enteric Phage Typing (IFEPT) (L. R. Ward, personal communication). Later, phage-typing schemes were developed for other frequently isolated serovars, such as *S.* Typhimurium ***(12)***, *S.* Enteritidis ***(13)***, and *Salmonella* Agona ***(14)***. So, *Salmonella* strains within a particular serovar may be differentiated into a number of phage types by their pattern of susceptibility to lysis by a set of phages with different specificity. Characterization based on the pattern of phage lysis of wild strains isolated from different patients, carriers, or other sources is valuable in epidemiological study. The phages must have well-defined propagation strains that allow reproducible discrimination between different *S.* Typhimurium strains. Different schemes have been developed in different countries. There is a high degree of correlation between the phage type and the epidemic source. Using a standardized phage-typing scheme in Europe, a dramatic shift in the epidemiology of *S.* Enteritidis in Western Europe could be monitored ***(15)***. Serovar Typhimurium cultures can be differentiated into numerous distinct variants by phage typing and other epidemiological typing methods. Some of these variants are frequently associated with disease in a single-host reservoir but are rarely cultured from other sources, indicative of a narrow host range. The most striking example of this apparent adaptation to one particular host reservoir are in serovar Typhimurium phage types (definitive type) DT2 and DT99 cultured from pigeons. Some of these DT2 strains from pigeon showed genomic rearrangements at rRNA operons, typical for host-adapted *Salmonella* ***(16)***. Serovar Typhimurium pigeon isolates are almost exclusively detected with disease in this reservoir (and not in other birds or mammals) since surveillance began some seven decades ago. Furthermore, DT2 and DT99 seem to be the only serovar Typhimurium variants to infect pigeons, as truly broad-host-range serovar Typhimurium variants, such as phage types DT49 and DT104, are very rarely isolated from this source. Apparently, host-adapted variants of serotype Typhimurium also infect ducks (DT8 and DT46) and wild birds (DT40), but these hosts are also susceptible to infection with true broad-host-range serovar Typhimurium variants ***(17)***.

1.2.2. Generalized Transduction

Studies during the past 10 years have demonstrated that the number of bacteriophages and bacteria in many natural ecosystems are very high. For instance, phage numbers often reach concentrations of 100 billion phages per milliliter of seawater ***(18,19)***. Today, many microbiologists believe that phages are a prime regulator of biomass and carbon and energy flow in the natural environment, in particular in aquatic habitats ***(20)***. Depending on the indicator strain, it is possible to isolate phages from nearly every wastewater sample. Bacterial cells can act as safe reservoir for phage genome, allowing them to remain part of ecosystem for periods greatly exceeding the infecting life of a phage. Because of the continuous induction of a fraction of these lysogenic populations, they act as continuous sources of phages as transducing phages. The remainder of the lysogenic population that is not induced can serve as excellent recipients for transduced genes because they are themselves immune to superinfection (e.g., ES18 can transduce in P22 lysogens). Most wildtype phages of *Salmonella* are able to cause generalized transduction ***(21)***. The mechanism of generalized transduction was discovered by Zinder and Lederberg in 1952 ***(22)***. The temperate short-tail phage PLT22 (Lilleengen phage type LT22, now designated P22), isolated from *S*. Typhimurium, is able to transduce DNA into smooth *Salmonella* strains of groups A, B, and D. Kuo and Stocker ***(23)*** reported that a temperate long-tail phage ES18 is also capable of transduction into smooth and rough strains of *Salmonella*. Several transducing phages related to P22, including MG40, L, PSA68, and KB1, have been reported as transferring chromosomal as well as plasmid-borne determinants ***(24)***. The isolation of a P22HT mutant, which can transduce with higher transducibility, made salmonella strain construction very easy ***(25)***.

1.2.3. Lysogenic Conversion

A second possible factor contributing to the temporal dominance of individual phage types may be the horizontal transfer of virulence genes. Hardt and co-workers screened a representative set of *S*. Typhimurium phage types isolated in Germany for the presence of strains that are lysogenized by a bacteriophage carrying *sopE*, a gene encoding a protein involved in bacterial invasion of intestinal epithelial cells ***(26)***. The *sopE* gene was only present in phage types DT204, DT204c, DT49, DT175, DT68, and DT29 ***(27)***. Phage types DT175 and DT68 were only rarely isolated in Germany. However, cattle-associated DT204c strains were the predominant multiple antibiotic resistant *S*. Typhimurium isolates in England and Wales in the 1980s ***(28–30)*** and were frequently isolated in Germany from humans and cattle in the 1990s. DT49 was the most commonly isolated *S*. Typhimurium strain from humans in England and Wales during the late 1980s ***(30)*** and was frequently isolated from cattle and poultry. DT204 was first isolated in 1972 from cattle and beef products in Germany and was resistant against sulfonamides and tetracycline (R-type SuT) ***(31)***. It was the phage type most frequently isolated from pigs and cattle in Germany during the 1970s. Multiple antibiotic resistant strains of DT204 (most commonly R-type CSSuT) spread rapidly among cattle populations after their emergence in Germany in 1974 and in England and Wales in 1977 ***(28,29,31)***. The finding that the *sopE* gene is mainly present in these epidemic cattle-associated phage types is intriguing because it suggests that this virulence gene may confer an advantage during circulation in the bovine reservoir ***(26)***. Lysogenic conversion of *S.* Typhimurium ATCC 14028 with *sopE*Φ resulted in a 2.5-fold increase in invasive ability compared with the wildtype phages ***(32)***, which may be sufficient to result in a small increase in transmissibility of *sopE*-positive strains and increase prevalence over time. Phage-mediated transfer of virulence factors may not be uncommon in serotype Typhimurium because putative virulence genes have been identified in the genomes of several of its prophages, including *nanH* and *sodCIII* in *FELS*-1, *gogB* in *Gifsy*-1, *sodCI* and *sseI* (*srfH*) in *Gifsy*-2, *sspH1* in *Gifsy*-3, and *sopE1* in SopEΦ ***(33–37)***.

So bacteriophages seem to mediate horizontal transfer of virulence functions among *Salmonella* strains in two different ways: by transduction and also by lysogenic conversion. The typing phages interfere with the prophages and/or cryptic phages of wild strains and so the complex genetic short-term evolution can be demonstrated in the laboratory. This is one reason for the successful application of phage typing in *Salmonella* epidemiology since the 1950s.

1.2.4. Phage Therapy and Food Sanitation

1.2.4.1. Phage Therapy

In the G. Eliava Institute of Bacteriophages, Microbiology and Virology, Tbilisi, Georgia, many *Salmonella* bacteriophages were collected for prophylaxis against salmonellosis ***(38)***. Institutes and factories in places like the Institute for Epidemiology and Microbiological in Gorki or Ufa were also producing these phage products for Soviet use. Liquid in bottles or capsules contained "Bacteriophagum salmonellae gr. ABCDE liquidum et siccum indumento acidoresistenti." These early studies were poor and uncontrolled. Today, phages of Stefan Slopek's group at the Bacteriophage Laboratory, Ludwik Hirszfeld Institute of Immunology and Experimental Therapy, have also been used for therapeutic treatment; 1307 patients with suppurative bacterial infections caused by multidrug-resistant bacteria of different species were treated with specific bacteriophages; this therapy was highly effective ***(39)***. Berchieri et al. have extended work on veterinary diarrhea to *S.* Typhimurium in poultry. With one strain of phage, mortality could be reduced from 60% in the phage free control group to 3%. The phage caused reduction of viable numbers of *S.* Typhimurium in the crop, small intestine, and cecum for up to 12 h after inoculation, with smaller reductions in bacterial numbers in the liver at 24 and 48 h after infection ***(40)***. Granted, many of these and other historic studies do not meet the current rigorous standards for clinical trials, and there still remain many questions that must be addressed before lytic phages can be widely used for therapeutic use. Nonetheless, it may be possible in future that a mixture of two or three genetically engineered phages with well-characterized phage–Salmonella–host interactions will be used. Sequencing will be a prerequisite for excluding virulence properties ***(41)***. Furthermore, the phage lysate will have to be free of endotoxin [lipopolysaccharide (LPS) and lipid A]. Phage therapy can be useful in modern clinical practice, because antibiotic usage has negative as well as beneficial effects ***(42,43)***:

1. Generalized non-specific antimicrobial activity that damages the normal microflora enabling colonization by opportunistic pathogens.
2. Side effects, such as allergy and toxicity, including effects of the immune system.
3. Selection of antibiotic resistant bacteria and enhanced rates of transfer in the absence of the normal flora.
4. Enhanced spread of fungal and yeast infections.

Therefore, phage treatment with genetically engineered phages could be helpful, for example in the treatment of healthy long-term carriers (like typhoid Mary) who excrete multi-resistant *S.* Typhi, but the pharmaceutical application of phages will have to be carefully tested before use.

1.2.4.2. Food Sanitation

Procedures that reduce surface contamination of carcasses by *Salmonella* and other enteric pathogens have been examined over many years. Lytic bacteriophages were applied to chicken skin that had been experimentally contaminated with *Salmonella*. Goode and colleagues have demonstrated that phages can be effective at reducing *S.* Enteritidis contamination of carcass surfaces ***(44)***. Leverentz et al (2001) found that a phage mixture reduced *Salmonella* populations by approximately 3.5 logs on honeydew melon slices stored at 5 °C and 10 °C and by approximately 2.5 logs on slices stored at 20 °C, which is greater than the maximal amount achieved using chemical sanitizers ***(45)***.

1.3. Phage Typing of S. Typhimurium

1.3.1. Lilleengen Scheme

The phage-typing system developed by Lilleengen in 1948 distinguishes 27 Lilleengen types among serotype Typhimurium isolates ***(12)***. The 12 phages were isolated from sewage, manure, and *S.* Typhimurium cultures. Some of the phages were adapted on suitable strains of *S.* Typhimurium (*see* **Table 1**; ***46***). The phage-typing scheme (*see* **Table 2**) is used mainly in Scandinavia and

Table 1
Type Strains and Phages of *Salmonella* Typhimurium Lilleengen Scheme

Phage number	Homologous test strain number	Origin of phage
30	135	Adapted from phage 14
39	331	Adapted from phage 30
36	409	Adapted from phage 7
34	111	Adapted from phage 7
31	125	Filtrate of sewage
33	23	Adapted from phage 22
8	119	Filtrate of phage contaminated *S.* Typhimurium culture (No. 113)
32	74	Filtrate of phage contaminated *S.* Typhimurium culture
37	306	Adapted from phage 4
4	22	Filtrate of phage contaminated *S.* Typhimurium culture (No. 52)
28B	193	Adapted from phage 28 A
2	100	Filtrate of sewage

has been used successfully in investigations of outbreaks of human infection ***(47,48)***. LT2, phage type 2 according to this scheme, was the first *Salmonella* strain with an edited genetic map (in 1965) and was the first *S.* Typhimurium strain for which the genome was sequenced. We have transferred the original phages and propagation strains from the Swedish Institute for Infectious Diseases Stockholm, Sweden (Ralph Wollin).

1.3.2. Felix and Callow Scheme

The phage-typing scheme of Callow distinguished 12 types of *S.* Typhimurium with 11 phages (*see* **Table 3**; ***49***). The origin of the phages is as follows: 1, 1a, and 1b are adaptations of one of the preparations used for *Salmonella* Paratyphi B (phage 3b) to adequate homologous strains

Table 2
Used Phage-Typing Scheme for *Salmonella* Typhimurium According to Lilleengen

	Typing phages											
Phage type	30	39	36	34	31	33	8	32	37	4	28B	2
1	SCL, CL	CL	CL	CL	CL	CL	CL	CL	CL	++n, SCL	–	–
2	CL	CL	CL	CL	CL	–	SCL	CL	CL	++n, SCL	–	–
3	SCL, CL	CL	CL	CL	CL	±sn	–	CL	CL	CL	–	–
4	CL	CL	SCL	SCL, CL	CL	–	–	CL	SCL	±m	–	–
5	CL	CL	CL	CL	CL	CL	CL	–	–	–	CL	–
6A	SCL, CL	CL	CL	CL	CL	CL	CL	++n, SCL	–	–	–	–
6B	SCL, CL	CL	CL	CL	CL	CL	CL	–	–	–	–	–
7	CL	CL	CL	CL	CL	–	SCL	–	–	–	–	–
8	++n	CL	CL	CL	++n	++n	SCL	–	–	–	–	–
9	++n	CL	CL	CL	++n	++n	–	–	–	–	–	–
10	+, ++sn	SCL	SCL	SCL	+, ++sn	+, ++sn	–	CL	±m	–	–	–
11	±sn	SCL	±m	±m	–	–	–	CL	–	–	–	–
12	±sn	++n	SCL	CL	SCL	±sn	–	++n	–	–	–	–
13	+sn	CL	++sn	–	++n	+sn	–	–	±sn	±sn	–	–
14	CL	CL	–	±m	±m	CL	–	–	–	–	CL	–

15	CL	CL	–	±m	±m	CL	–	–	–	–	±sn	–
16	CL	–	–	–	–	CL	–	–	–	–	–	–
17	–	–	–	–	–	–	–	–	–	–	CL	–
18	–	–	–	–	–	–	–	–	–	–	–	CL
19	CL	CL	CL	–	–	–	–	+n	–	–	–	CL
20	CL	–	CL	–	–	–	–	–	–	–	–	CL
21	CL	CL	CL	–	–	–	–	+n	–	–	–	–
22	CL	–	SCL, CL	–	–	–	–	–	–	–	–	–
23	CL	++m, SCL	++m, SCL	++m, SCL	++m, SCL	–	–	++m, SCL	–	–	–	–
24	SCL	CL	–	–	–	–	–	–	–	–	–	–
25	CL	CL	CL	–	–	–	CL	SCL	–	–	–	–
26	CL	CL	SCL	–	–	SCL	–	SCL	–	–	–	–
Ph 30	++s	–	–	–	–	–	–	–	–	–	–	–
Ph 36	–	–	++s	–	–	–	–	–	–	–	–	–

CL, confluent lysis; SCL, confluent lysis.

Table 3
Type Strains and Phages of *Salmonella* Typhimurium Felix and Callow Scheme

Phage number	Homologous test strain	Origin of phages
1	M 307	
1a	M 419	Adaptation of phage 3b from *Salmonella* Paratyphi B
1b	M 1414	
2	M 668	Isolated from feces
2a	M 154	Adapted from *S*. Typhimurium phage 2
2c	M 591	
2b	M 2540	Isolated from rough strain of *Salmonella* Enteritidis (Gärtner)
2d	M 1272	
3	M 298	
3a	M 736	Adapted from *S*. Typhimurium phage 2a
3b	M 435	
4	M 1166	A natural phage from a *S*. Typhimurium culture
5	M 4938	Anderson, Sept. 1963
35	M 3620	Anderson, Aug. 1963

of *S*. Typhimurium. Phage 2 was isolated directly from feces of a person infected with *S*. Typhimurium, phages 2a and 2c being its adaptations. Phages 3, 3a, and 3b are adaptations of phage 2a (adaptation of an adapted phage). Phages 2b and 2d were isolated from a rough culture of *S*. Enteritidis and phage 4 from a *S*. Typhimurium culture. All these preparations can be divided into three serological groups: group I—1, 1a, 1b, 2b, 2d, 3, 3a, and 3b; group II—2, 2a, and 2c; and group III—4.

On the basis of reactions with the aforementioned preparations, Felix and Callow ***(49)*** worked out and applied a scheme of phage typing containing 12 phage types, marked with the same symbols as the phage preparations. In 1963, a new phage was added in our laboratory, which enabled us to distinguish a new phage type 5. It was also observed that the strains of type 2 occasionally reacted with phage 4, which enables us to distinguish a new variation of type 2, namely 2var1. Additionally, we use the O1 phage. The Felix and Callow scheme with additional phages is in general use and is demonstrated in **Table 4**.

1.3.3. Anderson Scheme

Further investigations by Callow ***(49)*** led to an increase in the number of lysates up to 29 and the possibility of distinguishing 34 phage types. With the aid of her preparations, Callow was able to carry out a further division into types and variants of the strains classified according to the first scheme (*see* **Table 4**) and to distinguish 10 new types among strains untypeable by the old method. A large number of the Callow preparations (13 of 29) represent adaptations of phages isolated from lysogenic strains of *S*. Paratyphi B. They can be neutralized by antiserum against phage 3b for *S*. Paratyphi B or antiserum against phage 1 for *S*. Typhimurium. The revised scheme was adopted for routine use in the Enteric Reference Laboratory (today the Laboratory of Enteric Pathogens) in London in 1958; it rapidly showed that its sensitivity was of a high order, and since then it was progressively expanded by Anderson until it now distinguishes more than 300 types with 34 phages. The genealogy of the Anderson phages is described in **Fig. 1**. Phages released from the Anderson propagating strains (i.e., the BA strains) are designated by the name of their host strains with a preceding "P"; for example, PBA22 is the phage released from strain BA22, the propagating strain for Anderson phage A22. It was clear that definitive numerical designations must eventually be given to the many types of the new scheme. This had been done by Anderson et al. ***(50)***; they reported DT designation in addition to the provisional type (PT) designation that have been used for many years ***(50)***. It is impossible to describe the whole scheme with more than 300 phage types here in this chapter. Strains showing a lysis pattern that did not conform to any recognized definite or provisional phage types of *S*. Typhimurium were designated as "react but did not conform" (RDNC). Strains that did not react with any of the typing phages were designated as "untypeable." In 2002, this scheme was extended by Linda Ward, Laboratory of Enteric Pathogens, Health Protection Agency, Colindale, London, United Kingdom for untypeable strains by a further five phages of a supplementary scheme (add.1–3, add.10, and add.18 *see* **Fig. 2**) to distinguish DT193, DT194, DT195, DT208, and PTU302. The phage add.10var2 was added separately to determine PTU310. These additional phages were used for untypeable strains only, if all 30 phages of the classical Anderson set are negative.

1.3.4. Phage Typing in the Wernigerode Institute

The Central Laboratory for Phage Typing was founded by H. Rische in Wernigerode, East Germany on May 12, 1955. At that time, *S*. Typhi, *S*. Paratyphi A and B, *S*. Typhimurium, and some other pathogens (Shigella

Table 4
Used Phage-Typing Scheme for *Salmonella* Typhimurium According to Felix and Callow

	Typing phages													
Phage type	1	1a	1b	2	2a	2b	2c	2d	3	3a	3b	4	5	35
1	CL	CL	CL	CL	CL	CL	CL	CL	CL	+++	CL	–	CL	CL
1var2	CL	CL	CL	–	–	CL	–	CL	CL	CL	CL	–	CL	CL
1a	–	CL	CL	CL	CL	CL	CL	CL	CL	CL	CL	–	CL	CL
1avar2	–	CL	CL	–	–	CL	–	CL	CL	CL	CL	–	+++	CL
1b	–	–	CL	SCL	SCL	CL	SCL	CL	CL	SCL	CL	–	–	CL
2	–	–	–	CL	CL	–	CL	–	–	–	–	–	±	–
2var1	–	–	–	CL	CL	–	CL	–	–	–	–	+++	–	–
2a	–	–	–	–	CL	CL	CL	CL	–	–	–	–	±	CL
2b	–	–	–	–	–	CL	–	CL	–	–	–	–	–	SCL
2c	–	–	–	+++	+++	+++	+++	+++	–	–	–	–	–	OL
2c Phi 4	–	–	–	SCL	SCL	SCL	SCL	SCL	–	–	–	+++	–	+++
2d	–	–	–	–	–	–	–	OL	–	–	–	–	–	–
3	–	–	+++	–	–	–	–	–	CL	SCL	±	–	–	–
3a	–	–	–	–	–	CL	–	SCL	–	OL	+++	–	–	CL
4	–	–	–	–	–	–	–	–	–	–	–	++	–	–
5	–	–	–	–	–	–	–	–	–	–	–	–	OL	CL
35	–	–	–	–	–	–	–	–	–	–	–	–	±	CL

CL, confluent lysis; OL, opaque lysis; SCL, confluent lysis.

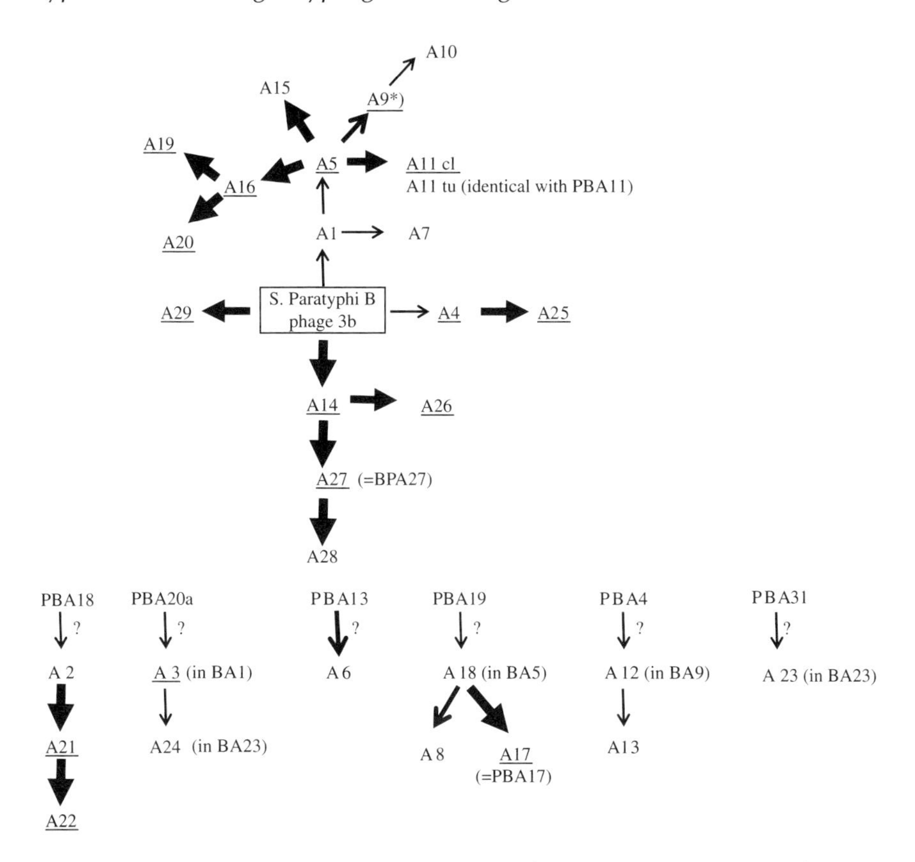

Fig. 1. Genealogy of Anderson typing phages (supplemented *see* **ref. *69***). *A9 was deleted.

sonnei, Staphylococcus aureus) were phage typed for epidemiological purposes. On October 3, 1965 the Central Laboratory for Phage Typing was integrated in the Institute for Experimental Epidemiology (IEE), Wernigerode. From the 1960s till now, the national epidemiological surveillance program at the Institute both in the former GDR and after reunification (the East German IEE was integrated into the Robert Koch Institute in 1991) for the whole of Germany was performed using a combination of the Felix/Callow and Lilleengen typing systems with 14 phages and 12 phages, respectively ***(51)***. We observed that different clones circulate in chicken flocks ***(52)***, and other clones of *S.* Typhimurium are frequently distributed in hospitals in Germany and Russia ***(53,54)***. Furthermore, the persistence of an epidemic type in different domestic animals was described (***31,55–58***). The better laboratory-

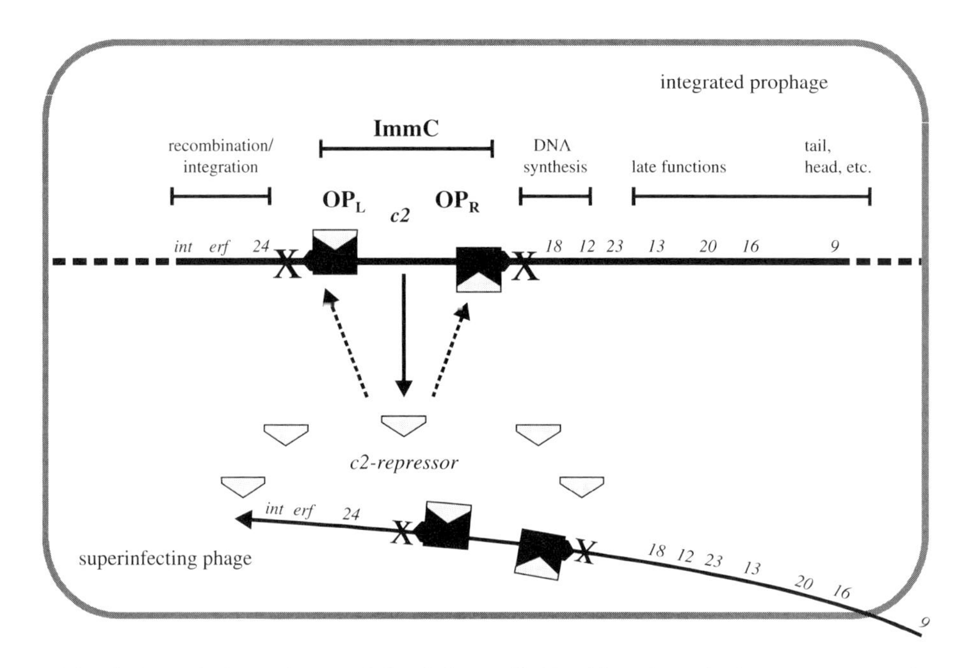

Fig. 2. Inoculation scheme of the *Salmonella* Typhimurium Anderson phage-typing plate.

based epidemiological analysis was helpful for control of salmonellosis ***(59)***. Since 1996, the extended phage-typing system of Anderson et al. ***(50)*** that distinguishes more than 300 definitive phage types (DT) (Robert Koch Institute and L. R. Ward, personal communication) has been used additionally for surveillance in Germany. This phage-typing system is used worldwide in Europe, the United States, and Australia. In most cases, the phage type is a stable property that provides an opportunity to follow the spread of a particular serotype Typhimurium clone within different host populations, food, feeds, and the environment over years or even decades ***(60,61)***. For example, molecular fingerprinting shows that the majority of multidrug-resistant DT104 isolates from Germany, Austria, the United Kingdom, the United Arab Emirates, the Philippines, and the Netherlands belong to a single clone that has spread pandemically during the 1990s ***(62)***.

The sensitivity, stability, and cheapness of phage typing make this a very attractive method for determining the spread of different, yet very closely related, serotype Typhimurium clones over time. Thus, phage typing in combination with other typing methods can be used to determine in which host reservoirs a particular serovar Typhimurium clone is associated with disease ***(63)***. Furthermore, laboratory-based surveillance is necessary for routine investigation of the two German *S*. Typhimurium live vaccine strains ***(64,65)***.

1.4. Terminology

The study of phage biology has generated a large number of descriptive terms that are commonly used in the literature. Several of these are provided:

1. Lytic growth: A bacteriophage lifestyle in which the phage DNA is replicated, phage proteins are synthesized, new infectious-phage particles are assembled, and progeny are released. This process is also called the lytic cycle. The host cell provides much of the raw materials and machinery for lytic growth and is often destroyed by lysis.
2. Lysogeny: A bacteriophage lifestyle in which the infecting-phage genome remains in a near-dormant state and is maintained within the host. This process is also referred to as lysogenic growth or the lysogenic cycle.
3. Pseudolysogeny: A form of unstable lysogeny in which the prophage is thought to remain extrachromosomal (non-integrated) to the host genome.
4. Virulent phage: A bacteriophage that is committed to lytic growth.
5. Temperate phage: A bacteriophage that can enter either lytic or lysogenic growth.
6. Prophage: A phage genome that is integrated in the host chromosome.
7. Lysogen: A cell carrying a prophage.
8. Induction: The entry of a lysogen into lytic phage growth.
9. Integration: The incorporation of a phage genome, usually by site-specific recombination, into the host chromosome to form a prophage.
10. Multiplicity of infection (MOI): The average number of phage particles that infect a single bacterial cell in a specific experiment.
11. Plaque: A visible circular zone of reduced growth on a lawn of cells; the result of the infection and lysis of a single bacterial cell on a bacterial lawn followed by multiple rounds of infection and lysis of the neighboring cells.
12. Superinfection immunity: A feature of a lysogen in which the cell is no longer sensitive to infection by phages similar or closely related to the prophage. This property is determined by the prophage and is not a normal host function.
13. Homoimmune phage: A phage that uses a homologous repressor/operator system to that of the resident prophage of a lysogen. Upon infection, the homoimmune phage genome is rendered inactive by the prophage repressor system and the lysogen will display superinfection immunity toward the infecting phage.
14. Heteroimmune phage: A phage that is unaffected by the resident prophage of a lysogen. The lysogen is susceptible to infection by the phage, and the infecting phage is considered heteroimmune with respect to the prophage.
15. Antirepressor: The gene product of the phage, which may inactivate the repressor, competes for the repressor-binding site on the DNA.

16. Host range: The spectrum of bacterial species on which a phage can form plaques. The host range can vary from very broad (many species) to highly specific (serovar till strain).

1.5. The Biological Base of Phage Typing

1.5.1. Host Controlled Modification and Restriction of Typing Phages

The DNA of the typing phages A1 and A7 share the same EcoRI digestion pattern (not shown). We suggested that their different plating efficiencies on various hosts, which are the basis of the phage-typing system, are caused by host-controlled modification and restriction. When phage A1 was propagated on host strain BA1 [= A1 (BA1)], it was significantly restricted by the strain BA7. When surviving phages A1 (BA1) were propagated on BA7, the plating efficiency of the resulting lysate A1 (BA1, BA7) was nearly the same on both strains. Therefore, strain BA7 seems to express at least one host-controlled modification/restriction system that is absent in strain BA1. BA1 has no such system, or it has a system that is (in addition to others) also expressed in BA7. Repropagation of the phages A1 (BA1, BA7) in BA1, yielding lysate A1 (BA1, BA7, BA1), renders A1 sensitive to BA7 restriction (*see* **Table 5**).

1.5.2. Adsorption Properties of Various Typing Phages and the Dependence on FhuA and TonB Proteins

Bacteriophages utilize for adsorption both LPS ***(66)*** and surface proteins ***(67)***. The bacterial receptor of the typing phages A8 and A18 (the latter being identical with phage ES18) is the FhuA protein, which is responsible for ferrichrome uptake ***(68)***. Another bacterial protein necessary for effective infection by these phages is the TonB protein. The S. Enteritidis typing phage H8 is a broad host range phage. It morphology and genome structure closely resembled those of bacteriophage T5 in the family Siphoviridae. H8 infects

Table 5
Efficiency of Plating of Anderson Phage A1 after Propagation of Different *Salmonella* Typhimurium Strains

	Strains	
Phages	BA1	BA7
A1 (BA1)	4.5×10^{10}	4×10^{4}
A1 (BA1, BA7)	1.5×10^{9}	3.6×10^{8}
A1 (BA1, BA7, BA1)	3×10^{10}	4.3×10^{6}

Table 6
Influence of FhuA and TonB to the Plating Efficiency of Anderson Typing Phages

		Anderson phages			Phages		
Strains	Relevant markers	A1	A8	A18	ES18	P22	Source
1. *fhu* mutants							
WR1173 (BA36)[a]	*fhu*A::MudJ	+	–	–	–	+	This paper
SL1027/23 (SL1027)[a]	*fhu*A	+	–	–	–	+	K. Hantke
WR1179 (BA36)[a]	*fhu*B::MudJ	+	+	+	+	+	This paper
2. *tonB* mutants							
SR1001 (enb-7)[a]	*ton*B	+	–	–	–	+	W. Rabsch
AIR36	*ton*B::MudJ	+	–	–	–	+	R. Tsolis
Control strains							
BA36	DT36 (Anderson typing system)	+	+	+	+	+	L. R. Ward
SL1027	*trp met*	+	+	+	+	+	B. A. D. Stocker
enb-7	*ent*-class II	+	+	+	+	+	J. B. Neilands

[a]Strain designations in brackets signify the parental strains.

different Salmonella serovars by initial adsorption to the outer membrane protein FepA (enterobactin receptor). The H8 infection is TonB-dependent ***(41)***. Phages A1 and P22 use the O12 polysaccharide antigen as a receptor. Therefore, *S*. Typhimurium strains carrying *fhuA* and *tonB* mutations appear resistant against A8 and A18, but sensitive to A1 and P22 (*see* **Table 6**).

1.5.3. Influence of Superinfection Exclusion Systems

Most Anderson typing phages belong to the P22-like phage family ***(69)***. Phage P22 has two different superinfection exclusion systems, *sieA* and *sieB*. Such protection systems may be present also in natural strains, because it has been shown that almost all natural isolates of *Salmonella* carry prophages, most of which belong to the P22 group. To study the influence of the superinfection exclusion systems on typing phages, strain BA36 was lysogenized with P22 wildtype (both *sie* systems are active) and with a P22 mutant defective in both *sie* systems. BA36 and its lysogenic derivatives were assayed with the Anderson typing phages. As summarized in **Table 3**, most typing phages (colored) were excluded by at least one of the *sie* systems of P22 wildtype (line 2), because the phages were inactive on BA36 (P22) but could effectively plate on the *sie*-defective mutant (line 3). Phages A12, A13, and 17 were suppressed on both lysogenic strains, indicating that they are homoimmune to phage P22 (*see* **Table 7**).

1.5.4. Influence of Various Immunity Systems

Most natural *Salmonella* strains harbor prophages, mainly of the P22 group ***(21)***. Their repressors may influence plating efficiencies of the likewise P22-related Anderson test phages, depending on homoimmunity or heteroimmunity to the resident prophage. *Salmonella* strain DB21 was lysogenized

Table 7
Influence of the P22 Superinfection Exclusion Systems (SIE) on Anderson Typing Phages

Salmonella indicator strain	Anderson typing phages (A)					
	A1–A7	A10, A 11	A12, A13	A14–A16	A17	A19–A35
BA36	+	+	+	+	+	+
BA36 (P22)	–	–	–	–	–	–
BA36 (P22 sieA, sieB)	+	+	–	+	–	+

Table 8
Influence of Different Prophages on the Phage Type

RKI number	Different prophages in *Salmonella* Typhimurium strains							Anderson phage type	Felix and Callow/ Lilleengen phage type
	Strain LT2 derivatives								
02-8516	LT2	G1 (+)	G2 (+)	G3 (−)	F1 (+)	F2 (+)	SΦ (−)	DT 4	1bvar2/2
02-8526	MA 4587	G1 (−)	G2 (−)	G3 (−)	F1 (+)	F2 (+)	SΦ (−)	DT 4	1b/2
02-8517	MA 6495	G1 (+)	G2 (+)	G3 (+)	F1 (+)	F2 (+)	SΦ (−)	DT 29	ut/n.c.
02-8501	MA 6502	G1 (−)	G2 (−)	G3 (+)	F1 (+)	F2 (+)	SΦ (−)	DT 29	ut/n.c.
02-8494	MA 6549	G1 (−)	G2 (−)	G3 (−)	F1 (+)	F2 (−)	SΦ (+)	DT 44	1var5/n.c.
	Strain SL1344 derivatives								
02-8508	MA 6118	G1 (+)	G2 (+)	G3 (−)	F1 (−)	F2 (−)	SΦ (+)	DT 44	1var5/n.c.
02-8500	MA 6244	G1 (−)	G2 (+)	G3 (−)	F1 (−)	F2 (−)	SΦ (+)	DT 44	1var5/n.c.
02-8507	MA 6497	G1 (−)	G2 (+)	G3 (+)	F1 (−)	F2 (−)	SΦ (+)	DT 29	ut/n.c.
	Strain ATCC14028 derivatives								
02-8511	NA 5958	G1 (+)	G2 (+)	G3 (+)	F1 (−)	F2 (−)	SΦ (−)	DT126	1var18/n.c.
02-8490	MA 5973	G1 (−)	G2 (+)	G3 (+)	F1 (−)	F2 (−)	SΦ (−)	DT126	1var18/n.c.
02-8487	MA 5975	G1 (−)	G2 (−)	G3 (+)	F1 (−)	F2 (−)	SΦ (−)	DT126	1var18/n.c.
02-8510	MA 6052	G1 (−)	G2 (−)	G3 (−)	F1 (−)	F2 (−)	SΦ (−)	DT1	1/1

G1 (Gifsy-1); G2 (Gifsy-2); G3 (Gifsy-3); F1 (Fels1); F2 (Fels2); SΦ (SopΦ).

with various P22-related phages representing different immunity classes. The resulting lysogenic derivatives were assayed with the Anderson test phages. We observed different phage type conversions ***(70)***.

1.5.5. Influence of Different Prophages on the Phage Type

There are a strain collection of N. Figueroa-Bossi which contain different prophages in different *S*. Typhimurium backgrounds ***(71)***. It was interesting to investigate phage type of these clones by our phage-typing sets. The results are described in **Table 8**. Phage Gifsy-3 has an influence on the phage type in all three *S*. Typhimurium strains. The sopE has also an influence in different *S*. Typhimurium backgrounds. On the other side, the phage Gifsy-1 has no influence in SL1344 and ATCC14028.

1.5.6. Lysis Variants With or Without Antirepressors

Figures 3–6 illustrated some examples of how homoimmune and heteroimmune phages can propagate in different backgrounds with or without antirepressor.

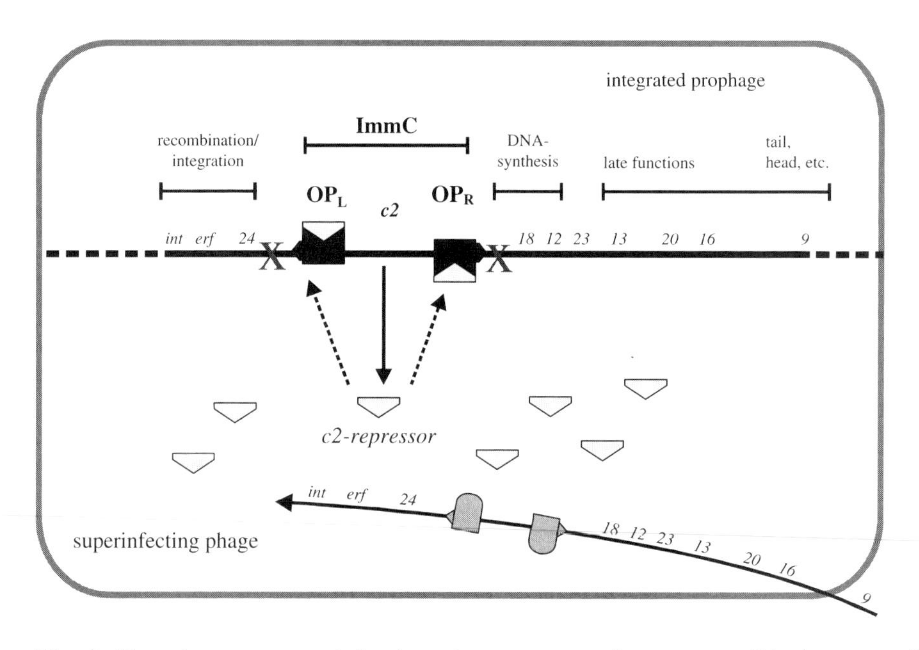

Fig. 3. Homoimmune superinfecting phage: propagation not possible because of blockage of operator-promoter sites OP_L and OP_R.

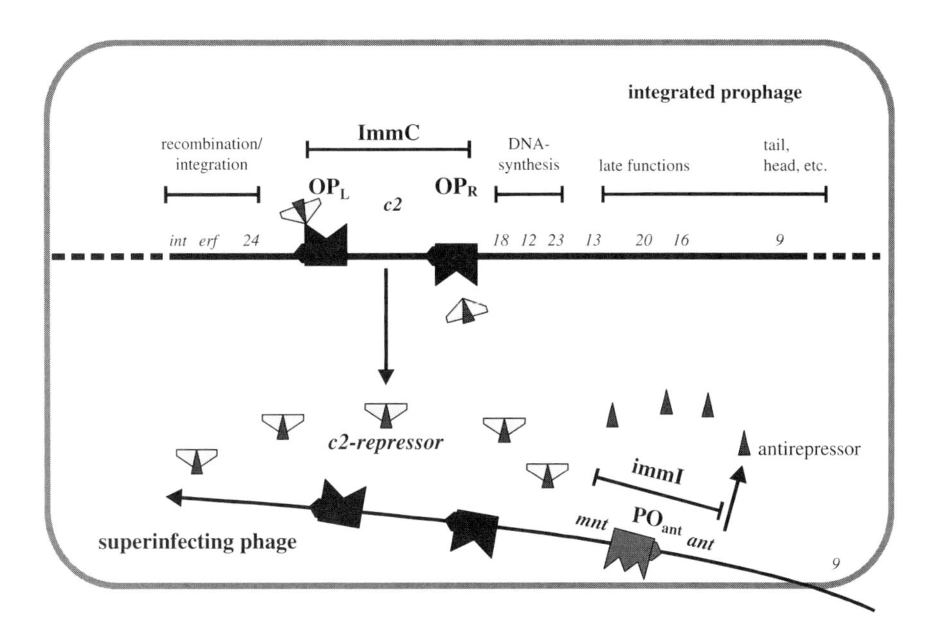

Fig. 4. Heteroimmune superinfecting phage: propagation possible.

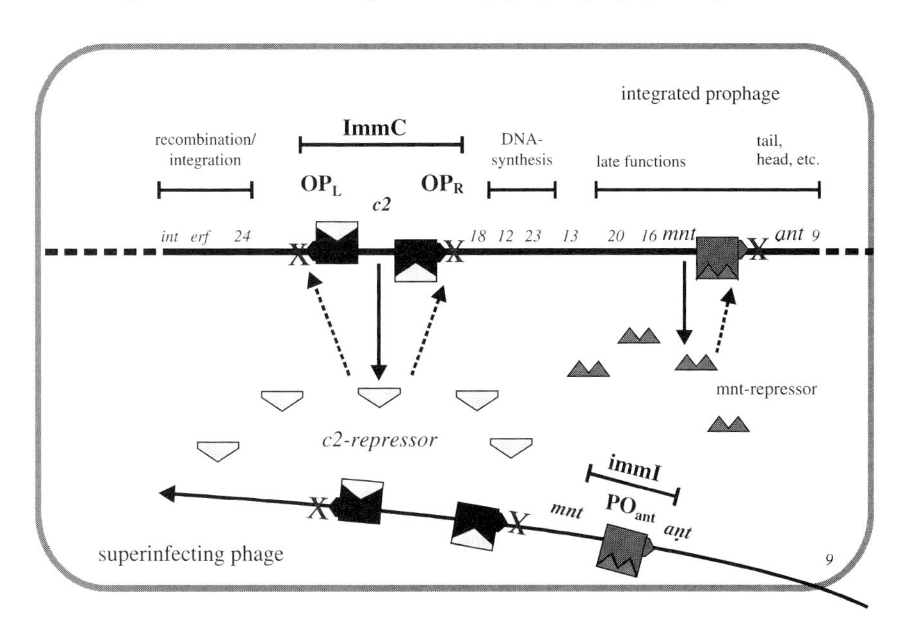

Fig. 5. Homoimmune superinfecting phage with expressed antirepressor: propagation possible.

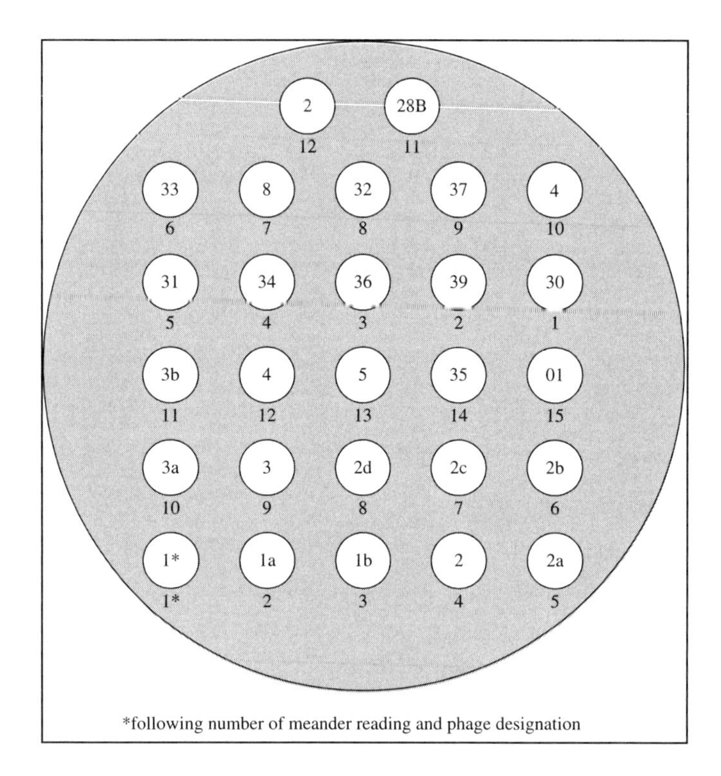

Fig. 6. Homoimmune superinfective phage with antirepressor and prophage with antirepressor and expressed *mnt* gene, which negatively regulates the antirepressor by binding downstream POant: propagation not possible.

2. Materials

Phage typing requires the use of plating media. For control of purity of *Salmonella* strains, Endo-Agar is recommended (*see* procedure above), but other plating media are also good tools for this purpose. A number of chromogenic *Salmonella* plating media are available on the market. Culture media have to be generally controlled before use. Control of color, pH, solidity, and microbiological efficiency are the main control measurements.

2.1. Endo-Agar

Endo-Agar differentiates bacteria producing acids from lactose from those that do not. In the metabolic pathway, ethanal (acetaldehyde) is formed, which reacts with sodium sulfite and fuchsine to form red colonies. The development of a metallic sheen occurs when the bacteria rapidly form ethanal. Bacteria that do not cleave lactose (e.g., most *Salmonella* subsp.) grow as colorless colonies. Different sizes of colonies may point to the presence of more than one *Salmonella* strain.

Endo-Agar is available as dry medium or as ready-to-use plates from various companies. The quality of these commercially available culture media has to be controlled before use, too.

1. Formula of Home-Made Endo-Agar ("Haus-Endo"): Endo-Agar (SIFIN, Berlin, Germany) for 1 L, meat extract (Merck, Berlin, Germany) 5.0 g, yeast extract (DIFCO/BD) 6.0 g, sterilization in a steamer (100 °C) for 90 min, decolorization by addition of 13.5 mL 10 % (w/v) sodium sulfite solution, and final pH 7.2–7.4.
2. Chromogenic Salmonella-Plating Media: For a clear differentiation of *Salmonella* spp. from other bacteria, the following newly developed chromogenic *Salmonella*-plating media are available:
   Salmonella-Ident-Agar (Heipha) detects all *Salmonella enterica* and *Salmonella bongori*. ASAP (AES-Chemunex) and SM ID2 (BioMerieux) detect all *S. enterica* and *S. bongori* except *Salmonella* Dublin. OSCM (Oxoid) detects all *S. enterica* and *S. bongori* except *Salmonella arizonae* and *Salmonella diarizonae*. Rambach-Agar (Merck, Heipha) detects most *Salmonella* subsp. *Enterica* serotypes but not all (e.g., *S.* Typhi, *S*, Paratyphi A, B, *Salmonella* Rostock, *Salmonella* Wassenar, *Salmonella* Marina, *Salmonella* Wayne) and not *S. arizonae* and *S. diarizonae*.
3. Enrichment Broth for Phage Typing ("Lysotypie-Bouillon"): Bacto Nutrient Broth (DIFCO/BD) 20.0 g, NaCl 8.5 g, distilled water 1 L, sterilization 121 °C, 15 min, and final pH 7.2–7.4.
4. Plating Medium for Phage Typing of S. Typhimurium (Anderson-Scheme) of S. Enteritidis and of other Salmonella spp. ("OX-3-plates")

   a) Basic Medium: Bacto Nutrient Broth (DIFCO/BD) 20.0 g, NaCl 8.5 g, Bacto Yeast extract (DIFCO/BD) 3.0 g, Caseinpeptone, pancreatic (Merck) 1.0 g, Bacto Peptone (DIFCO/BD) 2.0 g, Agar (Oxoid No. 3) 13.0 g, sterilization 121 °C, 15 min, and final pH 7.3–7.5.
   b) Supplements: Add the following supplements under sterile conditions: 0.1 M $CaCl_2$ 1.0 mL, 1.0 M $MgCl_2$ 1.0 mL, and trace-element solution according to Schloesser 5.0 mL. Control pH and pour into Petri dishes.
   c) Trace-element Solution According to Schloesser *(72)* $ZnSO_4 \cdot 7\ H_2O$ 1 mg, $MnSO_4 \cdot 4\ H_2O$ 2 mg, H_3BO_3 10 mg, $Co(NO_3)_2 \cdot 6\ H_2O$ 1 mg, $Na_2MoO_4 \cdot 3\ H_2O$ 1 mg, $CuSO_4 \cdot 5\ H_2O$ 0.005 mg, $FeSO_4 \times 7\ H_2O$ 0.7 g, EDTA 0.8 g, double distilled water 1 L, and sterilization by filtration (0.45 μm).

5 Plating Medium for Phage Typing of S. Typhimurium (Felix/Callow- and Lilleengen-Scheme); ("Blue-Plates")

   a) Basic Medium: Bacto Peptone (DIFCO/BD) 5.0 g, Bacto Yeast extract (DIFCO/BD) 5.0 g, Bacto Beef extract (DIFCO/BD) 1.0 g, Bacto Casamino Acids (DIFCO/BD) 5.0 g, Bacto Agar (DIFCO/BD) 17.0 g, distilled water 1 L, pH at this stage 7.7, sterilization 121 °C, 15 min, and final pH 7.2–7.4.

b) Supplements: Cool the basic medium to 60 °C and add freshly sterilized (30 min in a steamer) solutions of glucose and water blue (Aniline blue, water soluble) under sterile conditions: 24 mL Glucose (50 %) and 12 mL Water blue (10 %). Mix thoroughly, control the pH, and pour immediately into Petri dishes.

3. Methods

3.1. The Procedure of Phage Typing

The technique followed in the *Salmonella* phage typing (extended Anderson scheme) protocol from PHLS (London) is basically that described by Craigie and Felix ***(11)***. Most Reference Centers in other countries of the world using this method follow a closely similar technique, which, with only minor variations, is used for all *Salmonella* phage typing. The following description of the procedure is based on our laboratory experience over five decades.

It is recommended that received strains are subject to purity control by incubation on Endo-Agar plates. Morphological examination is followed by serological identification of the 4,[5],12.i:1,2 antigens. Rough-forms are excluded by agglutination with 0.85 % NaCl or 3.5 % NaCl with trypaflavin solutions. The O-antigens 4 and 5 as well as the H-antigens are examined. Special attention has to be paid to the first (specific) H-phase to diagnostically differentiate *S*. Paratyphi B (antigen formula 4,[5]12:b:1,2). If the strain is not available in the first phase, a swarming plate has to be prepared after SVEN GARD's protocol, which can then be transformed into a specific phase culture.

Cultures to be phage typed are inoculated by a sterile loop or glass stick into 4 mL enrichment broth to give a barely visible turbidity and are incubated without agitation at 37 °C until they reach a density of about 4×10^8 cells/mL (*see* **Subheading 2.1.5.**). The culture is inoculated onto a suitably marked Blue plate (Felix/Callow and Lilleengen) and Oxoid-3 plate (extended Anderson) using the standardized EnterNet protocol by flooding (*see* **Subheadings 2.1.6.** and **2.1.7.**). Excess culture is removed from the surface, and after drying the plate, the typing phages are spotted by a multipoint inoculator in a predetermined order corresponding to the marking of the plate. For Felix/Callow and Lilleengen, we use a 27 flexible loop multipoint inoculator (template, *see* **Fig. 7**, and after incubation the phage typing "Blue-plate," *see* **Fig. 8A**); for extended Anderson, we use a 35 flexible loop multipoint inoculator (*see* **Fig. 9**) (template, *see* **Fig. 2**, and after incubation the phage typing "OX-3-plate" overnight, *see* **Fig. 8B**). It is important when using multipoint inoculator that all loops are submerged in ethyl alcohol and flamed to prevent contamination of phage solutions with ethyl alcohol. Phage spots are dried, and the Blue plate and Oxoid-3 plate are incubated inverted for 5–18 h at 37 °C.

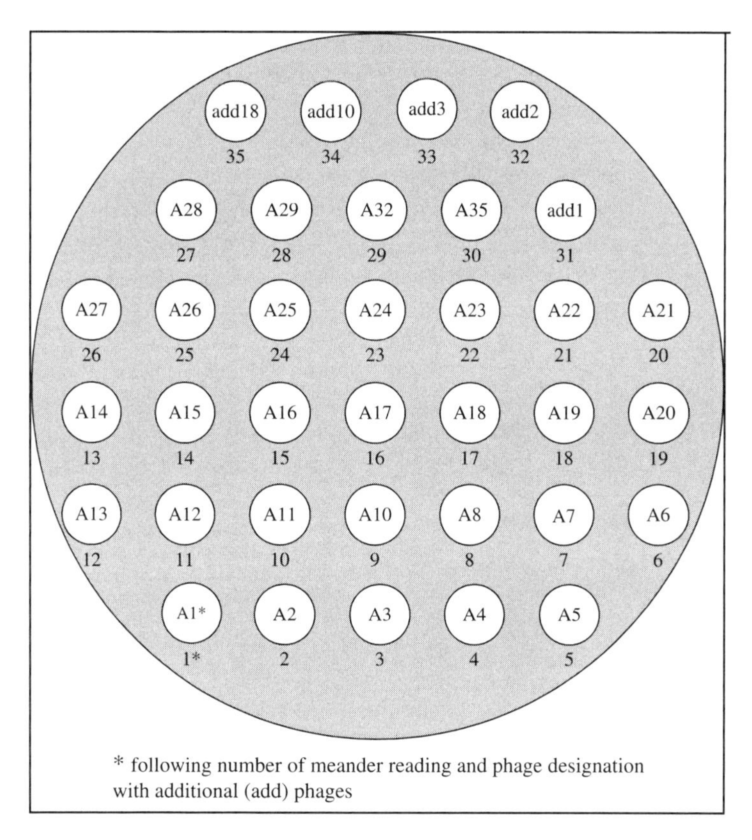

Fig. 7. Inoculation scheme of the *Salmonella* Typhimurium Felix/Callow and Lilleengen phage-typing plate.

Readings are carried out with a 10-times magnification hand lens through the bottom of the plates, using transmitted oblique illumination, or using stereo microscope at 10-times magnification with indirect illumination for the Oxoid-3 plate and direct light for the Blue plate. Transparent plastic Petri dishes are used. The lid needs to be removed for reading. The various degrees of lysis are recorded as shown in **Fig. 10**. The phage reactions are recorded on special phage-typing sheets for archiving.

At a given phage concentration, the amount of culture destroyed by a phage is obviously dependent on the size of the plaques; relatively few large plaques (often under 100) are needed to produce confluent lysis, whereas thousands of plaques of a minute-plaque phage may be necessary to produce lysis of a similar degree. The plaque sizes of typing phages may vary considerably,

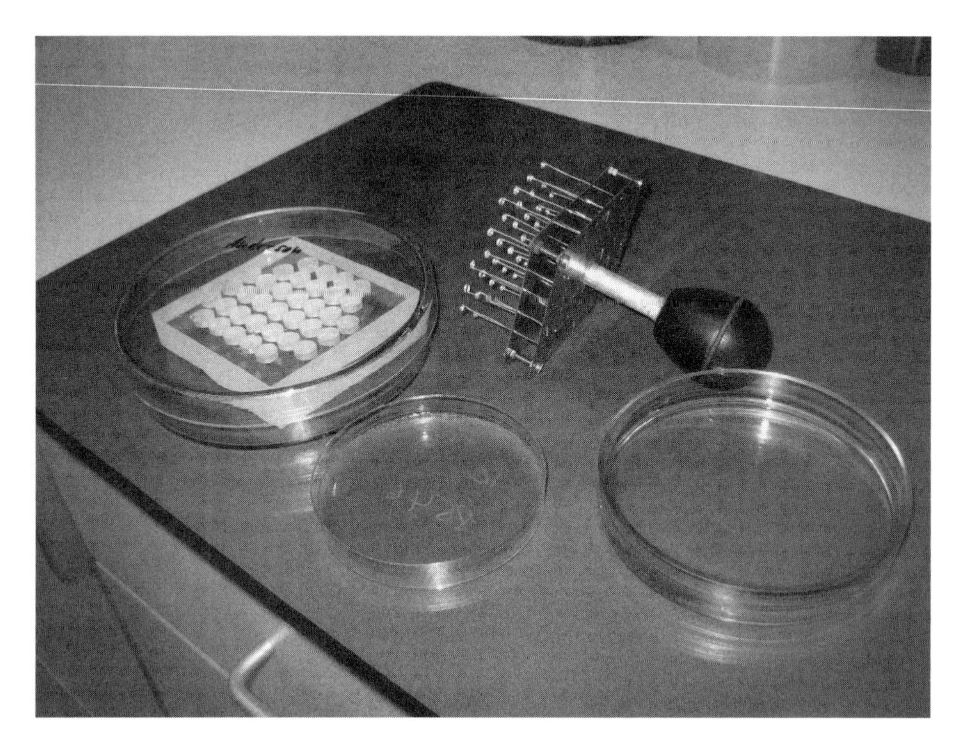

Fig. 8. (**A**) Felix/Callow and Lilleengen phage-typing plate (Blue plate) 1bvar.2/2 original sequenced LT2 (*see* **ref. 71**). (**B**) Extended Anderson phage-typing plate DT4 original sequenced LT2 (*see* **ref. 71**).

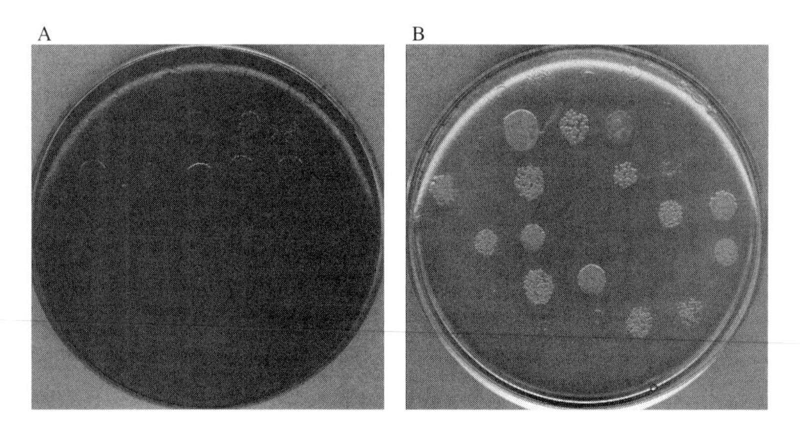

Fig. 9. Multipoint inoculator for *Salmonella* Typhimurium Anderson phages, template for phages (left), typing plate with agar, Petri dish with alcohol (right).

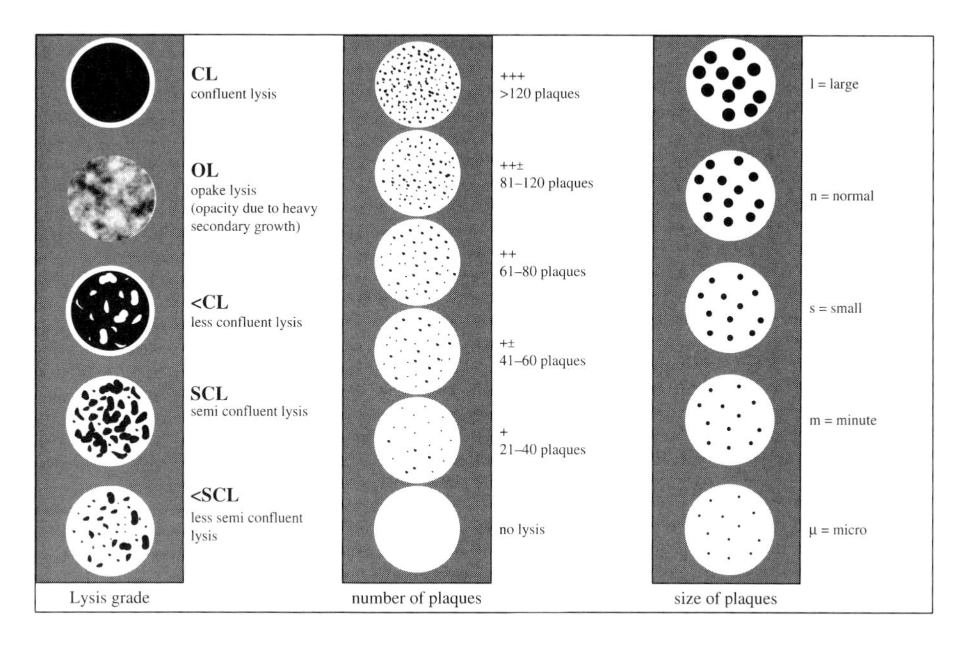

Fig. 10. Method of recording degrees of lysis on *Salmonella* phage-typing plate.

and this is one of the factors to be considered when deciding on the phage dilution to be used for routine typing, which is known as the routine test dilution (RTD).

The RTD is very important in this work because if the typing phages for any *Salmonella* serovar are used undiluted, they will usually lyse indiscriminately all phage types of the organism, so that no lytic patterns can be distinguished. This can be avoided by using the RTD, which is the highest phage dilution that will produce confluent lysis or a reaction approaching that order, on the type the phage is primarily intended to distinguish. The use of this dilution minimizes cross-reactions on other types and enables us to narrow the host specificity of the phages as far as is practicable. Thus, although the ideal of having a single-phage specific for each type can rarely be attained, the use of the RTD helps to make distinct the patterns on which type diagnosis depends.

3.2. Comparison of Different Phage Types of the Three Schemes

The phage-typing scheme of Anderson was developed because the proportion of strains that were untypeable by the Felix and Callow system increased. This was established by our Reference Center *(73)*. On the contrary, the combination of the Felix/Callow and Lilleengen systems gave good discrimination of some DT types, especially the use of Felix/Callow phages 2, 2a, 2c, and 4 and Lilleengen phages L30 and L33. In **Table 9**, examples show the discriminatory

Table 9
Comparison of *Salmonella* Typhimurium Phage Types by Anderson and Felix and Callow/Lilleengen

Anderson	Felix/Callow and Lilleengen
DT4	1b/2
DT4	1bvar2/2
DT4	1b/n.c.[a]
DT7	2d/ut
DT7	ut/Ph.30
DT7	ut/ut[b]
DT12	2/ut
DT12	2/Ph30
DT12	2var1/16
DT104	2/Ph30
DT104	2c/Ph30
DT104	ut/Ph30
DT104	ut/ut
DT120	2b/Ph30
DT120	2/Ph30
DT120	4/ut
DT120	ut/Ph30
DT120	5/26
DT120	ut/n.c.
DT120	ut/ut
DT186	2b/Ph30
DT186	35/Ph30
DT186	ut/Ph30
DT186	ut/n.c.
DT186	ut/ut

[a]Non-characteristic.
[b]Untypeable.

power of the combination of both schemes. This was shown also by phage typing of the old Demerec collection ***(74)***.

4. Notes

1. If the phage type of one strain is not clear, the 2-day-old Endo-Agar plate of this strain is reexamined for purity under the stereo microscope. Mixed cultures are purified by diluting an inoculation loop in 1.5 mL broth and spreading on five Endo-Agar plates while reducing cell densities subsequently. Thus, sufficient single colonies are generated to isolate the desired *S*. Typhimurium strain. If cultures are overgrown by swarming *Proteus* strains, Endo-Agar plates can be flooded with 96 % ethanol prior to spreading to suppress the swarming ability of *Proteus* spp. cultures. All other common protocols like *Salmonella* Ident Agar (Heipha) can also be applied to suppress the swarming ability.
2. Contaminated strains or mixed cultures are recognized by different colors as well as sizes and shapes of single colonies. There are on the Endo-Agar plate different colonies according their form, elevation, and margin, e.g., white circular convex, pink with and without turbid center, irregular flat colonies. Small convex colonies are often the best for typing. If the phage reactions are not clear, 10 colonies are picked and phage typed.
3. If the O1 phage is negative and no rough phage (London A59/6S.R.or rough phage Frankfurt/Main “Ffm”) reacts, the O1 phage is tested as concentrate (10^9 pfu/mL).
4. Logarithmic phase cells are needed for flooding the plate. It seems that overrepresentation of gene products around the chromosome is better for typing.
5. After flooding with the *Salmonella* suspension, the plates for typing have to be dry. If that is not the case, the phage drops can run into each other and give no clear phage reactions.
6. The template for the multipoint inoculator is used daily and filled up after sterilization. After 3 weeks’ usage, the phage titer may decrease sometimes and should be tested, for example, with DT36 reference strain.
7. It is important for using multipoint inoculator that all loops are submerged in ethyl alcohol because the loops contaminate quickly and bend easily, all loops are flamed to prevent contamination of phage solutions with ethyl alcohol, and all loops are cooled in air to prevent heat inactivation of phage particles giving inaccurate results.
8. New phage preparations have to be tested with the reference strain and positive and negative control strains.
9. Also RDNC (to any phage types) strains with untypical lysis pattern can be stable in outbreak investigations.
10. The comparison of phage typing and antibiotic resistance pattern results is very helpful. If outbreak strains show differences with both methods, the loss of an R-plasmid with restriction determinants can be the reason.

Acknowledgments

I thank Linda R. Ward, London, for supporting us with typing phages, Ralph Wollin, Stockholm, for sending the Lilleengen phages, N. Figueroa-Bossi for the well-characterized prophages in *S*. Typhimurium strains, Horst Schmieger, Munich, and Manfred Biebl, Regensburg, for the fruitful discussions, and Dianne Davos and P. H. Williams for critical reading. Formulas of plating media were developed or improved by the team of the nutrient media laboratory at the Institute in Wernigerode (Head: Rolf Reissbrodt). Furthermore, I thank H. Gattermann, V. Trute, S. Kulbe, and D. Busse of our laboratory for skillful technical assistance over many years.

References

1. Mead, P. S., Slutsker, L., Dietz, V., et al. (1999) Food-related illness and death in the United States. *Emerg. Infect. Dis.* **5,** 607–625.
2. Chalker, R. B. and Blaser, M. J. (1988) A review of human Salmonellosis: III. Magnitude of *Salmonella* infection in the United States. *Rev. Infect. Dis.* **10,** 111–124.
3. Voetsch, A. C., van Gilder, T. J., Angulo, F. J., et al. (2004) FoodNet estimate of the burden of illness caused by nontyphoidal *Salmonella* infections in the United States – Emerging Infections Program FoodNet Working Group. *Clin. Infect. Dis.* **15,** 38 Suppl. 3, S127–134.
4. Galbraith, N. S. (1961) Studies of human salmonelloses in relation to infection in animals. *Vet. Rec.* **73,** 1296–1303.
5. Rabsch, W., Altier, C., Tschäpe, H., and Bäumler, A. J. (2003) Foodborne *Salmonella* infections, in *Microbial Food Safety in Animal Agriculture*. Current Topics (Torrence, M. E. and Isaacson, R. E., eds.), Iowa State Press, pp. 97–107.
6. CDC (1996) *Salmonella* surveillance: annual tabulation summary, 1993–1995 US Department of Health and Human Services. CDC, Atlanta.
7. Sonnenschein, C. (1928) Bakteriendiagnose mit Bakteriophagen. *Deutsche Medizinische Wochenschr.* **25,** 1–4.
8. Marcuse, K. (1934) Über Typhusdiagnostik mit spezifischen Bakteriophagen (nach Sonnenschein). *Zentralbl. Bakteriol. Abt.i* **131,** 206–211.
9. Cragie, J. and Yen, C. H. (1938a) The demonstration of types of B. typhosus by means of preparations of Type II Vi-phage. 1. *Can. Public Health J.* **29,** 448–463.
10. Cragie, J. and Yen, C. H. (1938b) The demonstration of types of B. typhosus by means of preparations of Type II Vi-phage. 2. *Can. Public Health J.* **29,** 484–496.
11. Craigie, J. and Felix, A. (1947) Typing of typhoid bacilli with Vi bacteriophage. *Lancet* **i,** 823–827.
12. Lilleengen, K. (1948) Typing of *Salmonella* Typhimurium by means of bacteriophage. *Acta Pathol. Microbiol. Scand. Suppl.* **77,** 1–39.

13. Ward, L. R., de Sa, J. D., and Rowe, B. (1987) A phage-typing scheme for *Salmonella* Enteritidis. *Epidemiol. Infect.* **99,** 291–294.
14. Rabsch, W., Prager, R., Koch, J., et al. (2005) Molecular epidemiology of Salmonella enterica serovar Agona: characterization of a diffuse outbreak caused by aniseed-fennel-caraway infusion. *Epidemiol. Infect.* **133,** 837–844.
15. Fisher, I. S. and Enter-net participants (2004) Dramatic shift in the epidemiology of *Salmonella enterica* serotype Enteritidis phage types in Western Europe, 1998–2003 – results from the Enter-net international Salmonella database. *Eur. Surveill.* **9,** 43–45.
16. Helm, R. A., Porwollik, S., Stanley, A. E., et al. (2004) Pigeon-associated strains of Salmonella enterica serovar Typhimurium phage type DT2 have genomic rearrangements at rRNA operons. *Infect. Immun.* **72,** 7338–7341.
17. Rabsch, W., Andrews, H. L., Kingsley, R. A., et al. (2002) *Salmonella enterica* serotype Typhimurium and its host-adapted variants. *Infect. Immun.* **70,** 2249–2255.
18. Bratbak, G., Heldal, M., Norland, S., and Thingstad, T. F. (1990) Viruses as partners in spring bloom microbial trophodynamics. *Appl. Environ. Microbiol.* **56,** 1400–1405.
19. Procter, L. M., Okubo, A., and Fuhrmann, J. A. (1993) Calibrating estimates of phage-induced mortality in marine bacteria: ultrastructural studies of marine bacteriophage development from one-step growth experiments. *Microb. Ecol.* **25,** 161–182.
20. Thingstad, T. F. (2000) Elements of a theory for the mechanisms controlling abundance diversity and biogeochemical role of lytic bacterial virus in aquatic systems. *Limnol. Oceanogr.* **45,** 1320–1914.
21. Schicklmaier, P. and Schmieger, H. (1995) Frequency of generalized transducing phages in natural isolates of the *Salmonella* Typhimurium complex. *Appl. Environ. Microbiol.* **61,** 1637–1640.
22. Zinder, N. D. and Lederberg, J. (1952) Genetic exchange in Salmonella. *J. Bacteriol.* **64,** 679–699.
23. Kuo, T.-T. and Stocker, B. A. D. (1978) ES18, a general transducing phage for smooth and non-smooth *Salmonella* Typhimuirum. *Virology* **42,** 621–632.
24. Sanderson, K. E. (1972) Linkage map of *Salmonella* Typhimurium, edition IV. *Bacteriol. Rev.* **36,** 558–586.
25. Schmieger, H. (1972) Phage P22-mutants with increased or decreased transduction abilities. *Mol. Gen. Genet.* **119,** 75–88
26. Mirold, S., Rabsch, W., Rohde, M., et al. (1999) Isolation of a temperate bacteriophage encoding the type III effector protein SopE from an epidemic *Salmonella* Typhimurium strain. *Proc. Natl. Acad. Sci. USA* **96,** 9845–9850.
27. Hopkins, K. L. and Threlfall, E. J. (2004) Frequency and polymorphism of *sopE* in isolates of *Salmonella enterica* belonging to the ten most prevalent serotypes in England and Wales. *J. Med. Microbiol.* **53,** 539–543.

28. Wray, C., Beedell, Y. E., and McLaren, I. M. (1991) A survey of antimicrobial resistance in salmonellae isolated from animals in England and Wales during 1984–1987. *Br. Vet. J.* **147,** 356–369.
29. Threlfall, E. J., Ward, L. R., and Rowe, B. (1978) Spread of multiresistant strains of *Salmonella* Typhimurium phage types 204 and 193 in Britain.*Br. Med. J.* **2,** 997.
30. Threlfall, E. J., Frost, J. A., Ward, L. R., and Rowe, B. (1990) Plasmid profile typing can be used to subdivide phage-type 49 of *Salmonella* Typhimurium in outbreak investigations. *Epidemiol. Infect.* **104,** 243–251.
31. Kühn, H., Rabsch, W., Tschäpe, H., and Tietze, E. (1982) Characterization and epidemiology of a *Salmonella* Typhimurium epidemic strain. *Z. Ärztl. Fortbild. (Jena)* **70,** 607–610.
32. Ehrbar, K., Mirold, S., Friebel, A., Stender, S., and Hardt, W. D. (2002) Characterization of effector proteins translocated via the SPI1 type III secretion system of *Salmonella* Typhimurium. *Int. J. Med. Microbiol.* **291,** 479–485.
33. Fang, F. C., DeGroote, M. A., Foster, J. W., et al. (1999) Virulent *Salmonella* Typhimurium has two periplasmic Cu, Zn-superoxide dismutases. *Proc. Natl. Acad. Sci. USA* **96,** 7502–7507.
34. Figueroa-Bossi, N., Uzzau, S., Maloriol, D., and Bossi, L. (2001) Variable assortment of prophages provides a transferable repertoire of pathogenic determinants in *Salmonella. Mol. Microbiol.* **39,** 260–272.
35. Hardt, W. D., Urlaub, H., and Galan, J. E. (1998) A substrate of the centrisome 63 type III protein secretion system of *Salmonella* Typhimurium is encoded by a cryptic bacteriophage. *Proc. Natl. Acad. Sci. USA* **95,** 2574–2579.
36. Miao, E. A., Scherer, A., Tsolis, R. M., et al. (1999) *Salmonella* Typhimurium leucine-rich repeat proteins are targeted to the SPI1 and SPI2 type secretion systems. *Mol. Microbiol.* **34,** 850–864.
37. Brussow H., Canchaya C., and Hardt W. D. (2004) Phages and the evolution of bacterial pathogens: from genomic rearrangements to lysogenic conversion. *Microbiol. Mol. Biol. Rev.* **68,** 560–602.
38. Stone, R. (2002) Bacteriophage therapy: Stalin's forgotten cure, *Science* **298,** 728–731.
39. Weber-Dabrowska, B., Mulczyk, M., and Gorski, A. (2000) Bacteriophage therapy of bacterial infections: an update of our institute's experience. *Arch. Immunol. Ther. Exp. (Warsz),* **48,** 547–551.
40. Berchieri, A. Jr., Lovell, M. A., and Barrow, P. A. (1991) The activity in the chicken alimentary tract of bacteriophages lytic for Salmonella Typhimurium. *Res. Microbiol.* **142,** 541–549.
41. Rabsch, W., Wiley, G., Najar, F. Z., Kaserer, W., Schuerch, D. W., Klebba, J. E., Roe, B. A., Gomez, J. A. L., Schallmey, M., Newton, S. M. C., and Klebba, P. E. (2007) FepA-and TonB-dependent Bacteriophage H8: Receptor Binding and Genomic Sequence. *J. Bact.* (in press).

42. Chanishvili, N., Chanishvili, T., Tediashvili, M., and Barrow, P. A. (2001) Review: phages and their application against drug-resistant bacteria. *J. Chem. Technol. Biotechnol.* **76,** 689–699.
43. Hänggi, B. J. (2004) Die Phagentherapie und das Problem ihrer Verwirklichung – ein Beitrag zur gegenwärtigen Rückbesinnung auf ein medizinhistorisches Phänomen. Inauguraldissertation der Universität Bern, Schweiz.
44. Goode, D., Allen, V. M., and Barrow, P. A. (2003) Reduction of experimental Salmonella and Campylobacter contamination of chicken skin by application of lytic bacteriophages. *Appl. Environ. Microbiol.* **69,** 5032–5036.
45. Leverentz, B., Conway, W. S., Alavidze, Z., et al. (2001) Examination of bacteriophage as a biocontrol method for Salmonella on fresh-cut fruit: a model study. *J. Food Prot.* **64,** 1116–1121.
46. Kallings, L. O. and Laurell, A.-B. (1957) Relation between phage types and fermentation types of Salmonella Typhimurium. *Acta Pathol. Microbiol. Scand.* **40,** 328–342.
47. Lundbeck, H., Plazikowski, U., and Silverstolpe, L. (1955) The Swedish Salmonella outbreak of 1953. *J. Appl. Bacteriol.* **18,** 535–548.
48. Kallings, L. O., Laurell, A.-B., and Zetterberg, B. (1959) An outbreak due to Salmonella typhimurium in veal with special reference to phage and fermentation typing. *Acta Pathol. Microbiol. Scand.* **45,** 347–356.
49. Callow, B. R. (1959) A new phage-typing scheme for *Salmonella* Typhimurium. *J. Hyg., Camb.* **57,** 346–359.
50. Anderson, E. S., Ward, L. R., Saxe, M. J., and de Sa, J. D. (1977) Bacteriophage-typing designations of Salmonella Typhimurium. *J. Hyg. (Lond).* **78,** 297–300.
51. Kühn, H., Falta, R., and Rische, H. (1973) *Salmonella* Typhimurium, in *Lysotypie und andere spezielle epidemiologische Laboratoriumsmethoden* (Rische, H., ed.), VEB Gustav Fischer Verlag Jena, pp. 101–139.
52. Kohler, B., Vogel, K., Kuhn, H., et al. (1979) Epizootiology of *Salmonella* Typhimurium infection in chickens. *Arch. Exp. Veterinarmed.* **33,** 281–298.
53. Hasenson, L., Gericke, B., Liesegang, A., et al. (1995) Epidemiological and microbiological studies on salmonellosis in Russia. *Zentralbl. Hyg. Umweltmed.* **198,** 97–116.
54. Rabsch, W., Tschäpe, H., and Kühn, H. (1985) Phage type changes in multiple drug-resistant plasmids of S. Typhimurium strains from hospitals of different countries. *Z. Gesamte Hyg.* **33,** (1987) 264–266 (German).
55. Rabsch, W., Tschäpe, H., Tietze, E., and Kuhn, H. (1982) Characterization of individual *Salmonella* clones of an epidemic strain of the same phage- and biochemotype using plasmid determination. *Z. Gesamte Hyg.* **28,** 842–844 (German).
56. Böhme, G., Kühn., H., Tschape, H., and Rabsch, W. (1989) Epidemiology of salmonellosis. *Z. Gesamte Hyg.* **35,** 638–640 (German).
57. Jacob, W. K., Kuhn, H., Kurschner, H., and Rabsch, W. (1993) The epidemiological analysis of *Salmonella* Typhimurium infections in cattle – results of

lysotyping and biochemotyping in the region of East Thuringia from 1974 to 1991. *Berl. Munch. Tierarztl. Wochenschr.* **106,** 265–269 (German).
58. Kühn, H., Rabsch, W., and Liesegang, A. (1994) Current epidemiological status of salmonellosis of humans in Germany (Review). *Immun. Infekt.* **22,** 4–9 (German).
59. Aiogaily, Z., Anastassiadou, H., Barrow, P. A., et al. (1994) Control of Salmonella infections in animals and prevention of human foodborne Salmonella infections. *Bull. World Health Organ.* **72,** 831–833.
60. Ang-Kücüker, M., Tolun, V., Helmuth, R., et al. (2000) Phage types antibiotic susceptibilities and plasmid profiles of *Salmonella* Typhimurium and *Salmonella* Enteritidis strains isolated in Istanbul, Turkey. *Clin. Microbiol. Infect.* **6,** 593–599.
61. Rabsch, W., Tschäpe, H., and Bäumler, A. J. (2001) Non-typhoidal Salmonellosis: emerging problems. *Microbes Infect.* **3,** 237–247.
62. Helms, M., Ethelberg, S., and Molbak, K. (2005) DT104 Study Group. International Salmonella Typhimurium DT104 infections, 1992–2001. *Emerg. Infect. Dis.* (Jun) **11,** 859–867.
63. Rabsch, W., Andrews, H. L., Kingsley, R. A., Prager, R., and Tschäpe, H. (2002) *Salmonella enterica* serotype Typhimurium and its host-adapted Variants. *Infect. Immun.* **70,** 2249–2255.
64. Frech, G., Weide-Botjes, M., Nußbeck, E., Rabsch, W., and Schwarz, S. (1998) Molecular characterization of *Salmonella enterica* subspec. *enterica* serovar Typhimurium DT 009 isolates: differentiation of the live vaccine strain Zoosaloral H from field isolates. *FEMS Microbiol. Lett.* **167,** 263–269.
65. Rabsch, W., Liesegang, A., and Tschäpe, H. (2001) Laboratory-based surveillance of salmonellosis of humans in Germany – safety of *Salmonella* Typhimurium and *Salmonella* Enteritidis live vaccines. *Berl. Munch. Tierarztl. Wochenschr.* **114,** 433–437 (German) Reprints English from the author.
66. Hashemolhosseini, S., Holmes, Z., Mutschler, B., and Henning, U. (1994) Alterations of receptor specificities of coliphages of the T2 family. *J. Mol. Biol.* **240,** 105–110.
67. Luckey, M. and Nikaido, H. (1980) Diffusion of solutes through channels produced by phage lambda receptor protein of *Escherichia coli*: inhibition by higher oligosaccharides of maltose series. *Biochem. Biophys. Res. Commun.* **93,** 166–171.
68. Lucey, M. and Neilands, J. B. (1976) Iron transport in Salmonella Typhimurium LT2: prevention by ferrichrome of adsorption of bacteriophages ES18 and ES18h1 to a common cell envelope receptor. *J. Bacteriol.* **127,** 1036–1037.
69. Schmieger, H. (1999) Molecular survey of the Salmonella phage typing system of Anderson. *J. Bacteriol.* **181,** 1630–1635.
70. Rabsch, W., Mirold, S., Hardt, W. D., and Tschäpe, H. (2002) The dual role of wild phages for horizontal gene transfer among Salmonella strains. *Berl. Munch. Tierarztl. Wochenschr.* **115,** 355–359.

71. Bossi, L. and Bossi, N. F. (2005) Prophage Arsenal of Salmonella enterica Serovar Typhimurium, in *Phages – Their Role in Bacterial Pathogenesis and Biotechnology* (Waldor, M. K., Friedman, D. I., and Adhya, S. L., eds.), ASM Press Washington, D. C., pp. 165–186.
72. Schloesser, E. (1997) 3. Mikrobiologische Arbeitsmethoden, in *Methoden der Bodenbiologie*, 2nd edition, (Dunger, W. and Fiedler, H. J., eds.), Gustav Fischer Verlag Jena Stuttgart Lübeck Ulm, p. 92 (German).
73. Rabsch, W. 1995. Klassische epidemiologische Laboratoriumsmethoden, in *Salmonellosen des Menschen – Epidemiologische und ätiologische Aspekte. RKI-Schriften* 3/95, MMW Medizin Verlag München, pp. 118–134 (German).
74. Rabsch, W., Helm, R. A., and Eisenstark, A. (2004) Diversity of phage types among archived cultures of the Demerec collect of Salmonella enterica serovar Typhimurium strains. *Appl. Environ. Microbiol.* **70,** 664–669.

11

Salmonella Phages Examined in the Electron Microscope

Hans-W. Ackermann

Summary

Out of 177 surveyed bacteriophages, 161 (91%) are tailed and belong to the *Myoviridae, Siphoviridae*, and *Podoviridae* families (43, 55, and 59 viruses, respectively). Sixteen filamentous or isometric phages are members of the *Inoviridae, Leviviridae, Microviridae*, and *Tectiviridae* families (9%). Many tailed phages belong to established phage genera (P22, T1, T5, and T7), which are widespread in enterobacteria and other Gram-negatives of the Proteobacteria phylum.

Key Words: Dimensions; frequency; host range; *Myoviridae; Podoviridae; Siphoviridae.*

1. Introduction

The first *Salmonella* phage was reported as early as 1918 by Félix d'Hérelle *(1)*. *Salmonella* phages had always elicited much interest because of the medical and veterinary importance of their hosts, the ready availability of bacteriological media for phages and bacteria, and the use of phages in epidemiology (phage typing). The typing schemes for *Salmonella typhi, Salmonella paratyphi* B, and *Salmonella typhimurium* (*see* Chapter 9, this volume) are used worldwide. At least 983 enterobacterial phages were examined in the electron microscope (*2*; *see* **Note 1**). Of these, 906 are tailed and 77 are isometric or filamentous. No less than 177 phages infect salmonellas or are harbored by them. Reviews and classification schemes of enterobacterial phages have been published elsewhere *(3,4)*. Some *Salmonella* phages are famous. The best-known is phage P22, which has been the object of numerous morphogenetic and

From: *Methods in Molecular Biology, vol. 394: Salmonella: Methods and Protocols*
Edited by: H. Schatten and A. Eisenstark © Humana Press Inc., Totowa, NJ

genetic studies and is remarkable for its converting and transducing abilities. Phage O1 is well known as a diagnostic tool for *Salmonella* identification ***(3)*** because it is almost *Salmonella* specific and lyses most *Salmonella* strains. The Vi phages are particularly interesting because they (or are thought to be) specific for the Vi antigen of *S. typhi*. Phage Vi II is valuable because of its adaptability, which is the very basis of the *S. typhi* phage-typing scheme.

The label "*Salmonella* phage" does not mean that a phage is *Salmonella* specific. All enterobacteria are closely related. As a consequence, their phages are often polyvalent and infect hosts from several enterobacterial genera. Polyvalence is particularly frequent in phages of the *Escherichia–Klebsiella–Shigella* group and in plasmid-dependent filamentous or isometric phages. It is as if these bacteriophages do not recognize our bacteriological classifications. The term "*Salmonella* phage," though convenient, often means no more than a given phage has been propagated on a salmonella strain.

2. Materials

Medium-size centrifuges or ultracentrifuges, ammonium acetate buffer (1 M, pH 7.0), pointed Pasteur pipettes and tweezers, stain solutions (phosphotungstate, 2%, pH 7; uranyl acetate, 2%, pH 4.5), and support grids with carbon-coated Formvar films.

3. Methods

3.1. Electron Microscopy

Electron microscopy is the easiest, fastest, and least expensive way to identify novel *Salmonella* phages. Because enterobacterial phages are generally well known, novel phages can be instantly attributed to known species. The techniques of phage electron microscopy have been described elsewhere in detail ***(5)*** and may be summarized as follows.

3.1.1. Purification

Crude lysates always contain protein, DNA, and bacterial debris. They should never be examined. Purification is a must and easiest achieved by centrifugation. For example, 1 mL of filter-sterilized lysate is centrifuged for 1 h at 25,000 × *g* in a medium-size centrifuge. This is followed by two washings in ammonium acetate under the same conditions. Supernatants are discarded.

3.1.2. Staining

A drop of phage sediment is deposited on a grid and a drop of stain solution is added. After a minute, excess liquid is withdrawn with a filter paper.

3.1.3. Imaging and Magnification Control

The techniques of "wet" or darkroom photography are well known. They use fine-grain films, glossy high-contrast paper, developers, and fixers. Contrast and thus quality of prints can be improved dramatically by the use of graded filters. Digital electron microscopy obviates darkroom photography and facilitates archiving and exchanging micrographs. Magnification is controlled with the aid of catalase crystals or T4 phage tails.

3.1.4. Problems

Some phages are damaged during purification and lose their tails, simulating isometric viruses. Phage heads may be deformed and appear rounded. The exit of DNA is no problem because preparations usually contain intact particles. Much more serious is that digital electron microscopy is apparently not ready for virology; for example, I have not yet seen, even in prospects of electron microscopic companies, completely satisfactory digital images of viruses.

3.2. Salmonella *Phages by Family*

A total of 177 phages were surveyed, 161 of which were tailed (91%). They represented all three families of tailed phages and 20 morphotypes. Many *Salmonella* phages belong to established phage genera (P22, T1, T5, and T7). These genera are widespread in enterics and often occur in a wide range of α-, β-, and γ-proteobacteria, e.g., acinetobacters, pseudomonads, rhizobia, and vibrios *(3,6)*. Established genera are indicated in bold face. A few morphotypes are specific to enterobacteria or perhaps even to salmonellas (e.g., Beccles, Jersey, ViII, and SasL6). Many phages cannot be attributed to genera at the present state of phage classification. The main reason is poor electron microscopy and lack of serological and genomic data. Tailed phage morphotypes are illustrated by scale drawings (*see* **Fig. 3.1.3**). Isometric phage heads are icosahedra. Selected morphotypes are illustrated by micrographs (*see* **Figs. 2–6**). **Tables 1–3** list tailed phages by family, genus, species, and dimensions. Filamentous and isometric phages of salmonellas belong to four families (*see* **Table 4**); they may just as well be considered as coliphages because of their host ranges.

Some cautionary remarks are needed:

1. The tables summarize the dismal and confusing state of phage nomenclature. Unfortunately, 90 years after the discovery of bacteriophages, it is too late to devise a coherent naming system for phages.
2. Phage dimensions are often approximations and are halfway accurate only for phages measured after calibration with catalase or T4 tails. There is much chaff in phage literature. Some investigators do not even name or measure their phages.
3. The taxonomical position of certain phage hosts (*Salmonella* Greenside, Maricopa, Poona, and Siegburg) is uncertain.

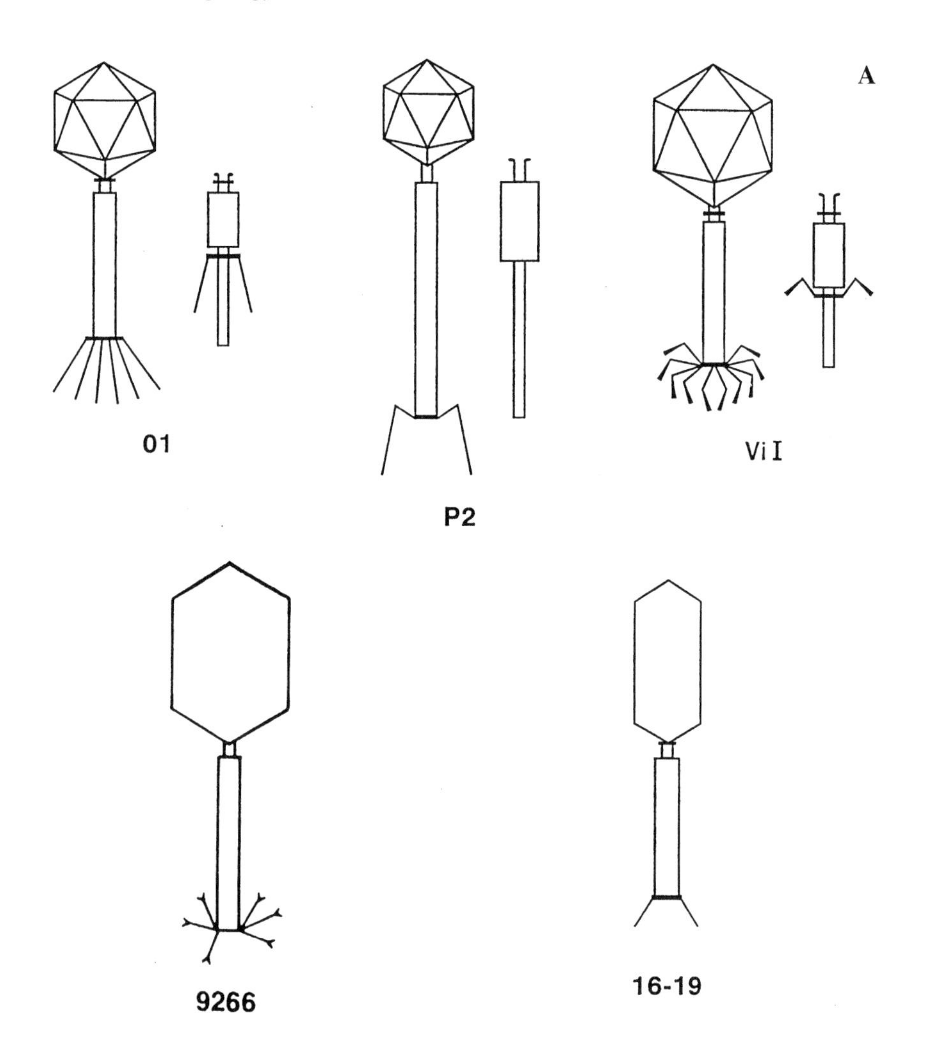

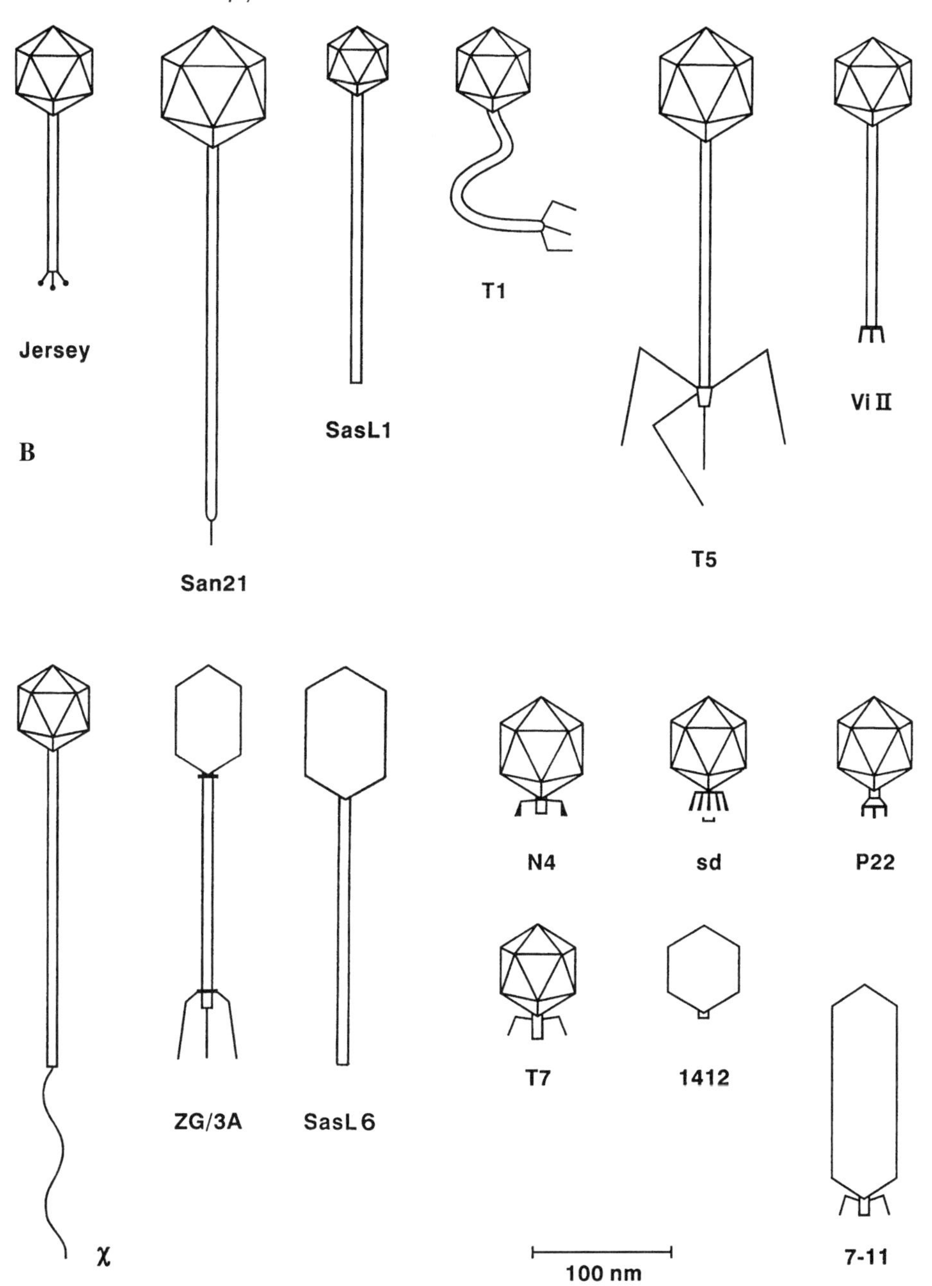

Fig. 1. Scale drawings of tailed *Salmonella* phages. **(A)** *Myoviridae*. **(B)** *Siphoviridae* and *Podoviridae*.

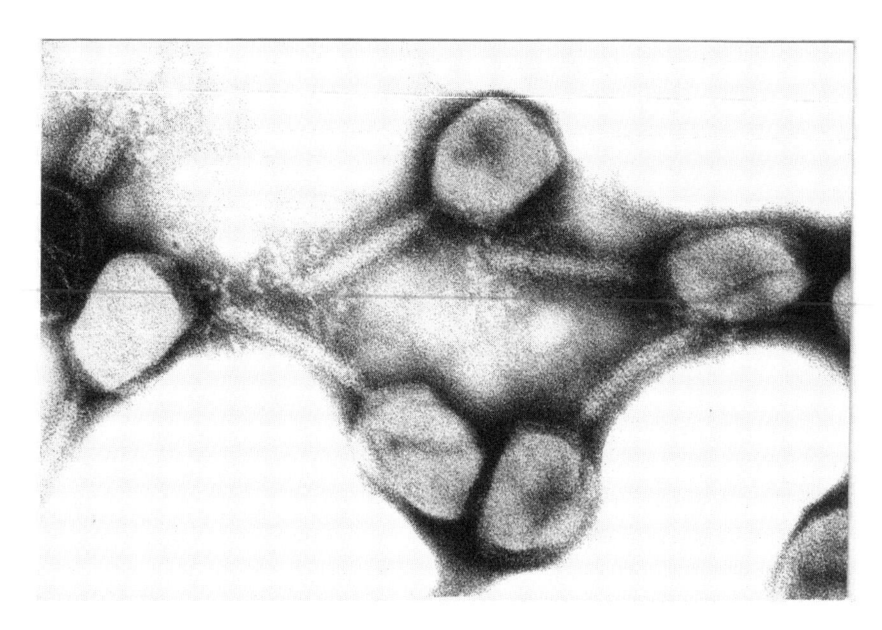

Fig. 2. Phage 9266Q, phosphotungstate (PT), ×297,000. Bar indicates 100 nm.

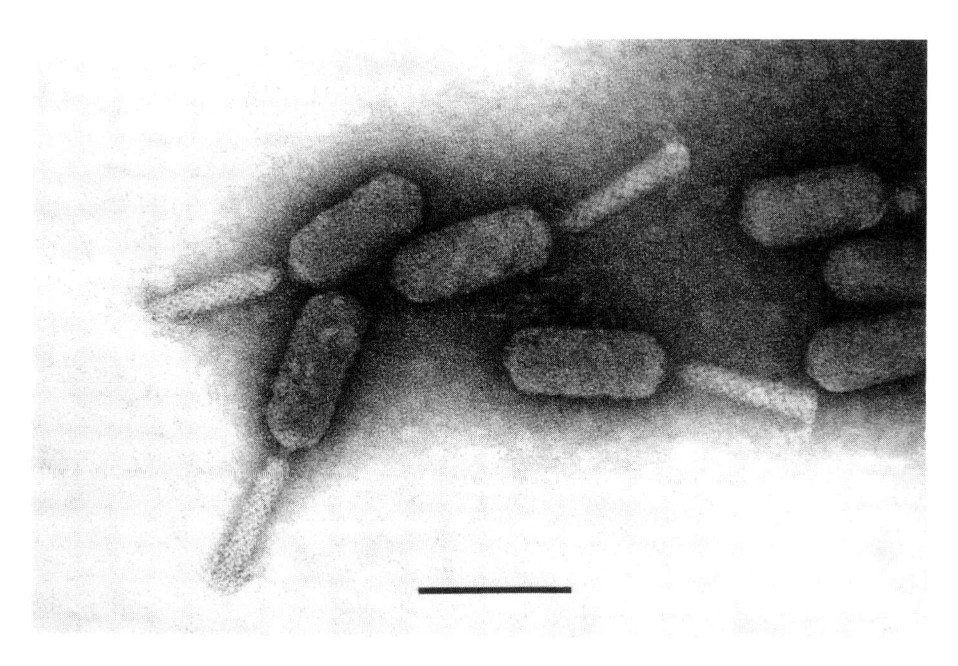

Fig. 3. Phages 16–19, uranyl acetate (UA), ×297,000.

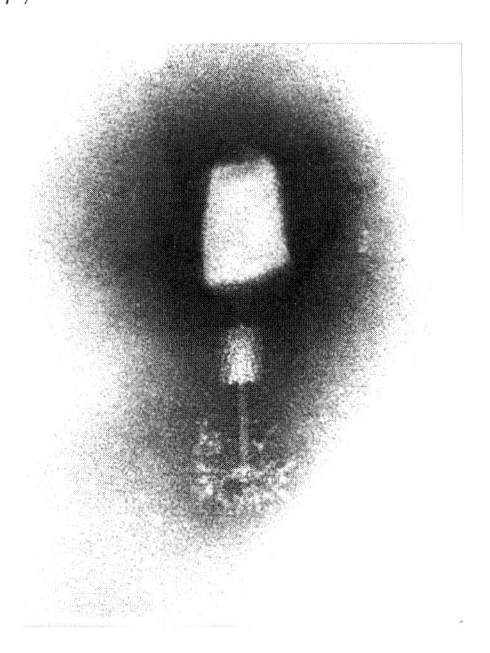

Fig. 4. Phage 9266Q with contracted sheath separated from tail fibers, phosphotungstate (PT), ×297,000.

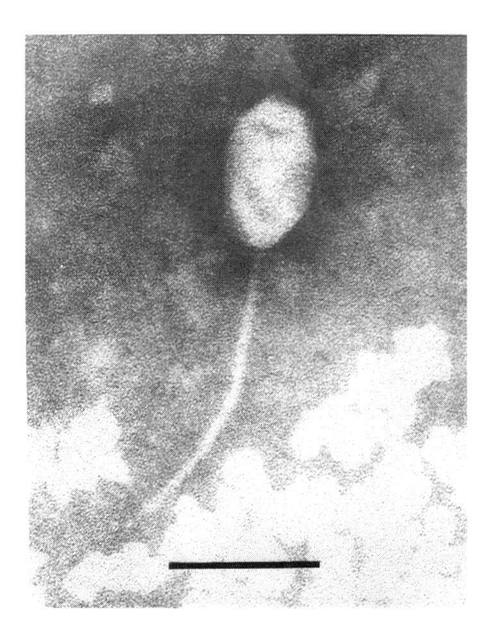

Fig. 5. Phage SasL6, phosphotungstate (PT), ×297,000; bar indicates 100 nm.

Table 1
Myoviridae

Genus or species	Phages	Usual host	Head (nm)[a]	Tail (nm)[a]	References
Beccles	Beccles, Taunton	*Salmonella paratyphi*	58	158 × 13	*(7)*
O1	O1	*Salmonella typhi*	*72*	*113 × 17*	***(7–9)***
O1	O2, O3, Dundee	*S. paratyphi*	76	105 × 15	*(7)*
O1	K	*S. typhi*	63–69	109 × 17	***(10)***
O1	P10	*Salmonella enterica* sv. Potsdam	55	95 × 12	***(11)***
O1	SPT1	*Salmonella typhimurium*	*73*	*113 × 17*	***(12)***
O1?	b, d, f, i, j, k, l	*Salmonella enterica* sv. Anatum	–	–	***(13)***
O1?	S1	*Salmonella enteritidis*	75	138 × 21	***(14)***
O1?	ϕ2	*S. typhi*	72	110 × 20	***(15)***
P2-like	Fels 2	*S. typhimurium*	55	110 × 20	***(16)***
P2-like	PSP3	*S. enterica* sv. Potsdam	–	–	***(17)***
P2-like	P3, P9a	*S. enterica* sv. Potsdam	55	118 × 12	***(11)***
P2-like	SopEϕ	*S. typhimurium*	58	133 × 19	***(18)***

ViI	ViI	*S. typhi*	92	108 × 17	***(15,19,20)***
ViI	San15, San17	*S.* spp.	*88*	*112 × 17*	***(21)***
ViI	Hei10	*S. enterica* sv. Heidelberg	*88*	*112 × 19*	***(22)***
ViI?	CT	*S. typhimurium*	93	109	***(23)***
9266	9266Q	*S. enterica* sv. Newport	*123 × 71*	*122 × 18*	***(24)***
16–19	16–19, 20.2, 36	*S. enterica* sv. Newport	*103 × 42*	*99 × 15*	***(25–28)***
16–19	966A, 449C		*104 × 43*	100 × 14	(***25–27***)
16–19	San11, San12		102 × 43	*104 × 17*	***(21)***
?	GI	*S. enteritidis*	79	125 × 21	***(14)***
?	GIII		66	152 × 17	
?	GVI		59	124 × 14	
?	GVIII		69	156 × 22	
?	Sab3, Sab5	*S. enterica* sv. Bareilly	42	220	***(28)***
?	–	*S.* sp. sv. Johannesburg	–	–	***(29)***

[a]Dimensions determined after catalase calibration are in italics. nm, nanometers; sp., spp., species; sv., serovar; –, no name or no dimensions.

Table 2
Siphoviridae

Genus or species	Phages	Usual host	Head (nm)[a]	Tail (nm)[a]	References
Jersey	Jersey, 1, 2, 3a, 3aI, 1010	*Salmonella paratyphi*	68	116 × 8	***(7)***
Jersey	Hei2, Hei3	*Salmonella enterica* sv. Heidelberg	*58*	*115 × 8*	***(22)***
Jersey	San3, San7, San8	spp.	*58*	117 × 8	***(21)***
Jersey?	MB78	*Salmonella typhimurium*	60	90 × 10	***(30)***
Jersey?	Sab1, Sab2, Sab4	*Salmonella enterica* sv. Bareilly	76	104	***(28)***
San21	San21	spp.	*84*	*260 × 8*	***(21)***
SasL1	SasL1, SasL2, SasL3, Sas34, SasL5	*S. enterica* sv. Senftenberg	*56*	*190 × 7*	***(4,31)***
T1-like	Hei5, Hei8	*S. enterica* sv. Heidelberg	*60*	150 × 8	***(22)***
T1-like	C557	*S. enterica* sv. Newport	64	150 × 8	***(27)***
T1?	S1BL	*Salmonella typhi*	–	–	***(15)***
T5-like	San 2, 12 other phages	spp.	*82*	*204 × 7*	***(21)***

T5-like	G5	*Salmonella enteritidis*	–	–	***(14)***
T5-like	ϕ1	*S. typhi*	–	–	***(15)***
ViII	ViII	*S. typhi*	67	144	***(19,32,33)***
χ	χ	ssp.	*68*	*227*	***(34,35)***
χ	IRA	*S. typhimurium*	50	280	***(36,37)***
χ	San1	ssp.	*65*	*227 × 11*	***(21)***
ZG/3A	San27	ssp.	72 × 54	179 × 9	***(21)***
Hl9-J (?)	SasL6	*S. enterica* sv. Senftenberg	100 × 50	200 × 7	***(4,31)***
?	c, g, h	*S. enterica* sv. Anatum			***(13)***
?	C236, C625, C699	*S. enterica* sv. Newport	70	195 × 8	***(27)***
?	ES18	*S. enterica*	56	210	***(38)***
?	San24	spp.	*66*	*175 × 9*	***(21)***
?	P22-1	*S. typhimurium*	65?	180 × 10	***(39)***
?	P22-3, P22-11	*S. typhimurium*	69	290	***(40)***
?	P22-12	*S. typhimurium*	80	188	***(40)***
?	–	*S. enterica* sv. Johannesburg	–	–	***(29)***

[a]Dimensions determined after catalase calibration are in italics. nm, nanometers; sp., spp., species; sv., serovar; –, no name or no dimensions.

Table 3
Podoviridae

Genus or species	Phage	Usual host	Head (nm)[a]	Tail (nm)[a]	References
N4-like	Hei9	*Salmonella enterica* sv. Heidelberg	*58*	13	***(22)***
	San10	ssp.	*61*	*11*	***(21)***
	ViIII, ViIV, ViV, ViVI, ViVII	*Salmonella typhi*	*57*	13 × 8	***(19)***
	ε15	*S. enterica* sv. Anatum	65	10	***(41)***
P22-like	P22	*Salmonella typhimurium*	65	18	***(39–42)***
	Hei1, Hei6, Hei7, Hei11	*S. enterica* sv. Heidelberg	*57*	*13*	***(22)***
	L	*S. typhimurium*	60	Short	***(43)***
	LP7	S. sp. sv. Greenside	–	–	***(44)***
	MG40	*S. typhimurium*	50	Short	***(45)***
	PSA68	*S. typhimurium*	60	Short	***(46)***
	P22a1, P22-4, P22-7, P22-11	*S. typhimurium*	80	8	***(40)***
	2, 4, 8, 28b, 30, 31, 32, 33, 34, 36, 37, 39	*S. typhimurium*	50	Short	***(47)***
	3b, B.A.O.R., Worksop	*Salmonella paratyphi*	65	20 × 8	***(7)***

	27	*S. enterica* sv. Schwarzengrund	60	8	***(41)***
	1,37	*S.* sp. sv. Poona	60	9	***(41)***
	1(40)	*S.* sp. sv. Johannesburg	59	6	***(29,41)***
	$1,42_2$	*S.* sp. sv. Maricopa	61	9	***(41)***
	6,14(18)	*S.* sp. sv. Siegburg	60	7	***(41)***
	14(6,7)	spp.	60	8	***(41)***
	ε34	spp.	60	8	***(41)***
P22?	G4	*Salmonella enteritidis*	72	23	***(41)***
P22?	P4, P9	*S. enterica* sv. Potsdam	55	15 × 9	***(11)***
T7-like	SP6	*S. typhimurium*	50-60	–	***(48–50)***
	ST64B, ST64T	S. typhimurium	63	10 × 7-8	***(51)***
7–11	7–11, 40.3	*S. enterica* sv. Newport	*154 × 40*	*12 × 8*	**(25–27,52)**
	SNT-3	*S. typhimurium*	142	19	***(23)***
?	a, e	*S. enterica* sv. Anatum	60	Short	***(13)***
?	GIII	*S. enteritidis*	65	28	***(14)***
?	M	*S. typhi*	49–54	15–20	***(10)***
?	SNT-1	*S. typhimurium*	74	13	***(23)***
?	SNT-2	*S. typhimurium*	54	14	***(23)***
?	1412	*S. enterica* sv. Newport	61	9 × 3	***(52)***

[a]Dimensions determined after catalase calibration are in italics. nm, nanometers; sp., spp., species; sv., serovar ; –, no dimensions.

Table 4
Filamentous and Isometric Phages

Family	Nucleic acid	Phage	Host	References
Inoviridae	ssDNA	C-2	Ec-Sal-Se-P	***(53)***
		I_2-2	Ec-Sal	***(54)***
		SF	Ec-Sal	***(55)***
		tf-1	Ec-Sal-K	***(56)***
		X-2	Ec-Sal-Se	***(57)***
Leviviridae	ssRNA	Iα	Ec-Sal	***(54)***
		M	Ec-Sal	***(58)***
		PRR1	Ec-Sal-Ps-Vi	***(59)***
Microviridae	ssDNA	φX174, 1φ7, 1φ1, 1φ3, 1φ9	Ec-Sal-Shig	***(60)***
		φR	Ec-Sal	***(61)***
Tectiviridae	dsDNA	PRD1	Ec-Sal-P-Ac-Ps-Vi	***(62)***
		PR5	Ec-Sal-Ps	***(63)***

Ac, *Acinetobacter*; Ec, *Escherichia coli*; K, *Klebsiella*; P, *Proteus*; Ps, *Pseudomonas*; Sal, *Salmonella*; Se, *Serratia*; Shig, *Shigella*; Vi, *Vibrio*.

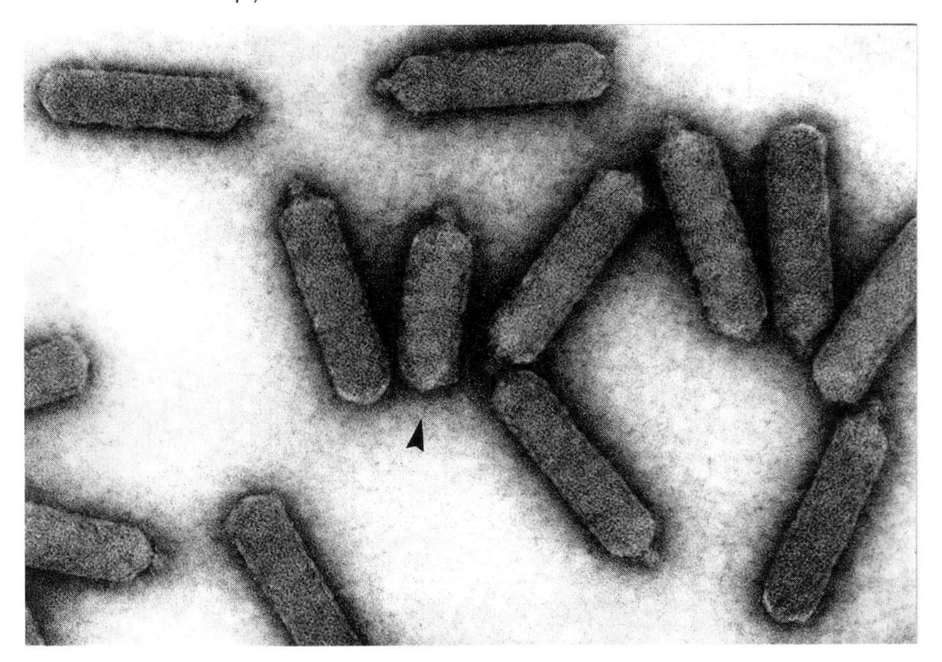

Fig. 6. Phages 7–11, uranyl acetate (UA), ×297,000. The arrow indicates an abnormally short head.

3.2.1. Myoviridae (43 Phages, 24 %)

1. Phages Beccles and Taunton are parts of the international *S. paratyphi* B phage typing set (7, *see* also Chapter 9, this volume). They tend to produce tails with abnormally long cores and several tail sheaths. The phages closely resemble coliphage P2 but differ from P2-like phages by their longer tails. They seem to represent a separate species of this genus which evolved by tail elongation.
2. Species O1 is characterized by a tail sheath with a criss-cross pattern. During contraction, the sheath separates from the base plate, leaving an empty space between sheath and base plate. Phages have a collar and six straight tail fibers, which are, in the quiescent state, folded along the tail. Although phage O1 itself is known for its *Salmonella* specificity, phages of this type are observed in many enterobacteria, especially in the *Salmonella–Serratia* division. They also occur in other proteobacteria, e.g., *Aeromonas, Pseudomonas*, rhizobia, and vibrios ***(3,6)***. Some members of the group produce particles with abnormally long tail cores and sheaths ***(3)***.
3. P2-like phages are characterized by a relatively small head, short kinked tail fibers, and a tail sheath that, upon contraction, becomes loose and slides down the tail core. Phages of the P2 type, commonly associated with *Escherichia coli*, are widespread in γ-proteobacteria and occur, for example, in *Haemophilus, Pasteurella*, and *Pseudomonas* ***(3,6)***.

4. Phage ViI is noted for infecting Vi-antigen-positive strains of *S. typhi* and *Citrobacter*. The phage is characterized by a large head, a collar, and numerous (12?) short, possibly kinked tail fibers. Phages of this type occur in salmonellas, *E. coli, acinetobacters*, and *rhizobia* ***(3)***.
5. Phage 9266Q (*see* **Figs. 2** and **4**) was isolated from sewage on a coli strain ***(24)***, had apparently died, and was recently revived by propagation on *Salmonella newport*. Phage 9266Q superficially resembles T4-like phages by its elongated head but differs from them in several aspects. It is slightly larger and has no collar, no base plate, and no long kinked tail fibers; instead, it has a set of 12 (?) feathery tail fibers of 30 nm in length. Consequently, true "T4-like phages" should only be diagnosed after visualization of tail fibers.
6. Phage 16-19 (*see* **Fig. 3**) and its relatives belong to a very rare morphotype with characteristically narrow heads. Tail fibers are short and straight. The DNA of 16-19 was found to contain 5-methylcytosine and glucose ***(25,26)***.

3.2.2. Siphoviridae (59 Phages, 33 %)

1. The Jersey species is a small group of phages characterized by short, rigid tails with six (?) club-shaped spikes. Several of these phages are parts of the *S. paratyphi* B typing set ***(26)***. These phages have been found in salmonellas only and may be *Salmonella* specific.
2. San21 is one of the largest siphoviruses reported in the literature. Besides its size, it is morphologically unremarkable. No similar phages have been found in other enterobacteria.
3. SasL1 is of very indistinctive morphology. It is considered as the representative of an independent species because it does not correspond to other well-studied enterobacterial phages.
4. T1-like phages have characteristic extremely flexible tail with conspicuous transverse striations and short kinked terminal fibers. Tails of phosphotungstate-stained particles are often curled around phage heads; tails of uranyl acetate-stained phages are more rigid and often appear straight. T1-like phages are usually considered as coliphages but are frequently observed in other enterobacteria (shigellas and klebsiellas). They have not been found outside enterobacteria ***(3,6)***.
5. T5-like phages are recognizable by their large heads and rigid tails provided with three long kinked fibers and one short fiber. Although commonly considered as coliphages, they infect salmonellas, klebsiellas, *Proteus*, and even *Vibrio cholerae* ***(3,6)***.
6. ViII is characterized by a rigid tail provided with a base plate and six spikes. The species seems to be small and restricted to salmonellas. ViII has been adapted to over 80 biovars of *S. typhi*. Its host range variants constitute the phage-typing scheme of *S. typhi* (*see* Chapter 9, this volume).
7. The χ (chi) species is characterized by a thick, rigid tail with conspicuous transverse striations and a normally coiled, single long tail fiber. Phages use this fiber to

adsorb to the flagella of motile bacteria and are thus flagella specific. Heads often show capsomers. Phage χ itself is polyvalent and infects salmonellas, *E. coli*, and serratias. Similar phages have been found in enterobacters, *Proteus*, and serratias *(3)*.

8. The ZG/3A species is named after a coliphage and characterized by a moderately elongated head. Particles have collars and terminal and subterminal tail fibers. Although essentially associated with *E. coli* and shigellas, ZG/3A-like phages occur in citrobacters and serratias *(3)*.
9. The SasL6 species is morphologically unremarkable. Particles (*see* **Fig. 5**) have a longer head than the ZG/3A type and a short tail fiber. SasL6 was assigned to the H19-J species of coliphages *(4)*, but this may have to be revised.

3.2.3. Podoviridae (59 Phages, 33 %)

1. N4-like phages constitute a poorly defined, probably heterogeneous group. They are found in *E. coli*, klebsiellas, and salmonellas, perhaps also in acinetobacters and rhizobia *(3,6)*. The morphological diagnosis rests on a weak criterion, namely the observation of tail fibers forming two small bushels or spheres at each side of the tail. N4 is a coliphage. A possible member of the group, *Salmonella* phage ε15, has no sequence relationships with N4 *(64)*. The N4-like Vi phages lyse *S. typhi* with Vi antigen and may be capsule specific.
2. P22-like phages are identified by their prominent baseplates with six spikes (three in profile). Phage heads often show capsomers. Phages are temperate and associated with *S. typhimurium*. They are known for their transducing and converting ability. P22-like phages are common in salmonellas and *Proteus* bacteria and have been reported in marine vibrios *(3)*. Many P22-like *Salmonella* phages of **Table 3** may be reisolates or variants of P22. The recently characterized phage ε34 has sequence relationships with P22 (Kropinski, personal communication).
3. The sd species is very small and so far limited to two coliphages and one *Salmonella* phage. Its characteristic feature is the presence of 12 (?) short fibers *(3)*.
4. T7-like phages are characterized by tapering tails with no base plate and short tail fibers, which are normally folded against the tail and invisible. These phages are typically found in *E. coli* and shigellas but can be expected in any enterobacteria. They are also widespread in acinetobacters, pseudomonads, rhizobia, and vibrios *(3,6)*.
5. The 7-11 species is characterized by very long heads and tapering tails without base plates (*see* **Fig. 6**). The species has 13 members only but is found in other enterobacteria (*E. coli, Enterobacter, Erwinia, Levinea, Proteus*, and *Yersinia*) *(3)*. It has not been found elsewhere and may be enterobacteria specific. Podoviruses with such long heads are very rare in the bacteriophage world. Phage 7-11 produces several classes of head size variants and contains mannose associated with its DNA *(25,26)*.
6. Phage 1412 is of uncertain standing. It has a very short stubby tail and a head with a generally wavy outline and may represent a species of its own.

3.2.4. Filamentous and Isometric Phages (16 Phages, 9 %)

The *Inoviridae* are thin filaments of 800–1200 nm in length exemplified by coliphage fd. The isometric phages are members of the *Leviviridae* (ssRNA phages), *Microviridae* (ϕX174-type), and *Tectiviridae* families. The latter family is most interesting as it includes viruses with a protein shell, an inner, lipid-containing vesicle, and a tail-like tube that is produced during infection. None of these phages can be called a *Salmonella* phage in a strict sense. All are polyvalent and most of them (*Inoviridae, Leviviridae*, and *Tectiviridae*) are pilus-specific, plasmid-dependent phages. For example, the broad-host-range tectiviruses will infect and lyse any bacteria harboring plasmids of the P, N, or W incompatibility groups ***(62,63)***.

4. Notes

1. The general phage survey ***(2)***, completed on December 31, 2006, comprises over 5500 phages from articles, books, and theses. Known defective phages, particulate bacteriocins, and phages from congress abstracts and personal websites were omitted.

References

1. D'Hérelle, F. (1918) Technique de la recherche du microbe filtrant bactériophage (*Bacteriophagum intestinale*). *C.R. Soc. Biol.* **81,** 1160–1162.
2. Ackermann, H.-W. (2006) Frequency of morphological phage descriptions in the year 2005. *Arch. Virol.* **152**, 227–243.
3. Ackermann, H.-W. and DuBow, M. S. (1987) *Viruses of Prokaryotes*, vol. II. *Natural Groups of Bacteriophages*. CRC Press, Boca Raton, pp. 85–100.
4. Ackermann, H.-W., DuBow, M. S, Gershman, M., et al. (1997) Taxonomic changes in tailed phages of enterobacteria. *Arch. Virol.* **142,** 1381–1390.
5. Ackermann, H.-W. (2006) Basic phage electron microscopy, in *Bacteriophages: Methods and Protocols* (Kropinski, A. M. and Clokie, M., eds.), Humana, Totowa, NJ, in print.
6. Fauquet, C. M., Mayo, M. A., Maniloff, J., Desselberger, U., and Ball, L. A., eds. (2005) *Virus Taxonomy: VIIIth Report of the International Committee on Taxonomy of Viruses*. Academic Press/Elsevier, London, pp. 35–79.
7. Ackermann, H.-W., Berthiaume, L., and Kasatiya, S. S. (1972) Morphologie des phages de lysotypie de *Salmonella paratyphi B* (schéma de Felix et Callow). *Can. J. Microbiol.* **18,** 77–81.
8. Lindberg, A. A. (1967) Studies of a receptor for Felix O-1 phage in *Salmonella minnesota. J. Gen. Microbiol.* **48,** 225–233.
9. Ackermann, H.-W. and Berthiaume, L. (1969) Ultrastructure des phages de lysotypie des *Escherichia coli* 0127:B8. *Can. J. Microbiol.* **15,** 859–862.

10. Bliznichenko, A. G., Milyutin, V. N., Tokarev, S. A., and Kirdeev, V. K. (1972) Study of biological properties of two variants of Vi-typhoid phages, features of the structure and cycle of development (Russian). *Vopr. Virusol.* **17,** 448–450.
11. Nutter, R. L., Bullas, L. R., and Schultz, R. L. (1970) Some properties of five new *Salmonella* bacteriophages. *J. Virol.* **5,** 754–764.
12. Voelker, L., Sulakvelidze, A., and Ackermann, H.-W. (2005) Spontaneous tail length variation in a *Salmonella* myovirus. *Virus Res.* **114,** 164–166.
13. Popovici, M., Szégli, L., Soare, L., et al. (1975) Caractéristiques des phages lysogènes de *S. anatum. Arch. Roum. Pathol. Exp. Microbiol.* **34,** 223–230.
14. Slopek, S. and Krzywy, T. (1985) Morphology and ultrastructure of bacteriophages. An electron microscopical study. *Arch. Immunol. Ther. Exp.* **33,** 1–217.
15. Bradley, D. E. and Kay, D. (1960) The fine structure of bacteriophages. *J. Gen. Microbiol.* **23,** 553–563.
16. Yamamoto, N. (1969) Genetic evolution of bacteriophage. I. Hybrids between unrelated bacteriophages P22 and Fels 2. *Proc. Natl. Acad. Sci. USA* **62,** 63–69.
17. Bullas, L. R., Mostaghimi, A. R., Arensdorf, J. J., Rajada, P. T., and Zuccarelli, A. J. (1991) *Salmonella* phage PSP3, another member of the P2-like phage group. *Virology* **185,** 918–921.
18. Mirold, S., Rabsch, W., Rohde, M., et al. (1999) Isolation of a temperate bacteriophage encoding the type III effector protein SopE from an epidemic *Salmonella typhimurium* strain. *Proc. Natl. Acad. Sci. USA* **96,** 9845–9850.
19. Ackermann, H.-W., Berthiaume, L., and Kasatiya, S. S. (1970) Ultrastructure of Vi phages I to VII of *Salmonella typhi. Can. J. Microbiol.* **16,** 411–413.
20. Kwiatkowski, B. and Taylor, A. (1970) Two-step attachment of Vi-phage I to the bacterial surface. *Acta Microbiol. Pol. A.* **2,** 13–30.
21. Ackermann, H.-W. and Gershman, M. (1992) Morphology of phages of a general *Salmonella* typing set. *Res. Virol.* **143,** 303–310.
22. Demczuk, W., Ahmed, R., and Ackermann, H.-W. (2004) Morphology of *Salmonella enterica* serovar Heidelberg typing phages. *Can. J. Microbiol.* **50,** 873–875.
23. Williams, F. P. Jr. and Stetler, R. E. (1994) Detection of FRNA coliphages in groundwater: interference with the assay by somatic *Salmonella* bacteriophages. *Lett. Appl. Microbiol.* **19,** 79–82.
24. Ackermann, H.-W. and Nguyen, T.-M. (1983) Sewage coliphages investigated by electron microscopy. *Appl. Environ. Microbiol.* **45,** 1049–1059.
25. Moazamie, N., Ackermann, H.-W., and Murthy, M. R. V. (1979) Characterization of two *Salmonella newport* bacteriophages. *Can. J. Microbiol.* **25,** 1063–1072.
26. Moazamie-Shamloo, N. (1976) Étude de deux phages de *Salmonella newport.* Ph.D. thesis, Université Laval, Quebec, Canada, 152 p.
27. Petrow, S., Kasatiya, S. S., Pelletier, J., Ackermann, H.-W., and Péloquin, J. (1974) A phage typing scheme for *Salmonella newport. Ann. Microbiol.* **125A,** 433–445.

28. Jayasheela, M., Singh, G., Sharma, N. C., and Saxena, S. N. (1987) A new scheme for phage typing *Salmonella bareilly* and characterization of typing phages. *J. Appl. Bacteriol.* **62,** 429–432.
29. Girard, R. and Chaby, R. (1981) Comparative studies on *Salmonella johannesburg* bacteriophages: virulence and interactions with the host cell lipopolysaccharide. *Ann. Microbiol. (Paris)* **132B,** 197–214.
30. Joshi, A., Siddiqui, J. Z., Rao, G. R. K., and Chakravorty, M. (1982) MB78, a virulent bacteriophage of *Salmonella typhimurium. J. Virol.* **41,** 1038–1043.
31. Kumar, S., Sharma, N. C., and Singh, H. (1997) Isolation of *Salmonella senftenberg* bacteriophages. *Indian J. Med. Res.* **105,** 47–52.
32. Kwiatkowski, B. (1966) The structure of Vi-phage II. *Acta Microbiol. Pol.* **15,** 23–26.
33. Vieu, J. F. and Croissant, O. (1966) Lyophilisation du bactériophage Vi II de *Salmonella typhi. Arch. Roum. Pathol. Exp. Microbiol.* **25,** 305–318.
34. Meynell, E. W. (1961) A phage, $\phi\chi$, which attacks motile bacteria. *J. Gen. Microbiol.* **25,** 253–290.
35. Schade, S. Z., Adler, J., and Ris, H. (1967) How bacteriophage χ attacks motile bacteria. *J. Virol.* **1,** 599–609.
36. Adamia, R. (1999) General mechanisms of phage-host bacterial cells' genomes interactions. Ph.D. thesis, Academy of Sciences of Georgia, Tbilisi, 31 p. (abstract).
37. Adamia, R. S., Matitashvili, E. A., Kvachadze, L. I., et al. (1990) The virulent bacteriophage IRA of *Salmonella typhimurium*: cloning of phage genes which are proportionally lethal for the host cell. *J. Basic Microbiol.* **30,** 707–716.
38. Casjens, S. R., Gilcrease, E. B., Winn-Stapley, D. A., et al. (2005) The generalized transducing *Salmonella* bacteriophage ES18: complete genome sequence and DNA packaging strategy. *J. Bacteriol.* **187,** 1091–1104.
39. Yamamoto, N. and Anderson, T. F. (1961) Genomic masking and recombination between serologically unrelated phages P22 and P221. *Virology* **14,** 430–439.
40. Young, B. G., Hartman, P. E., and Moudrianakis, E. N. (1966) Some phages released by P22-infected *Salmonella. Virology* **28,** 249–264.
41. Vieu, J. F., Croissant, O., and Dauguet, C. (1965) Structure des bactériophages responsables des phénomènes de conversion chez les *Salmonella. Ann. Inst. Pasteur* **109,** 160–166.
42. Anderson, T. F. (1960) On the fine structure of the temperate bacteriophages P1, P2 and P22, in *Proceedings of the European Regional Conference on Electron Microscopy*, Delft 1960, vol. 2 (Houwink, A. L. and Spit, B. J., eds.), De Nederlandse Vereniging voor Elektronenmicroscopie, Delft, pp. 1088–1011.
43. Bezdek, M. and Amati, P. (1967) Properties of P22 and a related *Salmonella typhimurium* phage. I. General features and host specificity. *Virology* **31,** 272–278.
44. Kitamura, J. and Mise, K. (1970) A new generalized transducing phage in *Salmonella. Jpn. J. Med. Sci. Biol.* **23,** 99–102.
45. Grabnar, M. and Hartman, P. E. (1968) MG40 phage, a transducing phage related to P22. *Virology* **343,** 521–530.

46. Enomoto, M. and Ishiwa, H. (1972) A new transducing phage related to P22 of *Salmonella typhimurium. J. Gen. Virol.* **14,** 157–164.
47. Svenson, S. B., Lönngren, J., Carlin, N., and Lindberg, A. A. (1979) *Salmonella* bacteriophage glycanases: endorhamnosidases of *Salmonella typhimurium* bacteriophages. *J. Virol.* **32,** 583–592.
48. Butler, E. T. and Chamberlin, M. J. (1982) Bacteriophage SP6-specific RNA polymerase. Isolation and characterization of the enzyme. *J. Biol. Chem.* **257,** 5772–5778.
49. Dobbins, A. T., George, M., Basham, D. A., et al. (2004) Complete genomic sequence of the virulent *Salmonella* bacteriophage SP6. *J. Bacteriol.* **186,** 1933–1944.
50. Scholl, D., Kieleczawa, J., Kemp, P., et al. (2004) Genomic analysis of bacteriophages SP6 and K1-5, an estranged subgroup of the T7 supergroup. *J. Mol. Biol.* **335,** 1151–1171.
51. Mmolawa, P. T., Willmore, R., Thomas, C. J., and Heuzenroeder, M. W. (2002) Temperate phages in *Salmonella enterica* serovar Typhimurium: implications for epidemiology. *Int. J. Med. Microbiol.* **291,** 633–644.
52. Ackermann, H.-W., Petrow, S., and Kasatiya, S. S. (1974) Unusual bacteriophages in *Salmonella newport. J. Virol.* **13,** 706–711.
53. Bradley, D. E., Sirgel, F. A., Coetzee, J. N., and Hedges, R. W. (1982) Phages C-2 and J: IncC and IncJ plasmid-dependent phages, respectively. *J. Gen. Microbiol.* **128,** 2485–2498.
54. Coetzee, J. N., Bradley, D. E., and Hedges, R. W. (1982) Phages Iα and I2-2: IncI plasmid-dependent bacteriophages. *J. Gen. Microbiol.* **128,** 2797–2804.
55. Coetzee, J. N., Bradley, D. E., Hedges, R. W., et al. (1986) Bacteriophages Fo*lac h*, SR, SF: phages which adsorb to pili encoded by plasmids of the S-complex. *J. Gen. Microbiol.* **132,** 2907–2917.
56. Coetzee, J. N., Bradley, D. E., Hedges, R. W., Tweehuizen, M., and Du Toit, L. (1987) Phage tf-1: a filamentous bacteriophage specific for bacteria harbouring the IncT plasmid pIN25. *J. Gen. Microbiol.* **133,** 953–960.
57. Coetzee, J. N., Bradley, D. E., Du Toit, L., and Hedges, R. W. (1988) Bacteriophage X-2: a filamentous phage lysing IncX-plasmid-harbouring bacterial strains. *J. Gen. Microbiol.* **134,** 2535–2541.
58. Coetzee, J. N., Bradley, D. E., Hedges, R.W., Fleming, J., and Lecatsas, G. (1983) Bacteriophage M: an incompatibility group M plasmid-specific phage. *J. Gen. Microbiol.* **129,** 2271–2276.
59. Olsen, R. H. and Thomas, D. D. (1973) Characteristics and purification of PRR1, an RNA phage specific for the broad host range *Pseudomonas* R1822 drug resistant plasmid. *J. Virol.* **12,** 1560–1567.
60. Ilyashenko, B. N., Tikhonenko, A. S., Dityatkin, S. J., and Rudchenko, O. N. (1965) Biological properties of small enteric phages containing DNA. *Mikrobiologiia* **34,** 814–819.

61. Bradley, D. E. (1961) Negative staining of bacteriophage φR at various pH values. *Virology* **15,** 203–205.
62. Olsen, R. H., Siak, J.-S., and Gray, R. H. (1976) Characteristics of PRD1, a plasmid-dependent broad host-range DNA bacteriophage. *J. Virol.* **14,** 689–699.
63. Wong, F. H. and Bryan, L. E. (1978) Characteristics of PR5, a lipid-containing plasmid-dependent phage. *Can. J. Microbiol.* **24,** 875–882.
64. Kropinski, A. M., Kovalyova, I. V., Billington, S. J., et al. (2006) The genome of ε15, a serotype-converting, Group E1 *Salmonella enterica*-specific bacteriophage. *J. Bacteriol.* in press.

12

Applications of Cell Imaging in *Salmonella* Research

Charlotte A. Perrett and Mark A. Jepson

Summary

Salmonella enterica is a Gram-negative enteropathogen that can cause localized infections, typically resulting in gastroenteritis, or systemic infection, e.g., typhoid fever, in both humans and warm-blooded animals. Understanding the mechanisms by which *Salmonella* induce disease has been the focus of intensive research. This has revealed that *Salmonella* invasion requires dynamic cross-talk between the microbe and host cells, in which bacterial adherence rapidly leads to a complex sequence of cellular responses initiated by proteins translocated into the host cell by a type III secretion system (T3SS). Once these *Salmonella*-induced responses have resulted in bacterial invasion, proteins translocated by a second T3SS initiate further modulation of cellular activities to enable survival and replication of the invading pathogen. These processes contribute to *Salmonella* entry into the host and the clinical symptoms of gastrointestinal and systemic infection. Elucidation of the complex and highly dynamic pathogen–host interactions ultimately requires analysis at the level of single cells and single infection events. To achieve this goal, researchers have applied a diverse range of microscopical methods to examine *Salmonella* infection in models ranging from whole animal to isolated cells and simple eukaryotic organisms. For example, electron microscopy and confocal microscopy can reveal the juxtaposition of *Salmonella*, its products, and cellular components at high resolution. Simple light microscopy (LM) can also be used to investigate the interaction of bacteria with host cells and has advantages for live cell imaging, which enables detailed analysis of the dynamics of infection and cellular responses. Here we review the use of imaging techniques in *Salmonella* research and compare the capabilities of different classes of microscope to address specific types of research question. We also provide protocols and notes on several LM techniques routinely used in our own research.

Key Words: *Salmonella*; infection; imaging; microscope; wide-field microscopy; confocal laser scanning microscopy; fluorescent staining; live cell imaging.

From: *Methods in Molecular Biology, vol. 394: Salmonella: Methods and Protocols*
Edited by: H. Schatten and A. Eisenstark © Humana Press Inc., Totowa, NJ

1. Introduction

1.1. Salmonella *Infection*

The virulence of *Salmonella enterica* depends on their ability to enter and survive in host cells. At the initial site of infection in the gut, *Salmonella* invasion involves triggering prominent cellular responses including remodeling of the actin cytoskeleton and plasma membrane to produce "membrane ruffles" ***(1,2)***. *Salmonella* thus enters the host by hijacking cellular machinery mediating uptake of material into the cell. A preferential, though non-exclusive, route of entry for *Salmonella* is provided by the specialized antigen-transporting epithelial cells (M cells) present in the epithelium overlying gut-associated lymphoid tissues ***(3–6)***. Upon invasion of epithelial cells and subsequent entry into other cell types (including macrophages), the bacteria continue to modulate cellular mechanisms controlling membrane trafficking to survive and replicate within a specialized membrane-bound compartment, the *Salmonella*-containing vacuole (SCV). Understanding the mechanisms underlying the intimate cross-talk between *Salmonella* and mammalian cells allows us to elucidate aspects of both microbial pathogenesis and cell biology. Microscopy methods are central to research in this area because they uniquely allow researchers to study the interactions between bacteria and host cells at the single cell, and single event, level. There is insufficient space here to discuss in detail the mechanisms of *Salmonella*/host cell interactions, and such details are available in other reviews (***5***,***7–9***). Nevertheless, a brief introduction to these aspects of *Salmonella* pathogenesis is necessary before discussing the ways in which microscopy is contributing to research in this field.

The ability of *Salmonella* to invade cultured cells depends on a region of the chromosome termed *Salmonella* pathogenicity island 1 (SPI-1) ***(10)***. Genes within SPI-1 encode a type III protein secretion system (T3SS) homologous to those involved in the secretion of virulence factors by a number of animal and plant pathogens ***(11,12)***. Unusually, *Salmonella* also possesses a second functional T3SS encoded by SPI-2. SPI-2 plays a critical role in intracellular survival and the systemic phase of *Salmonella* disease ***(7,13)***.

The first identified targets of the SPI-1-encoded T3SS were the *Salmonella* invasion proteins SipA, SipB, SipC, and SipD. Sips B–D form a translocon required for transfer of other effector proteins. The inactivation of *sipB, sipC*, or *sipD* genes confers a profound defect in invasion of cultured cells ***(11,12)***, but the absence of SipA has a subtler effect on invasion that is only apparent in the very early stages of infection of cultured cells ***(14,15)***. Subsequent studies have shown that SPI-1 translocates into host cells at least 12 proteins that are encoded within SPI-1 or elsewhere and have begun to reveal the complexity of host cell subversion by *Salmonella* ***(2)***. For example, it is now known that pathogenic *Salmonella* strains directly regulate Rho GTPase activity within host cells, and thereby modulate actin cytoskeleton and pro-inflammatory cytokine production, via one or more SPI-1-translocated guanine nucleotide exchange factor (GEF) proteins (SopE and SopE2) and a GTPase-activating protein (SptP) ***(16)***. Among the other SPI-1 effectors, SipA induces actin bundling and contributes to invasion ***(14,15)***, and SigD/SopB is an inositol phosphatase that appears to be involved in the regulation of membrane trafficking in addition to indirectly modulating Rho GTPase activity ***(17–19)***. Unlike mutation of genes encoding components of the T3SS system or translocon, elimination of individual SPI-1 effector proteins has comparatively minor effects on *Salmonella* infection of cultured cells because these proteins have overlapping functions and none are individually essential for invasion.

Prominent morphological changes and redistribution of host cell proteins occur at the site of *Salmonella* infection of cultured cells ***(1,8)***. Morphologically similar responses are observed on infection of non-polarized and polarized epithelial cells in culture and intestinal epithelium in vivo. Studying infection of polarized epithelial cells allows additional pathophysiological responses to *Salmonella* infection to be examined in addition to invasion itself, including regulation of ion permeability ***(20–22)***, tight junction modulation ***(23–25)***, cytokine production ***(26,27)***, and neutrophil transmigration ***(28,29)***.

Once *Salmonella* enter the cell, the activity of SPI-1 diminishes and expression of a second T3SS, SPI-2, discovered by signature-tagged mutagenesis, becomes more important, as shown by its critical role in the systemic phase of infection ***(13,30)***. Although mutants lacking SPI-1 T3SS are impaired in intestinal invasion, they retain significant virulence and are equivalent to the parent wild-type (WT) strain if administered intraperitoneally (i.p.). By contrast, SPI-2 mutants exhibit much greater attenuation of virulence whether delivered by oral or i.p. routes ***(13)***. Although SPI-2 has been regarded as being most critical for systemic infection, it has also recently been shown to be expressed in the intestine prior to invasion ***(31)*** and to be required for complete virulence in animal models of enterocolitis ***(32,33)***. As with SPI-1, the most critical SPI-2 secreted proteins form a translocon, whereas over 20 effector proteins directly affect cellular processes ***(7)***. *Salmonella* lacking functional SPI-2 T3SS or translocon are able to enter cells but are subsequently attenuated for survival and replication. Cellular responses occurring after infection with WT *Salmonella* that are curtailed upon infection with SPI-2 mutants include formation of *Salmonella*-induced filaments (Sifs), avoidance of

NADPH oxidase-dependent killing, modulation of SCV membrane dynamics, and association of F-actin and cholesterol around the SCV *(7)*. Infection of cells with strains lacking specific SPI-2 effector proteins, or expression of SPI-2 effectors in cells, has helped elucidate the likely functions of at least some of these proteins *(7)*.

1.2. Applications of Microscopical Methods to Study Salmonella *Infection*

1.2.1. Introduction to Microscopy Techniques

Although we are focusing here primarily on light microscopy (LM) techniques, including conventional phase-contrast, fluorescence, and confocal microscopy, these are just some of the wide range of microscopy techniques that have been applied to study bacterial morphology and the processes involved in infection. Alternative methods such as electron microscopy (EM) provide higher resolution and are particularly useful for studying the morphology of bacterial surface structures. EM has, for example, been used to examine the molecular architecture of T3SS of *Salmonella* and other pathogens ***(34–36)*** and the expression of surface structures (pili, flagella, etc.) ***(37–40)***. Atomic force microscopy (AFM) is capable of providing high-resolution images of surface structures ***(41)*** and, although it has yet to be used much to study *Salmonella* infection, has recently been employed to study the morphology of protein secretion and translocation pores of *Escherichia coli* ***(42,43)***. Although protocols for such methods are not given here, we will briefly discuss examples of the use of EM and refer the reader to other papers for technical aspects of the application of EM to study aspects of bacterial infection ***(44,45)***.

LM techniques are more frequently used to study the interaction of bacteria with cells because they are more readily accessible, and although they fall short of some other techniques in terms of resolution, they are very flexible and can be used to study dynamic processes occurring in living cells. We will discuss some technical aspects of these methods later but will first give a brief introduction to the types of imaging techniques available and how they have been applied to study *Salmonella* infection. Although we are concentrating on the most widely used and more widely available techniques, it is likely that other techniques will become more commonplace in *Salmonella* research. For example, multiphoton (MP) microscopy has distinct advantages for studying fluorescent entities deep within tissue and therefore has potential applications in studying infection of intact tissues. At the other extreme, AFM provides detailed information about surface structures and is increasingly being used to study living cells, so also has potential to study bacterial cell interactions.

Although use of AFM and MP microscopy in microbiology research has been somewhat limited to date, these techniques have been used in several studies of bacterial biofilms ***(46–50)***. Many researchers use a combination of techniques to address specific questions, reflecting the fact that different techniques have distinct advantages for specific purposes. Having said that, most researchers will not have the luxury of access to a vast range of microscopy systems, and here we intend to concentrate primarily on imaging techniques that are now fairly widely available, namely wide-field microscopy (WFM) and confocal laser scanning microscopy (CLSM).

It is worth briefly explaining some principles underlying these techniques to understand their relative merits for certain applications. The reader requiring more details is referred to other papers dealing with these issues in greater depth ***(51–56)***. WFM can also be referred to as conventional LM and in its simplest form involves a standard upright or inverted microscope (often with fluorescence capabilities) to which a camera (e.g., charge-coupled device [CCD] camera) is added to enable simple image acquisition. The term "wide-field" refers to the fact that light is detected from different focal planes at the same time so, in the case of fluorescence microscopy, the resulting images will include both "in-focus" and "out-of-focus" fluorescence. WFM imaging systems are highly adaptable; for example, they may incorporate shuttering, focus drives, and filter changers to enable the user to automate rapid switching between imaging parameters. These capabilities, together with the high sensitivity of camera systems, which allow rapid image acquisition and minimization of light exposure, make WFM a popular choice for live cell imaging.

Since its commercialization in the 1980s, confocal microscopy has become a fairly standard research tool, the most commonly encountered class being CLSM, which usually builds up an image of fluorescence (or reflectance) by scanning a laser point by point over a field of view; less frequently it is a line that is scanned rather than a point. The emitted light follows a reverse path (is "descanned") and passes through an aperture (pinhole) before detection using a detector, typically a photomultiplier tube (PMT). In allowing light from only one plane of focus to reach the detector, the confocal aperture enables discrimination of objects at different depths of focus, effectively enhancing axial resolution and thereby giving CLSM its major advantage over WFM. Most confocal laser scanning microscopes allow simultaneous detection of different fluorophores by selectively directing different wavelengths of emitted light to different detectors. The ability to rapidly switch between excitation wavelengths using an acousto-optic tuneable filter (AOTF) simplifies separation of fluorophore signals and is a standard feature of most modern confocal laser scanning microscopes.

When considering CLSM and WFM, we are discussing systems somewhere near two extremes in fluorescence imaging capabilities; with CLSM providing the best axial resolution achievable with commonly available systems but at the cost of sensitivity, speed, greatly increased risk of photodamage, and price. The other principle advantage that CLSM has over WFM is that its use of point-scanning allows flexibility over magnification (by "zooming" on a subregion of the specimen) and also makes it inherently applicable to techniques that require selective illumination of defined areas (photobleaching, photoactivation, and uncaging). Selective photobleaching can, for example, be used to study mobility of green fluorescent protein (GFP)-tagged cellular proteins in the fluorescence recovery after photobleaching (FRAP) technique, which has found widespread application in cell biology. For example, FRAP has been used to study turnover of actin and α-actinin during enteropathogenic *E. coli* (EPEC) infection *(57)*. While addition of photobleaching lasers to WFM and other types of confocal (e.g., spinning-disk confocals) can also enable such techniques on these systems, the flexibility of point-scanning systems usually gives them an advantage in this respect. However, the overriding advantages of WFM in other respects (speed, sensitivity, low photodamage, and cost) means that the most straightforward advice we can give is that if neither the enhanced resolution or functionality of CLSM are required, then WFM is most likely a better option.

Up until now, we have emphasized increased axial resolution as a major advantage of CLSM, but it should always be remembered that this is accompanied by a relative lack of sensitivity due to the large proportion of photons that are discarded at the confocal aperture. Moreover, optimal resolution can be a disadvantage in some circumstances because simultaneous detection of fluorescent entities in a broad depth of field may accelerate data acquisition. Typically, CLSM systems will allow the user to open the confocal aperture to increase sensitivity, and this is an option well worth exploring whenever sensitivity is a greater concern than is optimized resolution, as is most often the case in live cell imaging. In emphasizing the low axial resolution of WFM systems, we have yet to consider improvements that can be made subsequent to image acquisition by image processing. The application of deconvolution algorithms to stacks of WFM images allows reassignment of "out-of-focus" light to its point of origin and can result in marked improvement of image clarity (*see* **Fig. 1**). WFM coupled with deconvolution can often perform at least as well as CLSM in resolving fluorescent structures, especially when these are of relatively low intensity and samples are relatively thin *(58)*. The

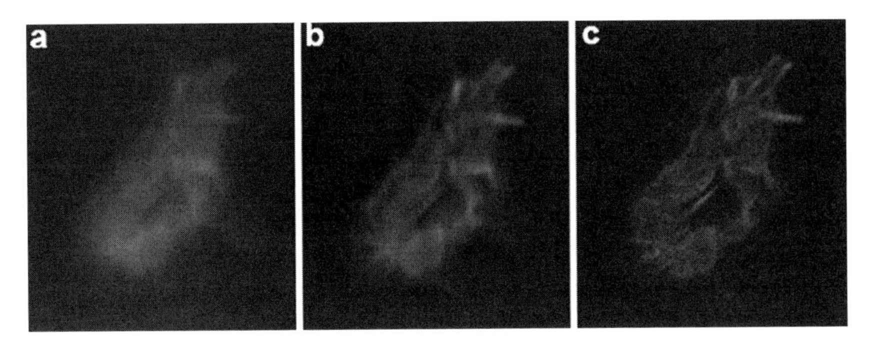

Fig. 1. Comparison of wide-field microscopy (WFM), WFM/deconvolution, and confocal laser scanning microscopy (CLSM) images of *Salmonella typhimurium*-infected Madin–Darby canine kidney (MDCK) cell stained with TRITC-phalloidin to localize F-actin accumulation in a *Salmonella*-induced membrane ruffle. Images were acquired from the entire cell at approximately 100-nm intervals and one selected for comparison. **(A)** Raw WFM image is blurred due to out-of-focus fluorescence. **(B)** After deconvolution using Volocity™ software (and a calculated PSF), a prominent increase in clarity is observed due to removal of out-of-focus fluorescence and increased spatial accuracy within this optical section. **(C)** A CLSM image (acquired with Leica SP2 AOBS system) of the same cell and same plane of focus as shown in **(A)** and **(B)**, or as near as it was possible to determine. The CLSM image shows similar "resolution" to the deconvolved WFM image in **(B)** apart from an arguably clearer delineation of some fine filamentous structures and a noticeably greater variation in signal intensity ("speckling") within actin structures, which may be indicative of the greater degree of noise expected in CLSM imaging. Field of view approximately 20 μm × 23 μm.

increased sensitivity of WFM systems also makes deconvolution an attractive option for live cell imaging if stacks of images can be acquired rapidly enough in relation to the movement of fluorescent entities to avoid introduction of artifacts. However, deconvolution can only be effectively applied where stacks of images at narrow focus increments are acquired in a way that optimizes sampling frequency (Nyquist sampling) *(54,58)*. For this reason, if the advantages of WFM are not marked, for example, where fluorescent labeling is bright and stable, high-resolution images can generally be obtained more simply using CLSM. It is also generally accepted that CLSM is likely to outperform WFM/deconvolution when thicker specimens with more complex distributions of fluorescence are to be imaged, because the contribution of "out-of-focus" fluorescence has a more profound effect on image resolution in such circumstances *(54,58)*. For this reason, CLSM has distinct advantages when studying *Salmonella* infection of polarized epithelia and intact tissues

1.2.2. Microscopical Localization of Salmonella in Cells and Tissues

Monitoring bacterial invasion and precise localization of *Salmonella* within cells and tissues is a routine requirement in *Salmonella* research. *Salmonella* invasion of epithelial cells in guinea pig ileum was first localized in transmission electron microscopy (TEM) studies ***(59)*** and subsequent studies built on these observations using a combination of LM and EM (scanning electron microscopy [SEM] and TEM) techniques ***(4,6,21,60)***. Examination of tissues by TEM alone is relatively laborious, and for this reason, most studies will examine a limited number of cells and thus potentially overlook significant events and give a rather limited view of *Salmonella* interactions ***(61)***. Therefore, where the optimal resolution of TEM is not required, methods that allow *en face* imaging of extensive areas of epithelium have a distinct advantage in facilitating observation of interactions of *Salmonella* with many cells. For example, CLSM has been used to localize *Salmonella* adhered to, and within, M cells in intact Peyer's patch tissue preparations, whereas parallel studies with SEM (sometimes examining the same cells previously imaged by CLSM) allowed surface morphology of *Salmonella*-infected epithelial cells to be examined ***(4,61,62)***. Similar techniques have also been applied to precisely localize infrequently encountered *Salmonella* within thick sections of intestine ***(63)*** and liver ***(64)***.

Identification of *Salmonella* using specific antibodies in conjunction with fluorescent-labeled second antibodies rapidly became the method of choice for localizing *Salmonella* in relation to cellular components or in specific cell types (***14,64–66***). Although immunolabeling of *Salmonella* is most often used in conjunction with WFM and CLSM, the technique would also be applicable to other techniques that have the potential to localize bacteria within intact tissues and other large samples, such as MP microscopy and deconvolution microscopy. Antibody staining has also been used to differentiate between, and quantify, internalized and external bacteria after infection of cells, exploiting the fact that bacteria within intact eukaryotic cells are inaccessible to externally applied antibodies unless the plasma membrane is permeabilized. This method is sometimes regarded as an alternative to the more commonly used, and arguably less laborious, gentamicin-protection assay of bacterial internalization. However, differential immunolabeling of adhered and invaded bacteria has some advantages, in that adherence and invasion are quantified in the same cells, it provides information on the frequently heterogeneous distribution of *Salmonella* within cells that cannot be determined by population-based assays, and it allows simultaneous monitoring of cell damage—cytotoxicity and cell loss being a potential source of serious artifacts in the gentamicin-protection assay. Different versions of the differential immunolabeling technique have been used. One uses transfer of cells to ice-cold phosphate-buffered saline (PBS) before application of ice-cold antibodies to label external *Salmonella* prior to fixation, and permeabilization with methanol to allow access of antibodies to internalized bacteria in a second round of labeling with a different fluorophore ***(22,67)***. In 2002–2003, we have adopted a variation of this technique that allows more flexibility with the timing of labeling. This protocol uses paraformaldehyde to fix, but not permeabilize, cells. The samples can then be labeled with antibodies to localize external bacteria, permeabilized with Triton X-100, then re-stained with antibodies and an alternative fluorophore to label all bacteria (described in **Subheading 3.3.2.**). In our hands each of these immunolabeling methods has proved more reproducible for assaying bacterial invasion than the gentamicin-protection assay, which is particularly difficult to apply to short infection times. Immunolabeling to discriminate external and internal bacteria can also be employed along with additional antibodies, or GFP expression, to localize cellular components or transfected cells, enabling quantification of invasion in transfected versus non-transfected cells and hence investigating the effect of protein expression or inhibition on invasion ***(68,69)***.

An alternative and simpler approach to localizing and enumerating bacteria is to pre-label them prior to infection studies. For example, bacteria can be fluorescently labeled with fluorescein derivatives such as fluorescein isothiocyanate (FITC) ***(70)*** or 5-chloromethylfluorescein diacetate (CMFDA) ***(71)***. Although such non-specific labeling has been used to monitor interaction of *Salmonella* with cells ***(70)***, it is important to consider the effect that the fluorescent molecules might have on surface properties or bacterial physiology. In all such techniques for labeling bacterial populations prior to experimentation, it is important to perform suitable controls to check for such effects.

Nowadays, the most commonly used techniques for labeling bacteria involve expression of GFP or alternative fluorescent proteins ***(18,68)***. GFP labeling has also been used in conjunction with immunolabeling to differentiate internalized bacteria from external ones using a simplified technique based on that described later (*see* **Subheading 3.3.2.**), only requiring fixation and one round of antibody labeling to localize external bacteria, GFP marking both external and internal bacteria ***(68)***. Alternative fluorescent proteins, such as GFP variants, might be used to simultaneously identify more than one bacterial population, e.g., distinct *Salmonella* strains during co-infection. It is important to consider the impact that high levels of GFP expression might have on bacterial fitness and to perform appropriate controls to test for this possibility. Indeed, some studies have highlighted adverse effects of plasmid expression of GFP on *Salmonella* infection *(72,73)*. In vivo studies using plasmid-encoded GFP expression are also likely to suffer from unpredictable loss of plasmid, so use of chromosomal insertion of GFP has been favored by some researchers *(74)*. Single-copy expression of GFP is also likely to have the additional advantage of decreasing the likelihood that GFP expression compromises the bacteria in some way *(74)*.

In addition to providing a method of detecting and enumerating *Salmonella*, GFP expression may be coupled to specific promoters to monitor expression of particular genes or sets of genes, either microscopically or through the use of flow cytometry or fluorescent-activated cell sorting (FACS). The "differential fluorescence induction" technique of Valdivia and Falkow *(75)*, which utilizes GFP and FACS, has allowed the isolation of promoters whose expression is dependent on particular environmental stimuli, such as low pH, or that are expressed during infection of specific cells or tissues *(75,76)*. The technique therefore has the potential to identify promoters and associated genes with differential expression in infected tissues at various stages of the infection, aiding our understanding of the temporal regulation of pathogenesis.

Single-copy *gfp* fusions have been shown to report gene induction in *Salmonella* as accurately as *lacZ* gene fusions and are being used successfully to monitor gene induction in vitro and during infection of mammalian cells *(74)*. The advantage this system offers is that through the use of flow cytometric analysis differential expression of the same gene in a genetically identical population can be recorded and hence facilitates analysis of variations in gene expression within populations and the effect this may have on infection.

An additional staining technique used in the microscopical analysis of *Salmonella* infection is live/dead staining, commercial kits of which are now available, e.g., the Live/Dead™ Baclight™ kit from Molecular Probes/Invitrogen. This method distinguishes "live" and "dead" bacterial cells using two readily distinguishable fluorophores with different permeabilities. One fluorophore is internalized by all cells, whereas the other can only be internalized by cells with compromised membranes, the "dead" population—perhaps more accurately referred to as the sick or dying population. Live/dead staining has so far been used to monitor *Salmonella* viability during infection of macrophages *(77)* but, along with an alternative fluorophore-labeling method that promises to allow monitoring of bacterial "vitality" *(71)*, has the potential to find further applications in *Salmonella* research. The "vitality" labeling method indicates activity of the electron transport chain by monitoring reduction of a compound to an insoluble fluorescent product and is also commercially available (RedoxSensor™ kit from Molecular Probes/Invitrogen).

1.2.3. Imaging Cellular Responses to Salmonella *Infection*

When microscopy methods were applied to study *Salmonella*-infected cells and tissues, it became clear that entry of *Salmonella* into epithelial cells required major changes in cell morphology. Formation of the characteristic *Salmonella*-induced "membrane ruffles" were studied extensively using TEM ***(23,59,60,78)*** and SEM ***(60,61,67,79)***, with fluorescence labeling and LM contributing toward determining how the redistribution of actin and various cytoskeletal proteins promoted formation of these ruffles ***(1,14,15,23,60)***. With the discovery of the SPI-1 T3SS, attention has turned to unraveling the roles of the various effector proteins in triggering cellular responses associated with bacterial pathogenesis.

Much of the work studying the mechanisms of *Salmonella* invasion has been done using cultured cells. The advantages of using such model systems, which are easy to set up, maintain, and genetically manipulate and analyze, are apparent. However, results obtained using non-epithelial and non-polarized epithelial cells (e.g., HeLa) should be regarded with caution as these cells do not always mimic the changes occurring in polarized epithelial cells and tissues during *Salmonella* infection. For example, requirements for specific Rho GTPases during *Salmonella* infection and the importance of each of the effectors SipA, SopA, SopB, SopD, and SopE2 during the same process have both been shown to be dependent on whether polarized or non-polarized cells are used ***(80,81)***. Furthermore, the relative ability of SPI-1 mutants to invade cells also differs between cultured cells and intestinal epithelial cells. Invasion of an *invA* mutant is reduced in comparison with WT in intestinal M cells, but the invasion defect is much more pronounced in cultured epithelial cells ***(3,62)***.

When studying *Salmonella* infection of polarized cells, CLSM has distinct advantages over WFM, which is often subject to "out-of-focus" light interfering with image resolution, even if deconvolution is applied. The redistribution of cellular proteins induced by bacterial infection has therefore been monitored by confocal microscopy in a large number of studies in both polarized epithelial cells ***(23,53,67,80,81)*** and in intact tissues ***(3,61,64,82,83)***. Elucidation of the infection process has been aided with the increasing sophistication of such systems. For example, confocal microscopy has allowed tight junction disruption in polarized epithelia to be examined and specifically to capture transverse (xz) images of living cells to monitor the loss of basolateral–apical diffusion barrier induced by *Salmonella* ***(24)***. In addition to greater resolution, CLSM offers the advantage of allowing events at the apical membrane of polarized cells to be distinguished from those occurring at a different depth in the cells. Examination of confocal optical sections at the apical surface of polarized MDCK cells and below this point led to the discovery that extensive constriction of cells at their apical poles during *Salmonella* infection contributes to gross changes in cell morphology and disruption of epithelial barrier function (*see* **Fig. 2**) *(23,24)*.

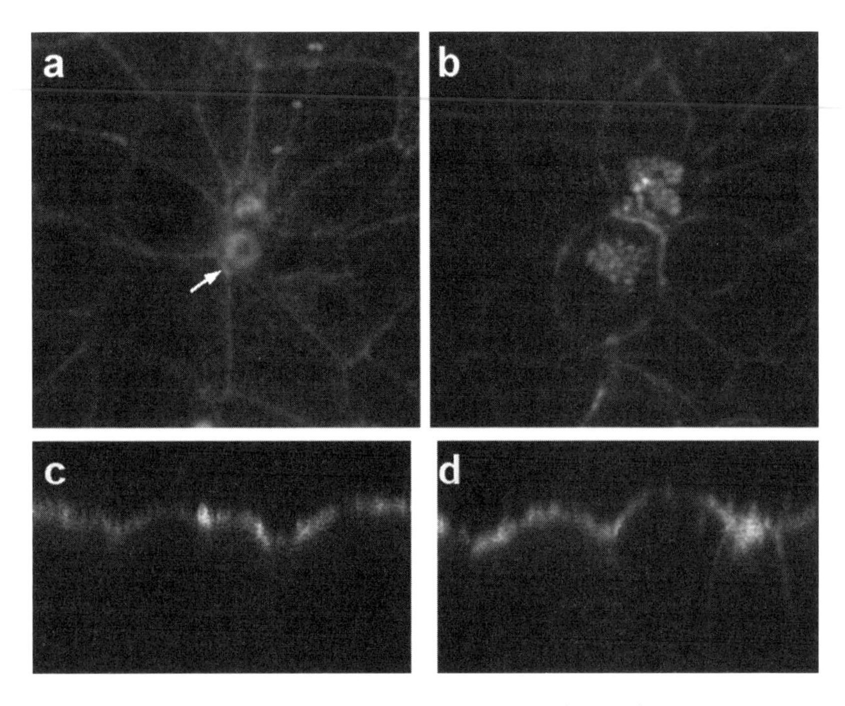

Fig. 2. Use of confocal laser scanning microscopy (CLSM) to study *Salmonella*-induced tight junction (TJ) modulation. **(A** and **B)** Madin–Darby canine kidney (MDCK) epithelial monolayers infected with green fluorescent protein (GFP)-expressing *Salmonella typhimurium* for 60 min, then labeling with phalloidin-TRITC to localize F-actin. CLSM images at the apical pole **(A)** reveal contraction of the peri-junctional actin ring in infected cells (arrow) and stretching of neighboring cells, whereas CLSM sections 4 μm below the apical portion **(B)** show more uniform cell profiles. Independently acquired TRITC and GFP images have been merged to show the relative position of bacteria and F-actin in single images. **(C** and **D)** x-z CLSM images of uninfected **(C)** and *S. typhimurium*-infected **(D)** MDCK monolayers following addition of BODIPY-sphingomyelin to the apical bathing medium. Fluorescent labeling is restricted to the apical membrane in control cells **(C)** due to the transmembrane diffusion barrier provided by intact TJs. Dysfunction of TJs in *Salmonella*-infected cells **(D)** allows migration of fluorescent lipid from apical to basolateral membrane compartments. Figure adapted from data first published in *Infection and Immunity* ***(24)***.

It is of course not an absolute requirement to use confocal microscopy for all fluorescence imaging, and indeed on many flatter cultured cells, excellent results can be obtained with conventional WFM ***(84)***. As discussed previously, WFM also has major advantages for live cell imaging which can address questions about the dynamics of *Salmonella* interactions with host cells, which cannot be studied easily, if at all, by static imaging techniques.

Understanding how *Salmonella* triggers cellular responses relies on studying highly dynamic processes. In the early 1990s, studies using a combination of fluorescence labeling, EM, and live cell imaging demonstrated that the triggering of membrane ruffling can occur within 1 min of *Salmonella* adherence to the surface of cells ***(60)***. Several studies have demonstrated the value of live cell imaging in determining the sequence of rapidly occurring events during *Salmonella* infection, which can be obscured by the non-synchronized nature of *Salmonella* interaction when restricting studies to the examination of cells fixed at discrete time points during infection ***(15,18,60)***.

Phase-contrast time-lapse microscopy of living cells infected with *Salmonella*, an example of which is shown in **Fig. 3**, revealed the effects of loss of SipA from *Salmonella. Salmonella sipA* mutants induce ruffles over a time course indistinguishable from WT, but after ruffle formation, they differ in their location relative to the ruffle, frequently moving to the periphery of the ruffle rather than staying centrally located ***(15)***. This behavior is associated with delayed entry into the ruffle and often detachment of the bacterium from the cell ***(15)***. This phenotype of the *sipA* mutant likely contributes to the decrease in invasion that is apparent only within the first few minutes of infection of cultured cells and was overlooked before the application of live cell imaging.

Imaging cellular processes in living cells can frequently reveal facets of bacterial interactions with host cells that would be impossible to determine by other means. This is especially evident when using specific fluorescent probes, e.g., calcium indicators, and GFP-PH and GFP-FYVE domains that specifically bind to certain phosphoinositides. Such probes enable monitoring of the redistribution of proteins and other cellular changes in real time and have revealed some of the dynamic processes occurring during *Salmonella* invasion. An example from our own research is the identification of cycles of PI(3)P generation on SCVs over several minutes, which were revealed by live cell imaging of *Salmonella* invading cells stably expressing GFP-FYVE ***(85)***. Without live cell imaging, the fact that a proportion of SCVs were labeled with PI(3)P would have led to the assumption that this indicated a temporary location rather than the repeated cycles of acquisition and loss identified in this

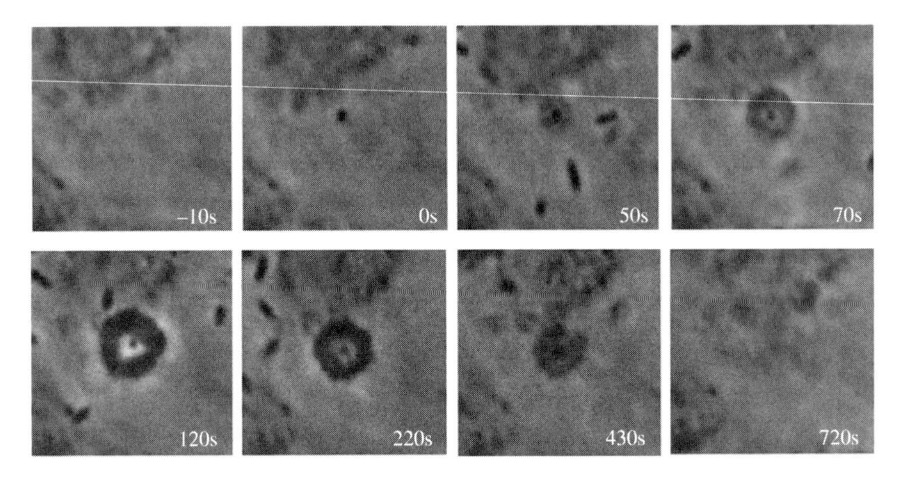

Fig. 3. Use of time-lapse phase-contrast microscopy to examine membrane ruffle propagation and development. Representative phase-contrast images of a membrane ruffle generated by wild-type *Salmonella typhimurium* (SL1344) in Madin–Darby canine kidney (MDCK) epithelial cells are shown at distinct points in ruffle development during a 20-min time course. Timestamps on each image indicate relative time compared with the first image in which this bacterium attached to cells (0 s). Similar time-lapse sequences have been used to analyze the kinetics of ruffle induction and dissipation, and of invasion ***(15,85)***. Time-lapse imaging also yields information on ruffle morphology, the position of bacteria relative to each developing ruffle, the intracellular location of invaded bacteria, and movement of intracellular proteins and structures.

study ***(85)***. Use of similar GFP probes also led to the finding that SigD/SopB contributes to PI(3)P generation on SCV membranes and to the formation of spacious phagosomes in infected cells ***(18)***.

Longer term live cell imaging can also be used to study later events occurring post-invasion, including SPI-2-mediated effects on intracellular trafficking, SCV development and maturation, and bacterial division. However, prolonging the time course over which images are acquired can be challenging as cells are more prone to damage during such experiments, especially from photodamage if fluorescent images are acquired. As discussed in **Subheading, 1.2.1.**, WFM has distinct advantages for live cell imaging, where the high-intensity excitation required for CLSM is much more likely to impair cellular processes. The additional challenges of prolonged imaging have, however, resulted in most studies of longer term infection being limited to static imaging techniques (***7,86–88***).

1.2.4. Imaging Secretion, Translocation, and Localization of Salmonella *Effector Proteins*

As we have already discussed, high-resolution microscopy methods (EM and AFM) have been used to study the architecture of the *Salmonella* T3SS machinery ***(34,35,89)*** and the translocation pores formed in the host cell membrane by an analogous secretion system expressed by *E. coli* ***(43)***. Fluorescence microscopy, including CLSM, has also been used to locate translocated effector proteins within infected cells using antibodies to the effector protein itself or, more commonly, epitope tags ***(84,90,91)***. Other methods have recently been devised to measure the rate of translocation of effector proteins that, due in part to the rather small numbers of molecules translocated into host cells, have proved difficult to localize by immunolabeling, especially in the early stages of infection. Quantitative immunolabeling of SipA within *Salmonella* after defined periods of interaction with host cells (determined by live cell imaging prior to fixation and labeling) has been used to estimate the rate of transfer of SipA into the host cell ***(92)***. A subsidiary technique measured, in transfected cells, recruitment of the GFP-InvB (the molecular chaperone of SipA) to the point of *Salmonella* infection as an indirect and semi-quantitative means of monitoring SipA translocation in real time ***(92)***. A further technique that might be applied to *Salmonella* has been demonstrated in a study quantifying secreted proteins IpaB and IpaC in *Shigella* using FLASH-tagging to fluorescently label the protein within bacteria and monitor its ejection during infection of host cells ***(93)***. Aside from these advanced techniques, fluorescence microscopy has also been used to monitor *Salmonella* effector translocation using a sensitive method employing tagging of effector proteins with TEM-1 β-lactamase. Treatment of cells with a fluorescent lactamase substrate CCF2/AM that is sequestered in the cytoplasm allows detection of TEM-1-tagged proteins translocated into the cells ***(81,94)***. This method effectively amplifies the signal from individual effector molecules as one protein molecule can produce many fluorescent molecules within the cell and thereby overcomes the problem of detection of potentially low numbers of effector molecules. Despite its inability to localize effector proteins (due to diffusion of fluorescent enzyme product within the host cell), this method has great potential as a sensitive single-cell assay of translocation.

As we have already alluded to, *Salmonella* effectors are often translocated in small amounts, making it difficult to use traditional localization techniques to identify their targets within infected cells. This is one reason why several studies have used the alternative approach of expressing effector proteins in cultured cells or model organisms. For example, transfection of cultured cells with GFP-tagged versions of effector proteins has facilitated identification of subcellular compartments, for which the effectors have specific affinity, and cellular activities that may be subverted by the effectors ***(66,84,87,88,90)***.

2. Materials

2.1. Preparing Cell Monolayers

1. Many of our *Salmonella* infection studies have utilized Madin–Darby canine kidney (MDCK) epithelial cells, and the details of culture methods given here are specific for these cells. Infection of other cells, both epithelial (e.g., Caco-2, HeLa, and Hep-2) and non-epithelial (e.g., macrophages, COS cells) can be studied using similar methods.
2. Cell culture medium: Minimum Essential Eagle's Medium, supplemented with 2 mM L-glutamine, 10% fetal calf serum, 1% non-essential amino acids, and 100 μg/mL kanamycin (*see* **Note 1**).
3. Glass coverslips: We generally use thickness 1 (130–170 μm) from, for example, VWR, Menzel-Gläser, or Corning (but *see* **Note 2**).
4. MDCK cells, like several other epithelial cell lines, e.g., Caco-2, can form functionally polarized monolayers when grown on permeable supports. The most commonly used permeable culture inserts include Transwell™ (Corning) and Anopore™ (Nunc). Epithelial cells able to form polarized layers should be grown on such permeable supports when the effects of *Salmonella* on certain epithelial cell properties, e.g., transport and barrier functions, are to be investigated.

2.2. Infecting Cell Monolayers for Immunofluorescence Microscopy

1. Luria-Bertani (LB) broth and LB agar plates, e.g., from Difco.
2. *Salmonella* strains grown overnight in LB broth, diluted 1:100 in fresh LB, and grown in a shaking incubator at 37 °C for 3.5 h (*see* **Note 3**).
3. Modified Krebs buffer: 137 mM NaCl, 5.4 mM KCl, 1 mM $MgSO_4$, 0.3 mM KH_2PO_4, 0.3 mM NaH_2PO_4, 2.4 mM $CaCl_2$, 10 mM glucose, and 10 mM Tris, adjusted to pH 7.4 at 37 °C with HCl (*see* **Note 4**).

2.3. Immunofluorescence Microscopy

1. PBS, pH 7.4.
2. Fixative: 2% paraformaldehyde in PBS (*see* **Note 5**).
3. Permeabilization buffer: 0.1% Triton X-100 in PBS.
4. Primary antibody: Either a specific monoclonal or polyclonal antibody generated against an antigen of interest, or an antibody fragment in some cases (*see* **Note 6**).
5. Secondary antibodies: Commercial fluorochrome-conjugated antibodies specific for the primary antibody (*see* **Note 7**).
6. Tetramethyl rhodamine isothiocyanate (TRITC)-conjugated phalloidin from Sigma (*see* **Note 8**).

7. Mounting media: We use Vectashield® mounting media for preservation of fluorescent labeling, usually employing the version supplied with 4′,6-diamidino-2-phenylindole (DAPI) (Vector Laboratories Inc.) (*see* **Note 9**).
8. Clear nail varnish.
9. Fluorescence microscope with appropriate fluorescence filters (*see* **Note 10**).

2.4. Wide-Field Imaging

1. Microscope system and hardware: Leica inverted microscope with Hamamatsu CCD camera and Prior Scientific hardware to control illumination and focus (*see* **Note 11**).
2. Image acquisition software: Openlab™ (Improvision) (*see* **Note 12**).
3. Deconvolution software: Volocity™ (Improvision) (*see* **Note 13**).

2.5. Confocal Laser Scanning Microscope

1. Confocal laser scanning microscope: Leica TCS SP2 AOBS attached to a Leica DM IRE2 inverted epifluorescence microscope (*see* **Note 14**).

2.6. Live Cell Imaging

1. Modified Krebs buffer (*see* **Subheading 2.2., step 3** and **Note 4**).
2. Coverslip chambers for 22 or 24 mm diameter coverslips (*see* **Note 15**).
3. Inverted microscope system: Leica DMIRB microscope with additional hardware (*see* **Notes 11** and **12**).
4. Microscope incubation chamber from Solent Scientific (*see* **Note 16**).

3. Methods

3.1. Preparing Cell Monolayers

1. Place glass coverslips between two pieces of tissue and spray with ethanol to sterilize. When dry, use forceps sterilized in ethanol to transfer the coverslips into the appropriate well plate (*see* **Note 17**).
2. Seed coverslips in well plates (or filters as appropriate) with epithelial cells. The number of cells used depends on the experimental requirements (*see* **Note 18**).
3. Incubate at 37 °C in humidified atmosphere with 5 % CO_2 until required confluence reached—generally 2–4 days.

3.2. Infection of Cell Monolayers for Immunofluorescence Microscopy

The following protocol includes volumes and other details specific for infection of cells grown on 13-mm coverslips. These would need to be adjusted for cells grown on other substrates.

1. Wash cells three times with 1 mL warm modified Krebs buffer, with the final wash media remaining in the well. Incubation for 10–15 min at 37 °C allows equilibration of the cells in the buffer.
2. Add 50 μl of the log phase *Salmonella* culture to the wells to give a multiplicity of infection of approximately 50:1 (*see* **Note 19**).
3. Incubate cells for the required time at 37 °C (*see* **Note 20**).
4. To remove non-adherent bacteria, extract each coverslip from the well plate with forceps and wash, with moderate agitation, in a beaker of PBS.
5. Place the coverslip in a well of a 12-well plate filled with 1 mL 2% paraformaldehyde (PFA) and leave to fix for 45 min at 4 °C.

3.3. Fluorescence Microscopy

There are multiple protocols that may be employed for immunofluorescence microscopy depending on which components of the cells or bacteria are to be localized. We introduce two representative protocols that we routinely use, the first for localization of *Salmonella* and F-actin, the second to measure *Salmonella* invasion by separately labeling the adhered and entire bacterial populations associated with cells. In each case, volumes refer to labeling of cells on 13-mm coverslips; these should be increased as required for larger coverslips or permeable supports.

3.3.1. Staining F-Actin and Salmonella

1. After fixation in PFA for at least 30 min (overnight if convenient), wash the coverslips in a beaker of PBS and place in a well containing 1 mL 0.1% (v/v) Triton X-100 for 10 min to allow permeabilization (*see* **Note 21**).
2. Wash coverslip again in PBS and place in a well of 1 mL PBS, before removing to an empty well.
3. Incubate cells with 50 μL primary antibody, in our case goat anti-*Salmonella* CSA-1 antibody (Kirkegaard and Perry Laboratories) diluted in PBS, for 45 min at room temperature (RT). Replace the lid on the plate to prevent drying (*see* **Note 22**).
4. Add a little PBS to each coverslip to aid its removal from the well plate with forceps. Wash each coverslip in PBS and place in a well of 1 mL PBS, before removing to an empty well.
5. Incubate cells with 50 μL secondary antibody (we use FITC-conjugated anti-goat antibody from Sigma) and TRITC-conjugated phalloidin (Sigma) diluted in PBS, for 45 min at RT in the dark (replace lid on the plate to prevent drying, and cover with foil to limit photobleaching) (*see* **Note 22**).

6. Remove coverslip from the well plate, wash in PBS, wipe the back of the coverslip, and blot its edge with tissue to remove excess liquid.
7. Mount coverslips by placing cell side down on microscope slides with a small drop of mounting media (we generally use Vectashield® mounting media with DAPI; *see* **Note 9**). Blot any excess liquid with tissue.
8. Paint around the coverslip edge with a minimal amount of clear nail varnish and allow to dry for at least 10 min and/or store at 4 °C, to seal and hold the coverslip in place.
9. Examine cells using a fluorescent microscope using the appropriate filters and objective lens (*see* **Note 10**).

3.3.2. Differential Antibody Staining of Adhered/Invaded Salmonella *(*see ***Note 23****)*

1. After fixation in PFA, wash coverslip in beaker containing PBS and place in a well containing 1 mL PBS, before removing to an empty well.
2. Incubate cells with 50 μL primary antibody (anti-*Salmonella* CSA-1 antibody from Kirkegaard and Perry Laboratories) diluted in PBS, for 45 min at RT. As no permeabilization step has been performed, only the *Salmonella* adhered to the cell surface are accessible to antibodies at this stage. Place the lid on the well plate to prevent drying.
3. Remove coverslip from the well plate, wash in PBS, and place in a well of 1 mL PBS, before removing to an empty well.
4. Incubate cells with 50 μL secondary antibody (we use FITC-conjugated anti-goat antibody from Sigma) diluted in PBS, for 45 min at RT. During this incubation and all subsequent steps, replace lid on the plate to prevent drying, and cover with foil to protect fluorophores from bleaching.
5. Remove coverslip, wash in beaker containing PBS, and place in a well containing 1 mL 0.1 % (v/v) Triton X-100 for 10 min to permeabilize the plasma membrane.
6. Remove the coverslip from the well plate, wash in PBS, and place in a well of 1 mL PBS, before placing in an empty well.
7. Incubate cells again with 50 μL anti-*Salmonella* antibody (as in step 2) diluted in PBS, for 45 min at RT. This time both adhered and invaded *Salmonella* will be accessible to antibodies allowing enumeration of all *Salmonella* associated with cells.
8. Remove coverslip from the well plate, wash in PBS, and place in a well of 1 mL PBS, before removing to an empty well.
9. Incubate cells with 50 μL secondary antibody conjugated to a different fluorophore (we use TRITC-conjugated anti-goat antibody from Sigma) diluted in PBS, for 45 min at RT. Using a different fluorophore at this stage to label the total population allows the adhered bacteria to be visualized as a separate population, allowing measurements of adhered and invaded *Salmonella* to be made.

10. Remove the coverslip, wash in PBS, and wipe the back of the coverslip and blot edge with tissue to remove excess liquid.
11. Mount coverslips cell side down on microscope slides using a drop of mounting media (we use Vectashield® mounting media with DAPI, *see* **Note 9**). Any excess liquid is blotted with a piece of tissue.
12. To seal and hold the coverslip in place, paint edges of coverslip with a minimal amount of clear nail varnish and leave to dry for 10 min and/or store at 4 °C.
13. Examine cells with a fluorescent microscope using the appropriate filters and objective lens (*see* **Note 10**).

3.4. Wide-Field Microscopy and Deconvolution

Although this protocol describes acquisition of stacks of wide-field images at small focal increments suitable for deconvolution (as illustrated in **Fig. 1**), in many cases, there will be no need to acquire a series of images at different focal planes, and in this case, images can be acquired without the use of automated imaging components. The protocols for WFM and CLSM (*see* **Subheading 3.5.**) refer to acquisition of images shown in **Fig. 1**, where MDCK cells were grown on glass coverslips and infected with *Salmonella typhimurium* for 15 min followed by fluorescent labeling of F-actin, *Salmonella*, and DNA (as described in **Subheading 3.3.1.**). Imaging was performed using the system described in **Note 11**. Although this necessarily refers to a specific imaging system, we have tried to make the advice given as general as possible.

1. Manually select appropriate fluorescent filter cube, in this case a triple dichroic and emission filter set (460/20, 520/35, and 600/40) mounted in filter cube (supplied by Chroma Ltd).
2. Select 63× oil-immersion objective lens (PLApo; 1.32 NA, phase 3), and add a small drop of immersion oil (Leica) to coverslip.
3. Focus on cells using appropriate filter for TRITC excitation (555/25 nm filter; Chroma Ltd) using filter wheel controller.
4. Examine FITC and DAPI labeling having used filter wheel controller to switch excitation (484/15 and 395/10 nm filters, respectively; Chroma Ltd).
5. Locate suitable area for imaging and likely brightest level of focus within the specimen (*see* **Note 24**).
6. Select maximum camera resolution (1344 × 1024; i.e. no binning), and 12-bit depth using Openlab™ 4 software (*see* **Note 25**).
7. Collect images of TRITC, FITC, and DAPI, and adjust exposure times and (minimally) camera gain to select suitable exposure times for each fluorophore. Ideally, the image will have a large dynamic range, without saturation and with low background (*see* **Note 26**).

8. Manually focus through specimen with camera "live" and TRITC excitation selected to check that brighter objects are not present at other levels that would saturate camera.
9. Adjust focus to a suitable start point (*see* **Note 27**).
10. Open an existing Openlab automation previously used to acquire similar image stacks of TRITC, FITC, and DAPI.
11. Modify the automation by typing in appropriate exposure times for TRITC, FITC, and DAPI as previously determined.
12. Modify the automation with required focus steps (in this case 100 nm) to be taken after acquiring each pair of images and to set the number of images to be acquired (*see* **Note 28**).
13. After running the automation, image stacks can be saved in Improvision's LIF format, transferred over the network, and imported directly into Improvision Volocity™ software running on an off-line workstation. Alternatively, images can be exported as a series of TIFF files for import into alternative software (*see* **Note 29**).
14. Apply deconvolution algorithm to the image stack (usually after cropping to accelerate image processing) using point spread functions (PSFs) calculated for each fluorophore wavelength and objective lens used (*see* **Note 30**).

3.5. Confocal Laser Scanning Microscopy

Although we have provided a protocol for imaging triple-labeled *Salmonella*-infected MDCK cells using a Leica CLSM system (*see* **Note 14**) as used to generate **Fig. 1**, the general principles described are applicable to other CLSM systems.

1. Select 63× oil-immersion objective lens (PLApoBL; 1.4 NA) (*see* **Note 31**).
2. Examine the slide to locate suitable area for imaging (in this case matched to previously acquired WFM images). At this stage, it is worth taking care to ensure the image is as clear as expected by eye (*see* **Note 32**).
3. Select appropriate zoom factor as required (*see* **Note 33**).
4. The confocal aperture (or "pinhole") is normally left at the default setting (equivalent to 1 Airy Unit) to optimize resolution (*see* **Note 34**).
5. Select averaging (e.g., averaging four frames as in **Fig. 1**) to reduce the effect of detector noise (*see* **Note 35**).
6. Select appropriate laser power for each fluorophore to allow images of these brightly stained cells to be acquired at PMT setting of 500–600 V, above which more detector noise is apparent (*see* **Note 36**).
7. Using the TRITC setting, locate and select the top and bottom extremes of the cell. This allows a stack of images to be acquired at user-defined intervals within the selected limits (*see* **Note 37**).
8. After adjusting AOTF setting and PMT voltage for each fluorophore (TRITC, FITC, and DAPI) and phase contrast (or brightfield or DIC) if required, save these settings to allow sequential imaging of the three fluorophores (*see* **Note 38**).

9. Choose option for sequential imaging and import previously defined TRITC, FITC, and DAPI settings to allow switching between two excitation wavelengths during image capture (*see* **Note 39**).
10. Acquire image sequence and save as a stack of TIFF files (thc default save option), which can be easily exported to other software.

3.6. Live Cell Imaging—Time-Lapse Phase-Contrast Microscopy

Although this protocol refers to the phase-contrast imaging we routinely perform to examine propagation of membrane ruffles (*see* **Fig. 3**), the automated imaging can be modified to include fluorescence imaging (e.g., GFP and/or other fluorophores) or differential interference contrast (DIC) imaging (*see* **Note 40**). The live cell imaging described here can be performed using a reasonably low-cost system (*see* **Note 11**), but faster switching and focusing, and more sensitive cameras are also available.

1. Wash a 22 or 24 mm diameter coverslip on which cells have been grown (as described in **Subheading 3.1.**) twice with warm (37 °C) modified Krebs buffer in its well, and use forceps to place it in the lid of the coverslip holder (*see* **Fig. 4**). Screw the chamber onto the lid and add 1 mL modified Krebs buffer into the well (*see* **Note 41**).
2. Place the coverslip holder on the microscope stage, which is enclosed within microscope incubator that has been running for at least 30 min, and preferably longer, to allow required temperature to be achieved and stabilized.
3. Examine cells under phase-contrast optics and select a suitable area for imaging.

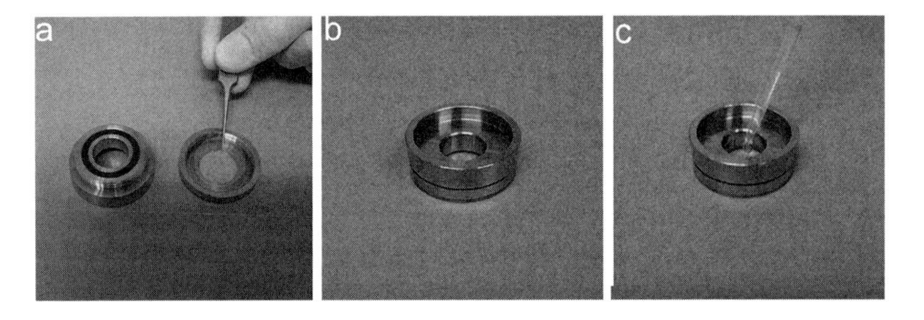

Fig. 4. Setting up the coverslip holder for live cell imaging. The coverslip holders we use are made in our workshop. **(A)** A 22-mm coverslip (cell side up) being positioned centrally within the chamber base using forceps. **(B)** The complete coverslip holder once the base has been screwed onto the lid and a seal formed between the coverslip and rubber O-ring in the underside of the upper part of the chamber. To the well is added 1 mL buffer, in this case modified Krebs buffer warmed to 37 °C **(C)**. Once on the microscope stage, log phase *Salmonella* are added to the well during image capture.

4. Open Improvision Openlab™ 4.0 software and a previously used automation for the acquisition of images using hardware described in **Note 11**. We typically set up the automation to capture images at three focal depths (at 1.5-μm increments to ensure data are acquired at correct level and allowing for the possibility of minor stage drift) at 10-s intervals over a 20- to 30-min time course. Run the automation prior to the start of imaging to allow for trouble-shooting and to optimize the quality of the images obtained, for example, by adjusting focus and lamp intensity.
5. Adjust automation to include specific exposure times and focus increments (as described in **Subheading 3.4.**) (*see* **Note 42**).
6. When ready to start imaging, begin the automation and then, after 1–2 min, add 50 μL log phase culture to the Krebs buffer in the coverslip holder (*see* **Note 43**).
7. Images can be exported for processing and analysis in the Openlab LIF format to an off-line Apple workstation running Openlab software or a PC running Volocity™ software. Alternatively, images can be exported as a Quicktime movie or as a stack of TIFF images for processing with alternative software.

4. Notes

1. We have mainly used media and serum from Sigma-Aldrich. Similar products are available from many suppliers.
2. We generally use 13 mm diameter coverslips when fluorescent labeling will be applied to fixed cells, and 22 or 24 mm diameter coverslips for live cell imaging. Coverslips are available in a variety of thicknesses but tend to vary somewhat within each batch (some suppliers may perform better than others in reproducibility). Most often, we use thickness 1 (130–170 μm) and do not go to any additional lengths to determine the precise thickness of individual coverslips, despite others reporting marked discrepancies in thicknesses of coverslips obtained from some suppliers. Microscope manufacturers generally specify 170 nm as the optimum coverslip thickness for performance of their objective lenses; so many researchers prefer to purchase thickness 1.5 (160–190 μm) as a result. It should, however, be remembered that many of these will be too thick to allow optimal resolution and that the additional distance from the coverslip to the upper parts of the cells grown on the coverslip should be taken into account. It is unlikely that small differences in coverslip thickness will noticeably affect image quality in most applications, but where optimal resolution is needed, for example, to ensure optimal precision in deconvolution, some researchers will check the thickness of all coverslips they use. However, coverslip thickness has much greater effect on performance of high NA dry and water-immersion lenses than oil-immersion lenses, and these typically have correction collars, which should be adjusted to the correct thickness of coverslip being used. It is also advisable to carefully adjust these collars while viewing the effects on image clarity to ensure correct positioning. When cells are destined to be fixed and labeled, it is possible to culture on multi-well slides (e.g., Lab-Tek), which provide a convenient means of

separately treating multiple cell samples on a single slide. At the end of the experiment, the chambers are removed and a coverslip placed onto the slide. Growing cells on coverslips is inconvenient for some live cell imaging applications as the coverslips need to be placed into a sealed holder before placing on the microscope stage (*see* **Subheading 2.6.**). Alternatives to this approach include plastic cell culture dishes with coverslip quality glass inserts (Matek). Cells can also be grown on conventional plasticware if low magnification is sufficient. High magnification and resolution can also be achieved with specialized chambers with optically clear plastic bases, although glass is preferable for DIC imaging.

3. Although we mostly use this method to obtain mid-log bacteria of optimized invasiveness, other growth parameters can also be used. Overnight cultures of *Salmonella* will invade cells suboptimally, whereas the period of growth of subcultured bacteria can be extended if late exponential/early stationary phase bacteria are required. Different dilutions of overnight cultures (from that of the 1:100 dilution we use) are preferred by some, e.g., 1:33 or 1:50.
4. We use Krebs as a convenient physiological buffer that avoids the requirement for CO_2 that comes with using bicarbonate-buffered media. Alternatives, which are especially useful for longer term infection studies, include conventional cell culture media or bicarbonate-free versions of these that incorporate alternative buffering to avoid the requirement for CO_2. For fluorescence microscopy, avoidance of Phenol Red (which is included in most cell culture media as a pH indicator) is advisable as this causes substantial background fluorescence.
5. PFA can cause some increase in background fluorescence. Many researchers routinely use a sodium borate treatment to quench this prior to labeling. PFA fixation works well for a large proportion of antigens, but in some cases solvent fixation, e.g., methanol, acetone, or mixtures of these, gives superior results. Solvent fixation also has the advantage that it permeabilizes cells and thus eliminates the requirement for this additional step.
6. For localization of more than one antigen, it is preferable to use primary antibodies from different sources, e.g., mouse and rabbit. If two antibodies from the same species have to be used, one or more may be conjugated directly to fluorophores to allow separate localization (commercial fluorophore labeling kits are available, such as the range of Zenon™ antibody labeling kits from Molecular probes/Invitrogen). Alternative approaches include blocking the available binding sites on one primary antibody with Fab fragments, e.g., those available from Jackson ImmunoResearch Laboratories, Inc., before application of a second primary from the same species.
7. Secondary antibodies are generally raised in larger animals, e.g., donkey, against immunoglobulin from species used to generate primary antibodies (typically mouse, rat, and rabbit) and are available in a wide variety of fluorophore conjugates (excited from UV through to red) from many suppliers, e.g., Jackson ImmunoResearch Laboratories, Inc., Molecular Probes/Invitrogen, Sigma. Some are specific

for individual classes of immunoglobulin and so can be used to detect monoclonal antibodies from the same species if they differ in antibody class. Although the methods described here use FITC and TRITC and give intense labeling, the use of these fluorophores is suboptimal and retained here for historical and economic reasons. Other alternative fluorophores, including Cy2, Cy3, Alexa Fluor® 488, 594, etc., have higher quantum yield (brightness) and improved photostability. If two antibodies from closely related species are used, e.g., mouse and rat, it may be necessary to use secondary antibodies absorbed with antibodies from the alternate species to avoid cross-reaction. However, this will generally decrease labeling intensity and, depending on class-specificity, may even eliminate it.

8. We generally use TRITC-phalloidin, but alternative fluorophore-conjugated versions, e.g., FITC and Cy5, are available from Sigma, Molecular Probes/Invitrogen, etc.
9. To mount cells and inhibit photobleaching, we routinely use Vectashield® supplied with DAPI, unless we specifically want to avoid an additional UV-excited dye interfering with labeling in that part of the spectrum. The main advantages of using this are that we avoid an additional DAPI-labeling step and also avoid preparing toxic solutions of DAPI or alternative DNA labels. Many researchers instead use Mowiol™ containing *n*-propyl gallate (to reduce bleaching) as mounting media, which unlike the Vectashield we use sets hard (though a hard-set version of Vectashield® is also available from Vector Laboratories). Alternatives available from other sources include SlowFade Gold™ and ProLong™ from Molecular Probes/Invitrogen. Anti-fade mountants can vary in their effectiveness against photobleaching and also in the initial intensity of fluorescence. The most effective in blocking bleaching generally decrease initial intensity to some extent. We have not performed a thorough side-by-side comparison of their properties. It is also important to bear in mind that mounting media may not be completely compatible with all fluorophores. For example, some users have reported decreased signal from Cy2 and other cyanine dyes over time when mounted in media containing phenylenediamine as an anti-fade.
10. This can be a relatively simple upright or inverted microscope with fluorescence capabilities. Usually, this will have separate fluorescence filter blocks for each fluorophore and be equipped with a range of objective lenses. The major suppliers include Leica, Olympus, Zeiss, and Nikon. All of these produce high-quality optical instruments and choice comes down to personal preferences and practicalities. Details of the Leica upright microscope we use for examining staining are available at http://www.bris.ac.uk/biochemistry/mrccif/techspecleicadm.html
11. The system we routinely use is based on a Leica DMIRB inverted microscope with Hamamatsu ORCA ER (12-bit CCD) camera and Prior Scientific filter wheel, shutters and motorized focus. Further details of the WFM system used are available at http://www.bris.ac.uk/biochemistry/mrccif/techspecopenlabsystem2.html. Alter-

native imaging systems and hardware options are available from other microscope manufacturers and imaging companies (e.g., *see* **Notes 10** and **12**).

12. The wide-field imaging system used to acquire images in **Fig. 1** is controlled by Improvision Openlab™4 software, which is a versatile Mac-based software from Improvision, which now also supplies a PC-based acquisition system, Volocity Acquisition. Alternative imaging systems, including integrated acquisition and deconvolution systems, are available from other companies, e.g., Applied Precision, Leica, Olympus, Universal Imaging (now part of Molecular Devices), Image Pro (Media Cybernetics), Kinetic Imaging (now part of Andor Technology), and Digital Pixel. Each of these has image processing and analysis capabilities. Some researchers prefer to build their own systems, and some freeware is available to control cameras and other hardware and/or undertake image processing and analysis. The most commonly used freeware is the Mac-based NIH image (its PC version from Scion imaging), functions of which have been incorporated in its Java-based successor, ImageJ, for which many plug-ins covering a wide variety of functions have been provided by users.
13. Deconvolution was (*see* **Fig. 1**) performed here using Improvision's Volocity™ software running on an off-line workstation that has relatively powerful three-dimensional (3D) rendering and image analysis capabilities. Other deconvolution software packages are available from several of the companies listed in **Note 12** as well as from others including Autoquant Imaging, Inc. We have not made careful comparison of these but have achieved very good results with the Deltavision™ system from Applied Precision and Autoquant software. Some freeware deconvolution software has also been made available, but we have not tested these.
14. We use one of a range of Leica CLSMs. The one used to acquire images in **Fig. 1** is a Leica TCS SP2 AOBS confocal laser scanning microscope attached to a Leica DM IRE2 inverted epifluorescence microscope, further details at http://www.bris.ac.uk/biochemistry/mrccif/techspecconfocal4.html. Other suppliers of confocal laser scanning microscopes include Zeiss, Olympus, and Nikon. Several faster-scanning confocal systems are also available from these suppliers, some of which utilize line-scanning instead of point-scanning. For rapid imaging of living cells, many researchers prefer the alternative confocal methodology available in the spinning-disk confocals (Yokogawa head), which offer improved light efficiency and often decrease photodamage but have somewhat reduced flexibility and axial resolution. Spinning-disk confocals are available from suppliers including Perkin-Elmer, Visitech, and Andor Technology. An alternative, simplified method of confocal imaging is provided by structured illumination systems involving WFM imaging of a specimen onto which a grid pattern is projected and moved during sequential image acquisition, e.g., Optigrid™ and Apotome™ Systems supplied by Improvision, Zeiss, etc.

15. The coverslip holders we use are made in our workshop (*see* **Fig. 4**), though commercial versions are also available. For alternatives to coverslip holders, refer to **Note 2**.
16. We have temperature control chambers from both Solent Scientific and Life Imaging Services. Although there are distinct advantages in sample stability (avoiding focus drift) by enclosing a large proportion of the microscope in a temperature-controlled environment, the advantages are more apparent with long-term imaging. Cells in dishes or in coverslip holders can also be maintained at 37 °C and perfused using on-stage heating devices as supplied by various companies, e.g., Harvard Apparatus, Scientifica, Life Imaging Services, and Bioptechs. If bicarbonate-buffered media are used, it is necessary to have a means of enriching the environment around the cells with CO_2. We use such a device from Solent Scientific. More sophisticated (and expensive) systems will also monitor and regulate CO_2 concentration.
17. For fluorescent labeling of fixed cells, we favor 13-mm coverslips in 12-well tissue culture plates; for live cell imaging applications, we grow cells on 22-mm coverslips in 6-well plates.
18. We typically seed 13-mm coverslips with 1 mL of 0.66×10^5 cells/mL or 1 mL of 0.33×10^5 cells/mL so that after 2 and 3 days, respectively, we obtain cells with approximately 80 % confluence. Coverslips of 22 mm are seeded with 2 mL of 1×10^5 cells/mL for 2 days or 2 mL of 0.5×10^5 cells/mL for 3 days, again for 80 % confluence. On Anopore™ permeable supports, to give fully polarized monolayers after 3–4 days, we seed with 0.5 mL of 1×10^6 cells/mL.
19. This gives high infection rate, which is ideal for some applications, e.g., live cell imaging, but decreased levels can be used, which may provide more realistic infection levels and are preferable for longer term infection studies. When comparing strains, it is important to check that growth characteristics are similar to ensure infection is comparable, i.e., similar numbers of bacteria at equivalent growth state. This can be measured by determining absorbance at 600 nm or by determining colony-forming units (cfu).
20. We routinely examine time points between 5 and 60 min. Longer term infection to study the fate of bacteria after invasion is usually performed with a short-term infection (e.g., 15 min) followed by washing, addition of gentamicin at 16 μg/mL, and incubation to the time point required.
21. Alternative detergents, e.g., saponin, are used in some protocols. Permeabilization is not required if solvent-fixation is employed.
22. It is important to stress that the labeling methods described here and in **Subheading 3.3.2** are minimalist protocols used for speed and convenience. We know the labeling will be bright, and we are not overly concerned with the presence of some non-specific labeling. This basic protocol does not include any steps to quench autofluorescence or to block non-specific binding. Several alternative modifications could be employed to decrease non-specific labeling, the most

common being to include bovine serum albumin or serum from unrelated species (preferably including the same species as that in which secondary antibodies were raised) before and/or during antibody incubations. For example, we have frequently used pre-blocking with 1:4 normal horse serum as well as addition of 5 % horse or sheep serum in the antibody diluent.

23. This is just one alternative method for differentially labeling adhered and invaded *Salmonella*, an alternative is the use of methanol fixation as discussed in **Subheading 1.2.2**. As with protocol in **Subheading 3.3.1.**, this is a "quick and dirty" protocol that is adequate for our requirements (*see* **Note 22**).
24. At this stage, it is worth taking care to ensure the image is clear (has high resolution) by eye. It was easy to do this with this specimen, as we are very familiar with the staining and clarity of image that should be obtained. An alternative is to use a standard and familiar slide. Commercial slides with cells labeled with multiple fluorophores are available from Molecular Probes/Invitrogen and other companies, but we use our own pre-prepared slides for this purpose. If the image is not optimal, it is worth checking that the coverslip correction collar (if applicable) is correctly adjusted and/or cleaning the coverslip and replacing the oil, as insufficient or contaminated oil is a common cause of poor image quality.
25. At this resolution, each pixel relates to a specimen area 100 nm × 100 nm and thus fulfills Nyquist sampling criteria for TRITC and FITC at least (theoretical optima approximately 115 and 102 nm, respectively). It is also possible to "bin" pixels (pooling signal detected by arrays of pixels; e.g., 2 × 2, 3 × 3) to increase sensitivity, decrease exposure time, and limit photodamage and phototoxicity. But in this case, we use maximum camera resolution as we wish to obtain the highest possible image resolution.
26. Selecting the color "look up table" (LUT) that shows adjustable range of intensities as red (top of range) and blue (bottom of range) is useful for this adjustment of camera parameters, as it becomes instantly apparent where an image approaches 0 and 4095 (limits of 12-bit range). Exposure time will need to be adjusted if pixels exceed these limits. The avoidance of "out-of-range" pixels is most important when applying deconvolution to image stacks, as any inaccuracies in measured intensity will compromise the operation of deconvolution algorithms. If deconvolution is not used, it may sometimes be necessary to allow some features to saturate in order to adequately image dimmer ones.
27. For optimal deconvolution, this will need to be a few microns beyond the region of the specimen of interest to optimize deconvolution of that part. The images shown in **Fig. 1** are from a stack of 65 images acquired at 100-nm steps, although this is likely to be more than is required to acquire high-quality deconvolved images. Where deconvolution is not required, an image stack need only cover actual limits of cells if volume data required, or possibly just one level of focus if that is all the experiment requires.

28. The automation starts at the current point of focus and moves focus by user-defined increments, in this case by controlling Prior ProScan™II motorized focus, a cost-effective option but by no means the fastest or most precise focus drive available. Faster stepping would be required to collect 3D data in living cells where movement of fluorescent entities during collection of image stacks would compromise the data, especially if deconvolution was to be applied.
29. This software allows 3D rendering as well as constrained iterative deconvolution. Because we have had some problems deconvolving stacks of maximum-size 12-bit images (due to complexity of the data processing and limits of computing power), a region was selected that included the required part of the cell and the image stack "cropped" to that region (approximately 400 × 300 pixels).
30. For images shown in **Fig. 1**, deconvolution parameters were set at 95 % confidence (requiring 14 iterations in this case). These options have usually given good results for this type of image stack but may need adapting for different ones. Measured PSFs may give more reliable results in some cases but have not been extensively examined by us.
31. This is one of a range corrected for chromatic aberration in the blue part of the spectrum.
32. As explained in **Note 24**, it is important to optimize image quality during visual inspection. It is useful to check that there is no additional factor, such as contamination of oil or incorrect objective setting that will compromise image quality. With thicker specimens, e.g., tissues that are more likely to be examined by CLSM than WFM, it can be much harder to determine if the image is optimal because of the level of out-of-focus fluorescence, so it may be worth checking this with a pre-prepared, familiar, and clear specimen such as a standardized slide.
33. In this case, to match the pixel size of WFM images collected with equivalent lens (100 × 100 nm), the scan size was set at the default size 512 × 512 and zoom factor adjusted to 4.65 (pixel size is displayed so it is easy to adjust zoom to appropriate level interactively). The same area as imaged by WFM can then be zoomed in on, and the field of view adjusted (by panning) as necessary. Although the pixel size of 100 × 100 nm met the requirements of Nyquist approaches for WFM in theory at least, CLSM could have facilitated an increase in resolution (approximately 1.3-fold), such that this pixel size may be slightly larger than the theoretical optima. It is, however, also worth bearing in mind that increasing zoom also increases the light dose per pixel so may be accompanied by increased photobleaching, and, in the case of live cell imaging, phototoxicity.
34. The confocal pinhole can be opened if required to increase sensitivity, which is especially useful for live cell imaging. Significant increases in signal can be achieved with moderate (often hardly detectable) loss of axial resolution. It is also worth considering the light-collecting properties of the objective lens when applications require optimal signal, e.g., live cell imaging. For example, signal intensity achieved with the oil-immersion lenses available on our CLSM system

decreases with increasing magnification such that users will often prefer to use a ×40 or ×63 objective lens and zoom rather than choosing ×100 objective.

35. Depending on the characteristics of the detector, and the amplification applied, there may be additional benefits in noise reduction to be gained from averaging more than four frames. It should be taken into consideration that more averaging comes at the expense of increasing the time taken to acquire stacks, in addition to increasing the light dose received by the sample, and consequentially risk of photobleaching. For live cell applications, averaging is often limited to fewer frames or completely avoided because of considerations of phototoxicity, as well as movement-based artifacts. Most CLSM systems will now allow line averaging rather than frame averaging, which is particularly useful where movement is likely to affect information within acquired images, as the time between repeated line scans is negligible compared with the frame interval.
36. Today's CLSM systems generally have AOTFs to allow rapid and independent adjustment of illumination intensity from individual laser lines. In the case of fixed (and anti-fade mounted) specimens, the laser power is less critical than for live cell imaging, where it is often necessary to use higher PMT voltages to enable laser power, and hence photobleaching/photodamage, to be minimized. As with WFM, we use a color LUT that clearly identifies pixels with intensities 0 or 255 (extremes of 8-bit range) by displaying them as green or blue, respectively. In the case of CLSM, there is less advantage to be gained by using 12-bit imaging because, in contrast to WFM, the number of photons detected for each pixel is usually rather small. The image in **Fig. 1** was acquired using the blue diode (405 nm), Ar laser (488 nm), and green HeNe laser (543 nm) with detection set at 411–482 nm for DAPI, 497–538 nm for FITC, and 555–690 nm for TRITC.
37. Of course, it is not always necessary to obtain image stacks; sometimes a single image or at most a small number of sections are all that are required if a limited part of the cell depth contains the required information. In the case of the image shown in **Fig. 1**, the full depth amounted to an approximately 6-μm stack and step size was set at 122-nm intervals (as close as possible to 100 nm with this focus device). The narrow focus increments were selected to optimize data acquisition (Nyquist sampling) and, in this case, allowed identification of the sections closest to those obtained from the equivalent WFM stack. More often, larger step sizes are selected, e.g., 500 nm, to limit the time required to obtain images, and thereby photobleaching, as well as to limit file size.
38. This approach has distinct advantages over acquiring images simultaneously, as it usually avoids the cross-talk between fluorophore signals that is almost inevitable if all three fluorophores are excited simultaneously. For example, DAPI emission spectrum overlaps with those of both FITC and TRITC, whereas the emission of FITC is also likely to contaminate the TRITC channel; the latter usually occurs when the labeling with green fluorophore is relatively intense.

39. In this case, the image stacks were acquired using "line-by-line" sequential imaging where the AOTF is controlled to rapidly switch between excitation lines for every line scanned such that the fluorophores are imaged at millisecond intervals and appear simultaneously on the display panel. Other parameters such as PMT settings, the wavelength range detected, and pinhole size can be adjusted between fluorophore settings only in the other, slower, and sequential modes, i.e. "between frames" and "between stacks" in the Leica confocal software.
40. DIC imaging affords enhanced contrast for some samples compared with phase contrast, although in our hands we have preferred images acquired by phase contrast to monitor ruffle propagation. Automated imaging of fluorescence, e.g., GFP, alongside phase contrast is somewhat more straightforward than with DIC, as the DIC optical components dramatically reduce detected fluorescence signal intensity. This can be avoided by using a motorized filter cube to switch between DIC and fluorescence filter blocks, although there will be an additional time delay introduced into the time course using his approach. The loss of sensitivity using DIC simultaneously with fluorescence is much greater than the small loss of light transmission caused by the phase ring within phase-contrast objectives.
41. The coverslip should be placed cell side up, centrally within the chamber lid to ensure a seal is made when the base of the chamber is screwed on (*see* **Fig. 1**). To check a seal has been made, when the buffer has been added, hold the chamber with one hand and with the other, use a piece of tissue to gently trace around the edge of the hole in the lid over which the coverslip is situated. After excess liquid has been absorbed, there should be no further liquid. If there is, remove the buffer from the central well, unscrew the chamber, and place the coverslip back in its well. Dry the chamber and repeat the setup. An alternative method of positioning the coverslip is to place the coverslip cell side down onto the rubber seal of the base and place the lid gently over the top. Whilst screwing the lid on, the back of the coverslip should be gently and evenly pressed down to hold it in position.
42. Live cell imaging automations can be customized to switch between, for example, GFP and phase-contrast imaging. This can be used to examine the dynamics of intracellular GFP-tagged probes or GFP-expressing bacteria alongside morphological responses to infection (e.g., ***85***). Automated shuttering of the light source is beneficial to decrease likelihood of photobleaching or phototoxicity compromising data acquisition. It is also normally advisable to select cells expressing low or medium levels of GFP-tagged protein as high levels of expression frequently interfere with normal cellular processes.
43. In order to compare one strain or growth condition with another, it is necessary for the cultures to have been incubated for the same length of time. Therefore, we set up cultures at 45-min intervals, and subsequently the imaging of each culture is timed to allow 20- to 30-min time courses to be acquired while also allowing time to set the system up for capturing the next time course.

Acknowledgments

We thank Alan Leard for assistance with microscopy, image processing, and protocols, and Alice Jarvie for images in **Fig. 1**. We also acknowledge all our colleagues who have contributed to the studies of *Salmonella* infection and the development of microscopical techniques discussed here. Although we have referred to imaging equipment and consumables that we have used in these studies, we do not intend to imply that these are superior to others available that we have not used. Inclusion of alternative sources of equipment and consumables is not exhaustive and also not intended as direct recommendation. Work in this laboratory is supported by Biotechnology and Biological Sciences Research Council (BBSRC) and Medical Research Council (MRC). The University of Bristol Cell Imaging Facility was established with funding from MRC.

References

1. Finlay, B. B., Ruschkowski, S., and Dedhar, S. (1991) Cytoskeletal rearrangements accompanying *Salmonella* entry into epithelial cells. *J. Cell Sci.* **99** (Pt 2), 283–296.
2. Galan, J. E. (2001) *Salmonella* interactions with host cells: type III secretion at work. *Annu. Rev. Cell Dev. Biol.* **17,** 53–86.
3. Clark, M. A., Hirst, B. H., and Jepson, M. A. (1998) Inoculum composition and *Salmonella* pathogenicity island 1 regulate M-cell invasion and epithelial destruction by *Salmonella typhimurium. Infect. Immun.* **66,** 724–731.
4. Clark, M. A., Jepson, M. A., Simmons, N. L., and Hirst, B. H. (1994) Preferential interaction of *Salmonella typhimurium* with mouse Peyer's patch M cells. *Res. Microbiol.* **145,** 543–552.
5. Jepson, M. A. and Clark, M. A. (2001) The role of M cells in *Salmonella* infection. *Microbes Infect.* **3,** 1183–1190.
6. Jones, B. D., Ghori, N., and Falkow, S. (1994) *Salmonella typhimurium* initiates murine infection by penetrating and destroying the specialized epithelial M cells of the Peyer's patches. *J. Exp. Med.* **180,** 15–23.
7. Waterman, S. R. and Holden, D. W. (2003) Functions and effectors of the *Salmonella* pathogenicity island 2 type III secretion system. *Cell Microbiol.* **5,** 501–511.
8. Patel, J. C. and Galan, J. E. (2005) Manipulation of the host actin cytoskeleton by *Salmonella* – all in the name of entry. *Curr. Opin. Microbiol.* **8,** 10–15.
9. Finlay, B. B. and Brumell, J. H. (2000) *Salmonella* interactions with host cells: *in vitro* to *in vivo. Philos. Trans. R. Soc. Lond. B Biol. Sci.* **355,** 623–631.
10. Hueck, C. J. (1998) Type III protein secretion systems in bacterial pathogens of animals and plants. *Microbiol. Mol. Biol. Rev.* **62,** 379–433.
11. Galan, J. E. and Collmer, A. (1999) Type III secretion machines: bacterial devices for protein delivery into host cells. *Science* **284,** 1322–1328.

12. Hueck, C. J., Hantman, M. J., Bajaj, V., Johnston, C., Lee, C. A., and Miller, S. I. (1995) *Salmonella typhimurium* secreted invasion determinants are homologous to *Shigella* Ipa proteins. *Mol. Microbiol.* **18,** 479–490.
13. Shea, J. E., Hensel, M., Gleeson, C., and Holden, D. W. (1996) Identification of a virulence locus encoding a second type III secretion system in *Salmonella typhimurium. Proc. Natl. Acad. Sci. USA* **93,** 2593–2597.
14. Zhou, D., Mooseker, M. S., and Galan, J. E. (1999) Role of the *S. typhimurium* actin-binding protein SipA in bacterial internalization. *Science* **283,** 2092–2095.
15. Jepson, M. A., Kenny, B., and Leard, A. D. (2001) Role of sipA in the early stages of *Salmonella typhimurium* entry into epithelial cells. *Cell Microbiol.* **3,** 417–426.
16. Galan, J. E. and Fu, Y. (2000) Modulation of actin cytoskeleton by *Salmonella* GTPase activating protein SptP. *Methods Enzymol.* **325,** 496–504.
17. Terebiznik, M. R., Vieira, O. V., Marcus, S. L., Slade, A., Yip, C. M., Trimble, W. S., Meyer, T., Finlay, B. B., and Grinstein, S. (2002) Elimination of host cell PtdIns(4,5)P(2) by bacterial SigD promotes membrane fission during invasion by *Salmonella. Nat. Cell Biol.* **4,** 766–773.
18. Hernandez, L. D., Hueffer, K., Wenk, M. R., and Galan, J. E. (2004) *Salmonella* modulates vesicular traffic by altering phosphoinositide metabolism. *Science* **304,** 1805–1807.
19. Norris, F. A., Wilson, M. P., Wallis, T. S., Galyov, E. E., and Majerus, P. W. (1998) SopB, a protein required for virulence of *Salmonella dublin*, is an inositol phosphate phosphatase. *Proc. Natl. Acad. Sci. USA* **95,** 14057–14059.
20. Bertelsen, L. S., Paesold, G., Marcus, S. L., Finlay, B. B., Eckmann, L., and Barrett, K. E. (2004) Modulation of chloride secretory responses and barrier function of intestinal epithelial cells by the *Salmonella* effector protein SigD. *Am. J. Physiol. Cell Physiol.* **287,** C939–948.
21. Finlay, B. B., Gumbiner, B., and Falkow, S. (1988) Penetration of *Salmonella* through a polarized Madin-Darby canine kidney epithelial cell monolayer. *J. Cell Biol.* **107,** 221–230.
22. Jepson, M. A., Lang, T. F., Reed, K. A., and Simmons, N. L. (1996) Evidence for a rapid, direct effect on epithelial monolayer integrity and transepithelial transport in response to *Salmonella* invasion. *Pflugers Arch.* **432,** 225–233.
23. Jepson, M. A., Collares-Buzato, C. B., Clark, M. A., Hirst, B. H., and Simmons, N. L. (1995) Rapid disruption of epithelial barrier function by *Salmonella typhimurium* is associated with structural modification of intercellular junctions. *Infect. Immun.* **63,** 356–359.
24. Jepson, M. A., Schlecht, H. B., and Collares-Buzato, C. B. (2000) Localization of dysfunctional tight junctions in *Salmonella enterica* serovar typhimurium-infected epithelial layers. *Infect. Immun.* **68,** 7202–7208.
25. Tafazoli, F., Magnusson, K. E., and Zheng, L. (2003) Disruption of epithelial barrier integrity by *Salmonella enterica* serovar typhimurium requires geranylgeranylated proteins. *Infect. Immun.* **71,** 872–881.

26. Elewaut, D., DiDonato, J. A., Kim, J. M., Truong, F., Eckmann, L., and Kagnoff, M. F. (1999) NF-kappa B is a central regulator of the intestinal epithelial cell innate immune response induced by infection with enteroinvasive bacteria. *J. Immunol.* **163,** 1457–1466.
27. Hobbie, S., Chen, L. M., Davis, R. J., and Galan, J. E. (1997) Involvement of mitogen-activated protein kinase pathways in the nuclear responses and cytokine production induced by *Salmonella typhimurium* in cultured intestinal epithelial cells. *J. Immunol.* **159,** 5550–5559.
28. Gewirtz, A. T., Simon, P. O., Jr., Schmitt, C. K., Taylor, L. J., Hagedorn, C. H., O'Brien, A. D., Neish, A. S., and Madara, J. L. (2001) *Salmonella typhimurium* translocates flagellin across intestinal epithelia, inducing a proinflammatory response. *J. Clin. Invest.* **107,** 99–109.
29. Hobert, M. E., Sands, K. A., Mrsny, R. J., and Madara, J. L. (2002) Cdc42 and Rac1 regulate late events in *Salmonella typhimurium*-induced interleukin-8 secretion from polarized epithelial cells. *J. Biol. Chem.* **277,** 51025–51032.
30. Hensel, M., Shea, J. E., Gleeson, C., Jones, M. D., Dalton, E., and Holden, D. W. (1995) Simultaneous identification of bacterial virulence genes by negative selection. *Science* **269,** 400–403.
31. Brown, N. F., Vallance, B. A., Coombes, B. K., Valdez, Y., Coburn, B. A., and Finlay, B. B. (2005) *Salmonella* pathogenicity island 2 is expressed prior to penetrating the Intestine. *PLoS Pathog.* **1,** e32.
32. Coburn, B., Li, Y., Owen, D., Vallance, B. A., and Finlay, B. B. (2005) *Salmonella enterica* serovar Typhimurium pathogenicity island 2 is necessary for complete virulence in a mouse model of infectious enterocolitis. *Infect. Immun.* **73,** 3219–3227.
33. Coombes, B. K., Coburn, B. A., Potter, A. A., et al. (2005) Analysis of the contribution of *Salmonella* pathogenicity islands 1 and 2 to enteric disease progression using a novel bovine ileal loop model and a murine model of infectious enterocolitis. *Infect. Immun.* **73,** 7161–7169.
34. Kubori, T., Matsushima, Y., Nakamura, D., Uralil, J., Lara-Tejero, M., Sukhan, A., Galan, J. E., and Aizawa, S. I. (1998) Supramolecular structure of the *Salmonella typhimurium* type III protein secretion system. *Science* **280,** 602–605.
35. Kubori, T., Sukhan, A., Aizawa, S. I., and Galan, J. E. (2000) Molecular characterization and assembly of the needle complex of the *Salmonella typhimurium* type III protein secretion system. *Proc. Natl. Acad. Sci. USA* **97,** 10225–10230.
36. Daniell, S. J., Kocsis, E., Morris, E., Knutton, S., Booy, F. P., and Frankel, G. (2003) 3D structure of EspA filaments from enteropathogenic *Escherichia coli. Mol. Microbiol.* **49,** 301–308.
37. Ramboarina, S., Fernandes, P. J., Daniell, S., Islam, S., Simpson, P., Frankel, G., Booy, F., Donnenbery, M. S., and Matthews, S. (2005) Structure of the bundle-forming pilus from enteropathogenic *Escherichia coli. J. Biol. Chem.* **280,** 40252–40260.

38. Baumler, A. J. and Heffron, F. (1995) Identification and sequence analysis of *lpfABCDE*, a putative fimbrial operon of *Salmonella typhimurium. J. Bacteriol.* **177,** 2087–2097.
39. Reed, K. A., Clark, M. A., Booth, T. A., et al. (1998) Cell-contact-stimulated formation of filamentous appendages by *Salmonella typhimurium* does not depend on the type III secretion system encoded by *Salmonella* pathogenicity island 1. *Infect. Immun.* **66,** 2007–2017.
40. Ginocchio, C. C., Olmsted, S. B., Wells, C. L., and Galan, J. E. (1994) Contact with epithelial cells induces the formation of surface appendages on *Salmonella typhimurium. Cell* **76,** 717–724.
41. Dufrene, Y. F. (2002) Atomic force microscopy, a powerful tool in microbiology. *J. Bacteriol.* **184,** 5205–5213.
42. Wang, H. W., Chen, Y., Yang, H., et al. (2003) Ring-like pore structures of SecA: implication for bacterial protein-conducting channels. *Proc. Natl. Acad. Sci. USA* **100,** 4221–4226.
43. Ide, T., Laarmann, S., Greune, L., Schillers, H., Oberleithner, H., and Schmidt, M. A. (2001) Characterization of translocation pores inserted into plasma membranes by type III-secreted Esp proteins of enteropathogenic *Escherichia coli. Cell Microbiol.* **3,** 669–679.
44. Knutton, S. (2003) Microscopic methods to study STEC. Analysis of the attaching and effacing process. *Methods Mol. Med.* **73,** 137–149.
45. Van Putten, J. P., Weel, J. F., and Grassme, H. U. (1994) Measurements of invasion by antibody labelling and electron microscopy. *Methods Enzymol.* **236,** 420–437.
46. Nunez, M. E., Martin, M. O., Chan, P. H., Duong, L. K., Sindhurakar, A. R., and Spain, E. M. (2005) Atomic force microscopy of bacterial communities. *Methods Enzymol.* **397,** 256–268.
47. Decho, A. W. and Kawaguchi, T. (1999) Confocal imaging of *in situ* natural microbial communities and their extracellular polymeric secretions using Nanoplast resin. *Biotechniques* **27,** 1246–1252.
48. Bryers, J. D. (2001) Two-photon excitation microscopy for analyses of biofilm processes. *Methods Enzymol.* **337,** 259–269.
49. Neu, T. R., Kuhlicke, U., and Lawrence, J. R. (2002) Assessment of fluorochromes for two-photon laser scanning microscopy of biofilms. *Appl. Environ. Microbiol.* **68,** 901–909.
50. Emerson, R. J. and Camesano, T. A. (2004) Nanoscale investigation of pathogenic microbial adhesion to a biomaterial. *Appl. Environ. Microbiol.* **70,** 6012–6022.
51. Stephens, D. J. and Allan, V. J. (2003) Light microscopy techniques for live cell imaging. *Science* **300,** 82–86.
52. White, N. S. and Errington, R. J. (2005) Fluorescence techniques for drug delivery research: theory and practice. *Adv. Drug Deliv. Rev.* **57,** 17–42.

53. Buda, A., Sands, C., and Jepson, M. A. (2005) Use of fluorescence imaging to investigate the structure and function of intestinal M cells. *Adv. Drug Deliv. Rev.* **57,** 123–134.
54. Jepson, M. A. (2006) Confocal or wide-field? A guide to selecting appropriate methods for cell imaging, in *Methods Express: Cell Imaging*, Scion Publishing Ltd (UK), pp. 17–48.
55. Pawley, J. (ed.) (1995) *Handbook of Biological Confocal Microscopy*. 2nd edn, Plenum Press New York.
56. Sheppard, C. and Shotton, D. (1997) *Confocal Laser Scanning Microscopy*, Bios Scientific Publishers, Oxford, UK.
57. Shaner, N. C., Sanger, J. W., and Sanger, J. M. (2005) Actin and alpha-actinin dynamics in the adhesion and motility of EPEC and EHEC on host cells. *Cell Motil. Cytoskeleton* **60,** 104–120.
58. Swedlow, J. R., Hu, K., Andrews, P. D., Roos, D. S., and Murray, J. M. (2002) Measuring tubulin content in *Toxoplasma gondii*: a comparison of laser-scanning confocal and wide-field fluorescence microscopy. *Proc. Natl. Acad. Sci. USA* **99,** 2014–2019.
59. Takeuchi, A. (1967) Electron microscope studies of experimental *Salmonella* infection. I. Penetration into the intestinal epithelium by *Salmonella typhimurium. Am. J. Pathol.* **50,** 109–136.
60. Francis, C. L., Starnbach, M. N., and Falkow, S. (1992) Morphological and cytoskeletal changes in epithelial cells occur immediately upon interaction with *Salmonella typhimurium* grown under low-oxygen conditions. *Mol. Microbiol.* **6,** 3077–3087.
61. Jepson, M. A. and Clark, M. A. (1998) Studying M cells and their role in infection. *Trends Microbiol.* **6,** 359–365.
62. Clark, M. A., Reed, K. A., Lodge, J., Stephen, J., Hirst, B. H., and Jepson, M. A. (1996) Invasion of murine intestinal M cells by *Salmonella typhimurium inv* mutants severely deficient for invasion of cultured cells. *Infect. Immun.* **64,** 4363–4368.
63. Monaghan, P., Watson, P. R., Cook, H., Scott, L., Wallis, T. S., and Robertson, D. (2001) An improved method for preparing thick sections for immuno/histochemistry and confocal microscopy and its use to identify rare events. *J. Microsc.* **203,** 223–226.
64. Richter-Dahlfors, A., Buchan, A. M., and Finlay, B. B. (1997) Murine salmonellosis studied by confocal microscopy: *Salmonella typhimurium* resides intracellularly inside macrophages and exerts a cytotoxic effect on phagocytes in vivo. *J. Exp. Med.* **186,** 569–580.
65. Salcedo, S. P., Noursadeghi, M., Cohen, J., and Holden, D. W. (2001) Intracellular replication of *Salmonella typhimurium* strains in specific subsets of splenic macrophages in vivo. *Cell Microbiol.* **3,** 587–597.
66. Salcedo, S. P. and Holden, D. W. (2003) SseG, a virulence protein that targets *Salmonella* to the Golgi network. *EMBO J.* **22,** 5003–5014.

67. La Ragione, R. M., Cooley, W. A., Velge, P., Jepson, M. A., and Woodward, M. J. (2003) Membrane ruffling and invasion of human and avian cell lines is reduced for aflagellate mutants of *Salmonella enterica* serotype Enteritidis. *Int. J. Med. Microbiol.* **293,** 261–272.
68. Jepson, M. A., Pellegrin, S., Peto, L., et al. (2003) Synergistic roles for the Map and Tir effector molecules in mediating uptake of enteropathogenic *Escherichia coli* (EPEC) into non-phagocytic cells. *Cell Microbiol.* **5,** 773–783.
69. Hardt, W. D., Chen, L. M., Schuebel, K. E., Bustelo, X. R., and Galan, J. E. (1998) *S. typhimurium* encodes an activator of Rho GTPases that induces membrane ruffling and nuclear responses in host cells. *Cell* **93,** 815–826.
70. Baumler, A. J., Tsolis, R. M., and Heffron, F. (1996) The *lpf* fimbrial operon mediates adhesion of *Salmonella typhimurium* to murine Peyer's patches. *Proc. Natl. Acad. Sci. USA* **93,** 279–283.
71. Freese, H. M., Karsten, U., and Schumann, R. (2006) Bacterial abundance, activity, and viability in the eutrophic River Warnow, northeast Germany. *Microb. Ecol.* **51**, 117–127
72. Knodler, L. A., Bestor, A., Ma, C., et al. (2005) Cloning vectors and fluorescent proteins can significantly inhibit *Salmonella enterica* virulence in both epithelial cells and macrophages: implications for bacterial pathogenesis studies. *Infect. Immun.* **73,** 7027–7031.
73. Wendland, M. and Bumann, D. (2002) Optimization of GFP levels for analyzing *Salmonella* gene expression during an infection. *FEBS Lett.* **521,** 105–108.
74. Hautefort, I., Proenca, M. J., and Hinton, J. C. (2003) Single-copy green fluorescent protein gene fusions allow accurate measurement of *Salmonella* gene expression in vitro and during infection of mammalian cells. *Appl. Environ. Microbiol.* **69,** 7480–7491.
75. Valdivia, R. H. and Falkow, S. (1996) Bacterial genetics by flow cytometry: rapid isolation of *Salmonella typhimurium* acid-inducible promoters by differential fluorescence induction. *Mol. Microbiol.* **22,** 367–378.
76. Bumann, D. (2002) Examination of *Salmonella* gene expression in an infected mammalian host using the green fluorescent protein and two-colour flow cytometry. *Mol. Microbiol.* **43,** 1269–1283.
77. Buchmeier, N. A. and Libby, S. J. (1997) Dynamics of growth and death within a *Salmonella typhimurium* population during infection of macrophages. *Can. J. Microbiol.* **43,** 29–34.
78. Francis, C. L., Ryan, T. A., Jones, B. D., Smith, S. J., and Falkow, S. (1993) Ruffles induced by *Salmonella* and other stimuli direct macropinocytosis of bacteria. *Nature* **364,** 639–642.
79. Reed, K. A., Booth, T. A., Hirst, B. H., and Jepson, M. A. (1996) Promotion of *Salmonella typhimurium* adherence and membrane ruffling in MDCK epithelia by staurosporine. *FEMS Microbiol. Lett.* **145,** 233–238.

80. Criss, A. K., Ahlgren, D. M., Jou, T. S., McCormick, B. A., and Casanova, J. E. (2001) The GTPase Rac1 selectively regulates *Salmonella* invasion at the apical plasma membrane of polarized epithelial cells. *J. Cell Sci.* **114,** 1331–1341.
81. Raffatellu, M., Wilson, R. P., Chessa, D., Andrews-Polymenis, H., Tran, Q. T., Lawhon, S., Khare, S., Adams, L. G., and Baumler, A. J. (2005) SipA, SopA, SopB, SopD, and SopE2 contribute to *Salmonella enterica* serotype typhimurium invasion of epithelial cells. *Infect. Immun.* **73,** 146–154.
82. Clark, M. A., Hirst, B. H., and Jepson, M. A. (1998) M-cell surface beta1 integrin expression and invasin-mediated targeting of *Yersinia pseudotuberculosis* to mouse Peyer's patch M cells. *Infect. Immun.* **66,** 1237–1243.
83. Jepson, M. A., Clark, M. A., Simmons, N. L., and Hirst, B. H. (1993) Actin accumulation at sites of attachment of indigenous apathogenic segmented filamentous bacteria to mouse ileal epithelial cells. *Infect. Immun.* **61,** 4001–4004.
84. Cain, R. J., Hayward, R. D., and Koronakis, V. (2004) The target cell plasma membrane is a critical interface for *Salmonella* cell entry effector-host interplay. *Mol. Microbiol.* **54,** 887–904.
85. Pattni, K., Jepson, M., Stenmark, H., and Banting, G. (2001) A PtdIns(3)P-specific probe cycles on and off host cell membranes during *Salmonella* invasion of mammalian cells. *Curr. Biol.* **11,** 1636–1642.
86. Brumell, J. H., Tang, P., Mills, S. D., and Finlay, B. B. (2001) Characterization of *Salmonella*-induced filaments (Sifs) reveals a delayed interaction between *Salmonella*-containing vacuoles and late endocytic compartments. *Traffic* **2,** 643–653.
87. Meresse, S., Unsworth, K. E., Habermann, A., Griffiths, G., Fang, F., Martinez-Lorenzo, M. J., Waterman, S. R., Gorvel, J. P., and Holden, D. W. (2001) Remodelling of the actin cytoskeleton is essential for replication of intravacuolar *Salmonella. Cell Microbiol.* **3,** 567–577.
88. Unsworth, K. E., Way, M., McNiven, M., Machesky, L., and Holden, D. W. (2004) Analysis of the mechanisms of *Salmonella*-induced actin assembly during invasion of host cells and intracellular replication. *Cell Microbiol.* **6,** 1041–1055.
89. Kimbrough, T. G. and Miller, S. I. (2000) Contribution of *Salmonella typhimurium* type III secretion components to needle complex formation. *Proc. Natl. Acad. Sci. USA* **97,** 11008–11013.
90. Brumell, J. H., Kujat-Choy, S., Brown, N. F., Vallance, B. A., Knodler, L. A., and Finlay, B. B. (2003) SopD2 is a novel type III secreted effector of *Salmonella typhimurium* that targets late endocytic compartments upon delivery into host cells. *Traffic* **4,** 36–48.
91. Brumell, J. H., Goosney, D. L., and Finlay, B. B. (2002) SifA, a type III secreted effector of *Salmonella typhimurium*, directs *Salmonella*-induced filament (Sif) formation along microtubules. *Traffic* **3,** 407–415.

92. Schlumberger, M. C., Muller, A. J., Ehrbar, K., et al. (2005) Real-time imaging of type III secretion: *Salmonella* SipA injection into host cells. *Proc. Natl. Acad. Sci. USA* **102,** 12548–12553.
93. Enninga, J., Mounier, J., Sansonetti, P., and Tran Van Nhieu, G. (2005) Secretion of type III effectors into host cells in real time. *Nat. Methods* **2,** 959–965.
94. Charpentier, X. and Oswald, E. (2004) Identification of the secretion and translocation domain of the enteropathogenic and enterohemorrhagic *Escherichia coli* effector Cif, using TEM-1 beta-lactamase as a new fluorescence-based reporter. *J. Bacteriol.* **186,** 5486–5495.

13

Analysis of Kinesin Accumulation on *Salmonella*-Containing Vacuoles

Audrey Dumont*, Nina Schroeder*, Jean-Pierre Gorvel, and Stéphane Méresse

Summary

Salmonella enterica is an intracellular bacterial pathogen that causes gastroenteritis and typhoid fever. Inside host cells, the bacterium is enclosed in a membrane bound compartment, the *Salmonella*-containing vacuole (SCV). Intracellular replication of *Salmonella* requires the translocation of effector proteins into the host cytosol. The SifA effector protein is important for the membrane stability of the SCV. Recently, we have shown that the *Salmonella sifA*$^-$ mutant presents on its vacuole an important accumulation of kinesin-1, a molecular motor involved in the plus-end-directed transport of various organelles. Kinesin-1 is not recruited on SCVs of mutants that do not translocate effector proteins. This indicates that SifA is a negative regulator of the recruitment of this molecular motor and reveals the existence of another effector that recruits kinesin-1. This chapter describes techniques that are used to screen by immunofluorescence microscopy the accumulation of kinesin-1 on strains of *Salmonella* carrying multiple mutations.

Key Words: *Salmonella typhimurium*; *Salmonella* mutant; *Salmonella*-containing vacuole; kinesin-1; immunolabeling.

*These authors contributed equally to this work.

From: *Methods in Molecular Biology, vol. 394: Salmonella: Methods and Protocols*
Edited by: H. Schatten and A. Eisenstark © Humana Press Inc., Totowa, NJ

1. Introduction

Salmonella enterica serovar Typhimurium (*S. typhimurium*) is a facultative intracellular bacterial pathogen responsible for gastroenteritis in humans and for a typhoid-like illness in certain mouse strains. *S. typhimurium* expresses a type III secretion system (TTSS-1) that translocates effector proteins, which enable the bacteria to invade epithelial cells by modulating the actin cytoskeleton ***(1)***. After its internalization, *S. typhimurium* resides in a compartment called the *Salmonella*-containing vacuole (SCV) ***(2)***. Both intracellular replication and bacterial virulence require the expression of a second TTSS (TTSS-2) and the translocation of its effectors inside the host cell cytosol ***(3)***.

De novo formed SCVs rapidly acquire early endosomal markers such EEA1 and transferrin receptor ***(4)***. These proteins are gradually removed and replaced by lysosomal membrane glycoproteins such as LAMP-1 ***(5–7)***. Infected cells are characterized by the enrichment of SCVs in lysosomal membrane glycoproteins and by the formation of tubular extensions of the vacuole called *Salmonella*-induced filaments ***(8)***. The formation of these tubular structures occurs along microtubules and requires both rab7 and the expression of the TTSS-2 effector SifA ***(9–11)***.

Microtubules use two families of molecular motors to drive transport in cells. Cytoplasmic dynein is the predominant minus-end-directed motor in higher eukaryotic cells, and kinesins are involved in the plus-end-directed transport of various organelles including mitochondria, lysosomes ***(12)***, the endoplasmic reticulum, and the Golgi complex ***(13,14)***. It has been reported that intracellular *S. typhimurium* causes a high accumulation of microtubules around the bacterial microcolonies ***(14)***. This microtubule bundling is accompanied by the recruitment of dynein and kinesin, and the inhibition of either molecular motor prevents bacterial replication ***(14)***.

The *S. typhimurium* mutant strain lacking the SifA effector looses its vacuolar membrane ***(6)***, indicating that SifA is involved in the dynamics of the SCV membrane. In addition, it has been shown that the *sifA*$^-$ mutant localizes to the cell periphery ***(15)***, in contrast to wild-type *Salmonella* that replicates in the perinuclear region of the cell. In view of the *sifA*$^-$ mutant phenotype and owing to the involvement of microtubules in the formation of *Salmonella*-induced filaments and bacterial replication, the question about the role of the microtubule molecular motors in the intracellular localization of the *sifA*$^-$ mutant arose.

The labeling of kinesin-1 after infection with a *Salmonella* strain lacking the SifA effector enabled to observe an unusually high accumulation of kinesin around the bacteria ***(15,16)***. However, this accumulation was not observed when infecting with a *Salmonella* strain that does not deliver TTSS-2 effectors in host cell. These results indicate that SifA is a negative regulator of the kinesin accumulation around the bacteria and that *S. typhimurium* translocates another effector, which is responsible for the recruitment of kinesin-1 on the SCV.

2. Materials

2.1. Preparation of Salmonella *Mutants*

2.1.1. Gene Disruption

1. Plamids pKD46 and pKD4 or pKD3 ***(16)***.
2. Primers for the gene you wish to disrupt.
3. Taq polymerase kit from Invitrogen.
4. Gel extraction kit from Qiagen.
5. Agarose from Invitrogen.
6. Luria–Bertani (LB) broth, Miller (Difco/BRL).
7. Ampicillin stock solution (100 mg/mL).
8. Arabinose from Sigma.
9. Deionized, sterile water.
10. Super optimal broth catabolite repression (SOC) medium (1 L): mix 20 g bacto-tryptone, 20 g bacto-yeast extract, 0.5 g NaCl, 2.5 mL KCL (1 M solution), and 1 L ddH_2O. Adjust pH to 7, autoclave to sterilize, and add 20 mL sterile glucose (1 M) before use.
11. Electrocompetent *Salmonella*.

2.1.2. P22 Transduction

1. LB medium.
2. Transducting broth: Prepare LB with E salt 1× and 0.2% glycerol.
3. E salts 50×: Mix 300 g citric acid monohydrate, 14.1 g $MgSO_4$, 1965 g $K_2HPO_4 \cdot 3H_2O$, and 525 g $NaNH_4HPO_4 \cdot 4H_2O$ in 1 L water. Be careful to add the different substances in the order dictated above.
4. P_{22} ***(17)***.
5. Chloroform.
6. LB-agar: LB medium supplemented with bacto-agar from Beckton-Dickinson.
7. Soft agar: Dilute one volume of LB-agar with three volumes of LB.
8. Antibiotics to select the transformed bacteria: stock of kanamycin solution (100 mg/mL) or chloramphenicol solution (100 mg/mL).
9. Ethylene glycol tetraacetic acid (EGTA) from Sigma.
10. Green plates: Prepare 1.3 g Aniline Blue and 12.4 g Alizarin Yellow in 200 mL water. Combine and autoclave with 8 g bacto-tryptone, 1 g yeast extract, 5 g NaCl, and 15 g LB-agar.

2.2. Cell Culture

1. HeLa cells (CCL-2); epithelial cells derived from human adenocarcinoma.
2. Cell culture medium: Dulbecco's Modified Eagle's Medium (DMEM) from Gibco/BRL, Invitrogen, supplemented with 10% fetal calf serum (FCS), 2 mM L-Glutamine from Gibco/BRL, and 1× nonessential amino acid from Gibco/BRL.
3. Trypsin–ethylediamine tetraacetic acid (EDTA) 1× in Hank's balanced salt solution (HBSS) from Gibco/BRL.
4. Phosphate buffered saline (PBS) 10× from Gibco/BRL. Prepare working solution by dilution of one volume with nine volumes of water.

2.3. Infection of Cells

1. LB medium supplemented with adapted antibiotics.
2. Earle's balanced salt solution (EBSS) from Gibco/BRL.
3. Gentamycin stock solution (10 mg/mL) from Sigma.
4. Cell culture medium.

2.4. Immunofluorescence

1. Microscope coverslips of 12 mm diameter from Marienfeld GmbH.
2. PBS.
3. Paraformaldehyde (PFA) 3% solution (100 mL): Heat 70 mL water to 70 °C. Add 10 mL 10× PBS and 3 g PFA (from TAAB laboratories). Stir until dissolved (use a stirring hot-plate in a fume hood). Add 100 μL 100 mM $CaCl_2$ and 100 μL 100 mM $MgCl_2$ while stirring. Allow to cool and fill up to 100 mL with water. Store at –20 °C in aliquots.
4. Solution of 1 M ammonium chloride (NH_4Cl).
5. Saponin from Sigma: 20% stock solution in water.
6. Whatmann paper (3 MM).
7. Parafilm.
8. Secondary antibodies from Jackson Immuno Research.
9. Horse serum from Sigma.
10. Solution of Mowiol 4-88 from Calbochiem (*see* **Note 1**).
11. Microscope slides (76 × 26 mm) from Menzel GmbH.

2.5. Absorption of Antibodies

1. Wild-type strain of *Salmonella* 12023.
2. 10 mM NaCl.
3. 10 mM KCl.
4. Acetone.
5. Sodium azide (NaN_3).

3. Methods

To analyze the involvement of translocated bacterial effectors in the modulation of the recruitment of kinesin-1, different *Salmonella* TTSS-2 effector mutants are prepared by the technique of gene disruption ***(16)***. However, because of the dominant activity of SifA on the negative regulation of kinesin-1 recruitment, the phenotype of these TTSS-2 effector mutants can only be observed on a *sifA*$^-$ genetic background. The absence of kinesin-1 around double mutant bacteria would demonstrate the involvement of this second effector in the recruitment of kinesin (*see* **Fig. 1**). To create *Salmonella* double-mutant strains, we used a technique based on the transduction by a lambda red bacteriophage to transfer mutations from strains to strains. After infection of epithelial cells with these different double mutants, the recruitment of kinesin-1 is analyzed by immunolabeling. However, because antibodies have the tendency to cross-react with bacterial epitopes, it is often necessary to

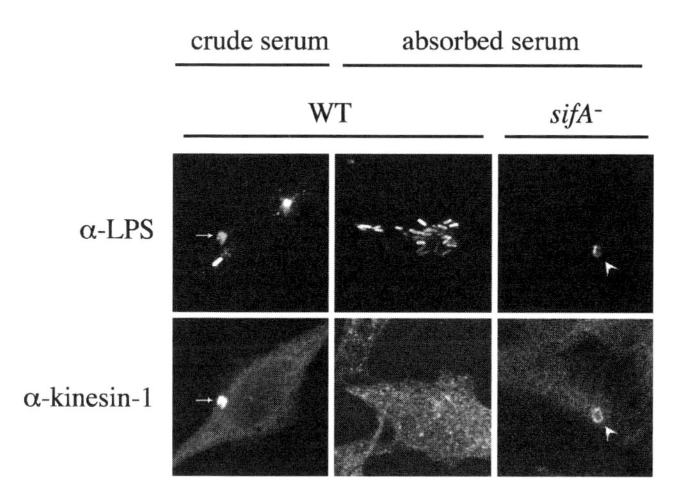

Fig. 1. Kinesin-1 immunolabeling of infected HeLa cells using crude rabbit serum or a serum previously absorbed with *Salmonella* acetone powder. HeLa cells were infected for 12 h with wild-type *Salmonella typhimurium* or a *sifA*$^-$ mutant strain and immunostained for lipopolysaccharide (LPS) and kinesin-1. Kinesin-1 was labeled using either absorbed or crude serum. A high accumulation of kinesin-1 was observed using purified serum around *sifA*$^-$ mutant bacteria (arrowhead) but not around wild-type *S. typhimurium*, indicating that the SifA effector is a negative regulator of the recruitment of this molecular motor. An unspecific labeling of the bacteria was often observed when using crude anti-kinesin-1 serum (arrow) due to the presence of contaminating anti-enterobacteria antibodies.

absorb them. This can be performed using acetone powder of *Salmonella* cells. Absorbed antibodies are then used to immunolabel infected cells.

3.1. Preparation of Salmonella Mutants

Salmonella mutants are obtained using the gene disruption technique developed by Datsenko and Wanner ***(16)***. This method requires a PCR amplification of an antibiotic cassette and a part of the gene of interest to disrupt using primers with extensions that are homologous to regions adjacent to the gene to be inactivated. The plasmid (pDK4) carrying a kanamycin resistance gene flanked by the flippase recombinase (FLT) recognition target (FRT) sites is used as template. The PCR product is then transformed in a *Salmonella* strain containing a plasmid expressing the lambda red recombinase, which is under control of an arabinose inducible promoter. The red recombinase mediates the recombination in the flanking homology regions of the target gene.

It is first necessary to prepare electrocompetent strain of wild-type *Salmonella* to introduce the plasmid expressing the lambda red recombinase.

3.1.1. Preparation of Electrocompetent Bacteria

1. Wild-type *Salmonella* are grown overnight in 10 mL LB at 37 °C under vigorous shaking. Bacteria are diluted 1:100 in 400 mL LB and incubated under the same conditions until the optical density (OD) reaches a value of 0.8 at 600 nm.
2. Bacteria are pelleted at 7500×g for 20 min at 4 °C. They are washed twice with 50 mL ice-cold water. In the last washing step, the bacteria are transferred in a 50 mL Falcon and pelleted. This pellet is resuspended in 800 μL cold water supplemented with 10% glycerol, aliquoted (50 μL), and stored at –80 °C (*see* **Note 2**).

3.1.2. Gene Disruption

1. Wild-type *Salmonella* (50 μL) are electroporated in a 0.2-cm chilled electroporation cuvette with the pKD46 plasmid (about 100 ng) containing the lambda recombinase and ampicillin resistance genes. A pulse at 25 μF/2.5 kV/200 ohm is applied. After electroporation, 350 μL warm SOC medium is added and the bacteria are incubated for 1 h at 37 °C. Then, bacteria are spread on LB-agar plates containing 100 μg/mL ampicillin, and these are incubated overnight at 37 °C.
2. The primers used for the gene disruption are designed to contain about 40 nucleotides that hybridize to the sequence of the gene you want to disrupt followed by the priming sites of the kanamycin resistance gene of the pKD4 plasmid. To disrupt SifA effector, we used the following primers: GCGCCCGCAGTTGAGATAAAAAGGGTCGATT-TAATCGTGTAGGCTGGAGCTGCTTC and for the reverse: GCCTGGCAA-GAGGTTACTCAGTAGGCAAACAGGAAGCATATGAATATCCTCCTTAG.

3. A PCR with about 50 ng pKD4 plasmid as template using the Taq DNA polymerase is performed according the supplier's protocol using the following program.
   5 min at 94 °C.
   Denaturation: 1 min at 94 °C.
   Annealing: 1 min at 55 °C (*see* **Note 3**).
   Extension: 2 min at 72 °C.
   Cycle 30 times back to **step 2**.
   10 min at 72 °C.

   This should amplify a fragment of about 1550 bp for the kanamycin cassette. To get enough PCR product to do the disruption of the target gene, four to six tubes of 50 μL PCR reactions are performed.
4. To remove the plasmid after amplification of the kanamycin cassette, the PCR is run on a 1% agarose gel and the 1550-bp amplified fragment is purified according to the protocol of the Gel extraction kit. After purification, the isolated DNA is quantified by measuring the OD at 260 nm.
5. The *Salmonella* strain carrying the pKD46 plasmid is grown in 10 mL LB medium supplemented with ampicillin (100 μg/mL), overnight at 30 °C, and under vigorous shaking.
6. The strain is subcultured using a dilution of 1:200 in 100 mL LB medium supplemented with ampicillin (100 μg/mL) and 10 mM arabinose, for the induction of the lambda red recombinase gene. This is then incubated at 30 °C for about 3.5 h until OD at 600 nm has reached a value of 0.6. Then, electrocompetent bacteria are prepared according to the protocol described above (*see* **Subheading 3.1.1.**).
7. To perform the disruption of the gene of interest by homologous recombination, the purified PCR fragment containing the kanamycin cassette is introduced by electroporation. To do so, the competent bacteria are electroporated with different concentrations of the PCR product, ranging from 250 to 800 ng/μL. SOC (1 mL) is added immediately, and the bacteria are then incubated 1 h at 37 °C under vigorous shaking. Hundred microliters of the transformed bacteria are spread on a LB-agar plate supplemented with 100 μg/mL kanamycin to select the strains containing the kanamycin resistance after recombination. As a negative control, the strain carrying the pKD46 plasmid grown in the absence of arabinose is used. The LB-agar plates are placed in an incubator at 37 °C overnight.
8. To purify the clones obtained, a single colony is picked and streaked on another LB-agar plate supplemented with 100 μg/mL kanamycin.
9. To check if the transformants have incorporated the kanamycin resistance by homologous recombination, a PCR to amplify the kanamycin gene is performed, using the same primers as in **step 2**. This should amplify a fragment of about 1550 bp. Wild-type *Salmonella* DNA is used as negative control.
10. To remove the pKD46 lambda red recombinase plasmid, which is temperature sensitive, the strain containing this plasmid is cured by cultivating it overnight at 43 °C in LB medium under vigorous shaking. The bacteria are then spread on a

LB plate without antibiotic and incubated at 37 °C overnight. Different colonies are tested for kanamycin resistance and ampicillin sensitivity to isolate the strain that does not contain the pKD46 plasmid (*see* **Note 4**).

11. To create a *Salmonella* mutant carrying several mutations, the method based on the P22 bacteriophage transduction is followed (*see* **Subheading 3.1.3.**).
12. The kanamycin gene incorporated by recombination is eliminated by using the pCP20 helper plasmid. This plasmid is an ampicillin- and chloramphenicol-resistant helper plasmid that expresses the FLP recombinase acting on the FRT sites flanking the resistance gene, this plasmid is also temperature sensitive. The pCP20 plasmid is transformed into the bacteria by electroporation.

    To check if the transformants have lost the kanamycin resistance, a PCR to amplify the kanamycin resistance cassette is performed using the same primers as in the previous PCRs. This should not give any amplification. As controls for this PCR, the mutant containing the kanamycin resistance and wild-type *Salmonella* DNA are used.
13. The strain containing the pCP20 plasmid is cured by cultivating it overnight in LB medium at 43 °C. The bacteria are then spread on a LB-agar plate without antibiotics and incubated at 37 °C overnight. Different colonies are tested for ampicillin sensitivity to isolate the strain that does not contain the pCP20 plasmid anymore.

3.1.3. P22 Transduction

Using the mutant bacteriophage strain P22 HT 105 int, a high-frequency transduction form of P22 which is specific for *Salmonella*, it is possible to transduce a mutation (antibiotic or auxotrophy linked), through the exchange of genetic material between two strains of *Salmonella*. This technique is used to create *Salmonella* mutants carrying several mutations.

The single mutant containing the kanamycin cassette (after **step 9**, *see* **Subheading 3.1.2.**) is used in this method allowing the selection of further mutants.

3.1.3.1. Preparation of P22 Lysates

The *Salmonella* donor strain is grown overnight in 1 or 2 mL LB medium at 37 °C under vigorous shaking. Hundred microliters of the overnight culture is diluted in 1 mL transducing broth, and 20 µL P_{22} phages at 10^9 plaque-forming unit (PFU)/mL are added. This is then incubated overnight under the same conditions. On the morning, there should be a clog floating in the tube indicating the lysis by the phages. At this time, the phage concentration is about 10^{11} PFU/mL. Twenty microliters of chloroform is added to kill the remaining bacteria. After vortexing, it is incubated 5 min at 37 °C under vigorous shaking. Bacteria are centrifuged 10 min at 4 °C full speed. The supernatant is saved and it is mixed with 20 µL chloroform.

3.1.3.2. P22 Titrations

Hundred microliters of an overnight *Salmonella* 12023 (wild-type) culture is mixed with 100 μL diluted phages and 3 mL soft agar. This is plated on a LB-agar plate. The plate is incubated overnight at 37 °C, and on the morning, the PFU is counted. Transducing phages are about one-third of the PFU's titer.

3.1.3.3. Transduction in the Recipient Bacteria

The recipient bacteria are grown overnight and subcultured at 1:100 during 4 h in transducing broth. Then, the bacteria (about 10^9 cells) are added to 100 μL of P22 phages diluted at 1:100 (about 10^8 PFU). This is incubated at 37 °C for 30 min under vigorous shaking. Hundred microliters of bacteria are then spread on a LB-agar plate with 100 μL 1 M EGTA, to dissociate the phages from the bacteria. The plates are incubated overnight at 37 °C and the clones are purified once on EGTA 10 mM–LB-agar plates (*see* **Note 5**).

3.2. Cell Culture

1. HeLa cells are routinely grown in 75-cm^2 flasks from Beckton-Dickinson in 15 mL cell culture medium. Cells are split every 3 days at a surface ratio of 1:4. For this, cells are washed once with 1× PBS and detached by incubating 4 min in the presence of 1 mL 1× trypsin–EDTA. The cells are resuspended in 9 mL cell culture medium, and 2.5 mL cells are added to 12.5 mL fresh culture medium.
2. For the infection of HeLa cells, the cells are grown on coverslips placed into a six-well plate. The coverslips must have been washed with acetone and ethanol and then dried and autoclaved. Four 12 mm diameter coverslips are placed per well, and 2 mL cell culture medium is added into each well. Resuspended cells are added into each well at a surface ratio of 1/10 (133 μL). The six-well plate is placed into the incubator for 18–24 h. The cells are then 40–60% confluent (*see* **Note 6**) and can be infected by *Salmonella*.

3.3. Infection of Cells

1. The day before the infection, *Salmonella* cultures are started in 1 mL LB medium supplemented with the appropriate antibiotics and grown overnight at 37 °C under vigorous shaking.
2. The overnight *Salmonella* culture is diluted at 1:33 in LB medium supplemented with the right antibiotics and incubated at 37 °C for 3.5 h under vigorous shaking. At this stage, the bacteria are in exponential growth phase and can efficiently infect cells.
3. The cell culture medium of the six-well plate is replaced by 1.5 mL warm EBSS. About 30 μL of the *Salmonella* subculture is added and the cells are incubated for

10 min at 37 °C. Invasive *Salmonella* are highly motile. The bacteria swimming quickly can be seen by phase contrast microscopy.
4. After infection, the EBSS is removed and the cells are washed twice (*see* **Note 7**) with cell culture medium supplemented with 100 μg/mL gentamycin. Then, the cells are incubated for 1 h at 37 °C in cell culture medium containing the same concentration of gentamycin. This antibiotic is used to kill extracellular bacteria.
5. The medium is removed and replaced by cell culture medium supplemented with 10 μg/mL gentamycin.

3.4. Absorption of Antibodies on Acetone Powder

Rabbit serum and mouse ascite fluid are commonly contaminated by antibodies that recognize dead *Salmonella* (*see* **Fig. 1**). This is possibly because of previous infection by enterobacteria. To resolve this problem and to be sure that the antibody used does not detect any bacterial antigens, it is often necessary to perform an absorption of the antibody. To do this, the antibody is incubated with *Salmonella* to eliminate any antibody specific for the bacteria.

3.4.1. Production of Acetone Powder

A 100-mL culture of wild-type S*almonella* is started and incubated overnight in LB medium at 37 °C under vigorous shaking.

The bacteria are pelleted and resuspended in 10 mM NaCl and 10 mM KCl. The cells are then centrifuged again and 50 volumes of acetone are added to the pellet. The suspension incubated on ice for 30 min. This is then centrifuged at 5000 × *g* for 10 min and the supernatant is discarded. Acetone is evaporated and the dry pellet is homogenized to powder.

3.4.2. Absorption

Acetone powder (10 mg) is added to 1 mL of 1:10 or 1:50 diluted antibody solution (pH 7). This is then incubated overnight at 4 °C. On the next day, it is centrifuged in an Eppendorf centrifuge for 15 min at maximum speed. The supernatant is filtered or spun in an ultracentrifuge at 100, 000 × *g* for 15 min in a TL-100 rotor.

The antibodies are then stored with 0.02% NaN_3 at 4 °C till further use.

3.5. Immunofluorescence

1. Before starting the immunofluorescence, the mounting medium has to be prepared: 6 g glycerol is placed in 50 mL canonical centrifuge tube, 1.4 g Mowiol 4-88 is added and stirred thoroughly, 6 mL distilled water is added and the solution is mixed for 1 h at room temperature, 12 mL 0.2 Tris, pH 8.5, is added, and then,

it is incubated at 50 °C for several hours with occasional stirring to dissolve the mowiol. Then, it is centrifuged at 5000 × *g* for 15 min to separate the solution from remaining mowiol crystals. The mowiol solution is stored at –20 °C in aliquots (1 mL). Once thawed, the mounting medium can be kept at room temperature for several weeks.

2. Sixteen hours post infection, the medium is removed. Cells are washed once with PBS and are fixed by adding 1 mL 3% PFA for 10 min at room temperature. Then, the cells are washed 3× with PBS. NH_4Cl (10–20 mM) is included in the last wash to saturate the free aldehyde groups. The cells can then be stored in PBS at 4 °C.
3. For the immunofluorescence, a solution of 0.1% saponin in PBS and a solution of 0.1% saponin and 10% horse serum in PBS for the dilution of the antibodies are prepared. Saponin is a mild detergent and permeabilizes the cells. This permeabilization is weak and saponin must therefore be included in all subsequent incubation buffers (*see* **Note 8**).
4. A moist chamber is prepared by placing wet Whatmann paper in a plastic box. A strip of parafilm is placed on the wet paper. The parafilm should be flat with no air bubbles underneath.
5. The rabbit antibody against kinesin-1 is diluted to the proper concentration in PBS/0.1% saponin/10% horse serum. Drops of diluted antibodies (50 µL) are placed on the parafilm. Coverslips are picked up, drained for excess buffer, and placed on the drop, cell side down. This is incubated for 30–60 min at room temperature (*see* **Note 9**).
6. The appropriate secondary antibodies are diluted in PBS/0.1% saponin/10% horse serum. Two beakers with PBS/0.1% saponin are prepared. Coverslips on which the primary antibodies were incubated are picked up and dipped in the two successive beakers. The excess buffer is drained and coverslips are placed on a 50µL drop of diluted secondary antibodies. This is then incubated for 30 min at room temperature.
7. Two beakers with PBS/0.1% saponin, one beaker with 1× PBS and one with water, are prepared. The microscope slides are washed with ethanol and 10 µL mounting medium per coverslip is dropped on it. Coverslips are washed in the successive beakers and the excess water is drained. The cell-free side is wiped and coverslips are placed, cells side down, on a drop of mounting medium. This is then dried for 60–120 min, in the absence of light.

4. Notes

1. Mowiol is carcinogenic; wear gloves and handle it under a fume hood.
2. Be careful to always keep the bacteria on ice and use a centrifuge cooled at 4 °C. Make sure that the water used is ice cold.
3. The annealing temperature varies according to the nucleotide sequence of the primer.
4. Usually, 80% of the clones are resistant to kanamycin and 20% sensible to ampicillin after clone purification.

5. After transduction, you have to check that you have eliminated P22 phages from your strains and that you have no prophages integrated in the bacterial chromosome. If so, your strain will be resistant to other P22 infection.

   To check that you have eliminated P22 phages from your strain, you have to look if it lyses or not on green plates. These plates are poorly buffered rich plates with an excess of glucose and two pH indicator dyes. The excess of glucose causes the strains to make more acid than usual. When a strain lyses, the acid released causes the plate to turn dark green. In addition, normal colonies will turn the plate dark after a sufficiently long period (more than 24 h). Therefore, green plates must be examined promptly. Single, phage-infected strains lyse and make small, dark-green colonies. Normal cells will make larger, light-green colonies. These colonies are phage-free but may contain lysogens.

   Strains may be cross-truck over phage on green plates. The portion of the streak before the phage will be healthy and light. After contacting the phage, normal cells will lyse, so the portion of the streak after the phage will be spotty and dark. Lysogens are immune to phage and will look healthy on both sides of the streak.
6. For immunofluorescences, it is important that the cells are not too confluent in order to be able to clearly visualize the phenotype and the specific immunolabelings.
7. Make sure to have removed all extracellular bacteria. If not, additional washes have to be carried out.
8. Solutions containing saponin should always be prepared shortly before use.
9. The incubation time may vary depending on the quality and specificity of the antibody used.

Acknowledgments

We thank Suzana Salcedo for critical review of this manuscript. A.D. was recipient of fellowships from the French Ministry of Research. This work was supported by institutional grants from the CNRS and INSERM and the Microban EU network n° MRTN-CT-2003-504227.

References

1. Galan, J. E. and Zhou, D. (2000) Striking a balance: modulation of the actin cytoskeleton by Salmonella. *Proc. Natl. Acad. Sci. USA* **97,** 8754–87561.
2. Garcia-del Portillo, F. and Finlay, B. B. (1995) The varied lifestyles of intracellular pathogens within eukaryotic vacuolar compartments. *Trends Microbiol.* **3,** 373–380.
3. Hensel, M., Shea, J. E., Gleeson, C., Jones, M. D., Dalton, E., and Holden, D. W. (1995) Simultaneous identification of bacterial virulence genes by negative selection. *Science* **269,** 400–403.
4. Steele-Mortimer, O., Méresse, S., Gorvel, J. P., Toh, B. H., and Finlay, B. B. (1999) Biogenesis of Salmonella typhimurium-containing vacuoles in epithelial cells involves interactions with the early endocytic pathway. *Cell Microbiol.* **1,** 33–49.

5. Méresse, S., Steele-Mortimer, O., Finlay, B. B., and Gorvel, J.-P. (1999) The rab7 GTPase controls the maturation of Salmonella typhimurium-containing vacuole in HeLa cells. *EMBO J.* **18,** 4394–4403.
6. Beuzon, C. R., Méresse, S., Unsworth, K. E., et al. (2000) Salmonella maintains the integrity of its intracellular vacuole through the action of SifA. *EMBO J.* **19,** 3235–3249.
7. Garcia-del Portillo, F. and Finlay, B. B. (1995) Targeting of Salmonella typhimurium to vesicles containing lysosomal membrane glycoproteins bypasses compartments with mannose 6-phosphate receptors. *J. Cell Biol.* **129,** 81–97.
8. Garcia-del Portillo, F., Zwick, M. B., Leung, K. Y., and Finlay, B. B. (1993) Salmonella induces the formation of filamentous structures containing lysosomal membrane glycoproteins in epithelial cells. *Proc. Natl. Acad. Sci. USA* **90,** 10544–10548.
9. Brumell, J. H., Goosney, D. L., and Finlay, B. B. (2002) SifA, a type III secreted effector of Salmonella typhimurium, directs Salmonella-induced filament (Sif) formation along microtubules. *Traffic* **3,** 407–415.
10. Brumell, J. H., Rosenberger, C. M., Gotto, G. T., Marcus, S. L., and Finlay, B. B. (2001) SifA permits survival and replication of Salmonella typhimurium in murine macrophages. *Cell Microbiol.* **3,** 75–84.
11. Stein, M. A., Leung, K. Y., Zwick, M., Garcia-del Portillo, F., and Finlay, B. B. (1996) Identification of a Salmonella virulence gene required for formation of filamentous structures containing lysosomal membrane glycoproteins within epithelial cells. *Mol. Microbiol.* **20,** 151–164.
12. Tanaka, Y., Kanai, Y., Okada, Y., et al. (1998) Targeted disruption of mouse conventional kinesin heavy chain, kif5B, results in abnormal perinuclear clustering of mitochondria. *Cell* **93,** 1147–1158.
13. Lippincott-Schwartz, J., Cole, N. B., Marotta, A., Conrad, P. A., and Bloom, G. S. (1995) Kinesin is the motor for microtubule-mediated Golgi-to-ER membrane traffic. *J. Cell Biol.* **128,** 293–306.
14. Guignot, J., Caron, E., Beuzon, C., et al. (2004) Microtubule motors control membrane dynamics of Salmonella-containing vacuoles. *J. Cell Sci.* **117,** 1033–1045.
15. Boucrot, E., Henry, T., Borg, J. P., Gorvel, J. P., and Meresse, S. (2005) The intracellular fate of Salmonella depends on the recruitment of kinesin. *Science* **308,** 1174–1178.
16. Datsenko, K. A. and Wanner, B. L. (2000) One-step inactivation of chromosomal genes in Escherichia coli K-12 using PCR products. *Proc. Natl. Acad. Sci. USA* **97,** 6640–6645.
17. Schmieger, H. and Buch, U. (1975) Appearance of transducing particles and the fate of host DNA after infection of Salmonella typhimurium with P22-mutants with increased transducing ability (HT-mutants). *Mol. Gen. Genet.* **140,** 111–122.

14

Magnesium, Manganese, and Divalent Cation Transport Assays in Intact Cells

Michael E. Maguire

Summary

Protocols are described for the measurement of radioisotopic cation influx and efflux in bacteria using *Salmonella* as a model system. Methods are discussed for both the use of primary radioisotopes for measurement, e.g., using $^{54}Mn^{2+}$ to measure Mn^{2+} influx, and the use of surrogate radioisotopes when the primary radioisotope is not available, e.g., use of $^{63}Ni^{2+}$ or $^{57}Co^{2+}$ in place of $^{28}Mg^{2+}$ to measure Mg^{2+} influx. Both vacuum filtration and centrifugation assays are described. In addition, the use and misuse of chelating agents is discussed.

Key Words: Transport; filter; filtration; centrifugation; chelators; binding; magnesium; manganese; iron; calcium; nickel; cobalt; *Salmonella*; bacteria; influx; efflux.

1. Introduction

1.1. Overview

Mg^{2+} and Ca^{2+} are fundamental constituents of all cells. They are present in millimolar concentration within cells. By contrast, there seems to be a general assumption that transition metal divalent cations, Fe^{2+}, Ni^{2+}, Co^{2+}, Zn^{2+}, Mn^{2+}, etc., are present only in trace amounts. In reality, in many cell types, the total concentrations of Zn^{2+} and Mn^{2+} can often approach millimolar, whereas iron is routinely sequestered in cells at substantial levels. Total cation concentration can be measured by atomic absorption or various mass spectrometric means. Of course, thermodynamically it is the free concentration of cation that is of interest. Unfortunately, the free concentration is much more difficult to measure.

From: *Methods in Molecular Biology, vol. 394: Salmonella: Methods and Protocols*
Edited by: H. Schatten and A. Eisenstark © Humana Press Inc., Totowa, NJ

Excellent dyes for Ca^{2+} are available and are applicable to a wide variety of cell types *(1,2)*. Dyes for other cations are less selective and must be carefully used *(3–6)*. Microelectrode measurement of cation concentrations is generally applicable only to large cells or purified reconstituted transporters. Thus, for the measurement of cation influx or efflux from whole cells, the investigator must most often rely on radioisotopes.

The mechanisms by which cells maintain cation homeostasis are only now being elucidated. Central to an understanding of cation homeostasis is the elucidation of the various transport processes by which cells mediate influx and efflux of cations. These can be readily determined using the methods outlined in this chapter. Our emphasis has been on the measurement of Mg^{2+} and Mn^{2+} influx in bacteria, but the methods are equally applicable to all cations and in any cell type that can grow in suspension culture. In addition, comments on the use of these methods to measure efflux from cells and cation uptake in nonsuspension cells and on the use of metal ion chelators are also included.

1.2. Choice of Assay Type

The principle of the assay involves the separation of the cation of interest in the extracellular compartment from that in the intracellular compartment. In addition, correction for or minimization of cation nonspecifically bound to the cell must be made. This separation is achieved usually by (1) filtration or centrifugation of the cells to separate them from the incubation medium and (2) the use of buffer and wash conditions to minimize nonspecific binding of cation to the cell surface.

Vacuum filtration of cells assumes that the cells are sufficiently hardy to withstand interaction with the filter itself, tolerant of rapid changes in buffer composition and temperature, and capable of withstanding the airflow and consequent mild drying action caused by vacuum filtration. Virtually, all bacteria and yeast as well as most Protista are quite compatible with filtration because of the strength of their cell wall. With these types of cells, the drying process after all liquid has passed through the filter does not generally lyse the cells and therefore does not present a problem. Filtration is rapid, mechanically simple, and does not require extensive and expensive apparatus.

A disadvantage to filtration however is that despite some drying inherent in the process, the cell wall retains considerable water content and therefore also retains anything dissolved in that content. Centrifugation through an oil solution (*see* **Subheading 4.**) can minimize this effect if it is a problem with some cells and is particularly useful with mammalian cells that are grown in suspension. It is also the optimal process for isolation of cells that are to be ashed for atomic absorption measurements of ion content because it removes most or all surface water. Its disadvantage is that the process is more cumbersome, sample processing is longer than for filtration, not as many samples can be processed in a given time period, and very rapid kinetic measurements cannot be made. Nonetheless, it is still relatively rapid, and the equipment necessary is generally already available in a laboratory.

1.3. Choice of Radioisotope

The choice of isotope can be quite simple. Simply use a radioactive isotope of the cation of interest, e.g., $^{45}Ca^{2+}$. However, the cation of interest may not have a radioactive isotope available. $^{28}Mg^{2+}$ is not readily available. Because prokaryotic Mg^{2+} transporters also transport either or both of Ni^{2+} and Co^{2+} *(7–9)*, $^{63}Ni^{2+}$ or isotopes of Co^{2+} can be used for study of Mg^{2+} transport. In other cases, the isotope may not be easy to work with. $^{42}K^{+}$ is available but it has a short half-life and its decay products are of extremely high energy and thus relatively dangerous. For most purposes therefore, most investigators would use $^{86}Rb^{+}$ instead. **Table 1** lists primary and surrogate radioisotopes of common divalent cations that may be of use in determining transport processes.

1.4. Genetics and Measurement of Transport Systems

Measurement of cation uptake in whole cells potentially and necessarily includes transport activity of more than one transport system. For example, *Salmonella enterica* serovar Typhimurium carries two distinct Mn^{2+} transport systems, MntH and SitABCD. Until one has characterized them as to their regulation and expression, measurement of Mn^{2+} uptake in whole cells could be measuring uptake through both systems or only one, depending on the experimental conditions. The genetics of the system, however, can be used to advantage to help solve this problem. Once a putative transporter has been identified, a knockout strain should be made. In the simplest case, transport should be measured in both the parental and knockout strains. The difference is presumed to be due to the transport system that was knocked out. This controls for the presence of other transporters capable of transporting the same cation. Of course, this assumes that knocking out the transporter in question does not alter transport by other systems or alter expression of other transport systems. Not infrequently, loss of one system will cause partial induction of another system in compensation. There are two primary ways around this. First, the transport system to be studied can be cloned into an inducible expression vector. The

Table 1
Isotopes Useful for Measurement of Cation Flux

Radioisotope	Useful for	Half-life	Decay particles and relative energy[a]	Liquid scintillation counting[b]	
				Window	Efficiency
$^{28}Mg^{2+}$	Mg^{2+}	21 h	3β, 2γ, very high	0–1000	NA[c]
$^{45}Ca^{2+}$	Ca^{2+}	165 days	β, high[d]	0–750	100
$^{54}Mn^{2+}$	Mn^{2+}	312 days	γ, Auger[e], low	0–400	35
$^{55}Fe^{2+}$	Fe^{2+}, Fe^{3+}	2.7 years	Auger, low	0–400	35
$^{59}Fe^{2+}$	Fe^{2+}, Fe^{3+}	44.6 days	β, γ, very high	0–1000	100
$^{57}Co^{2+}$	Co^{2+}, Mg^{2+}	271.7 days	Auger, γ, high	0–800	80
$^{60}Co^{2+}$	Co^{2+}, Mg^{2+}	5.27 years	β, γ, very high	0–1000	100
$^{63}Ni^{2+}$	Ni^{2+}, Mg^{2+}	100 years	β, low	0–600	85
$^{64}Cu^{2+}$	Cu^{2+}	12.7 h	Auger, $β^+$, β, high	0–1000	70
$^{65}Zn^{2+}$	Zn^{2+}	245 days	$β^+$, γ, very high	0–1000	90–100

[a] Low energy is considered to be roughly equivalent to ^{3}H, ^{14}C, or ^{35}S or a β-particle of less than 200 Kev, thus requiring minimal if any shielding. High energy is either a β-particle > 200 Kev, e.g., ^{32}P or a γ emission of < 200 Kev, requiring shielding. Very high energy is a γ emission of > 200 Kev regardless of other emissions and usually requires heavy shielding.

[b] Approximate. Based on standard Beckman Corporation instruments.

[c] Not applicable. Because of its extremely high energy, $^{28}Mg^{2+}$ is best counted by Cerenkov radiation by placing the sample in a scintillation vial containing water, not scintillation fluid. Efficiency is dependent on the type of counter and detection crystal and must therefore be individually determined. In addition, ^{28}Mg decays into ^{28}Al, which has a half-life of about 2 min. Because most cells can transport Mg^{2+}, but not Al^{3+}, uptake into cells separates these two isotopes. This raises the curious specter that if one counts experimental samples immediately after separation, the two radioisotopes are not yet back in a decay equilibrium and experimental counts will actually increase for a few minutes until equilibrium is reestablished. In practice, wait 15 min after separation to begin counting $^{28}Mg^{2+}$.

[d] Radioactive Ca^{2+} is by far the most dangerous isotope listed in this table because most Ca^{2+} taken up into the body ends up in bone for very long periods and thus has the potential to cause mutations in bone marrow cells. Other isotopes have relatively short transit times through the body and are not concentrated in bone with intimate access to marrow cells. Thus, their potential for causing damage is orders of magnitude lower.

[e] Auger electron, detected similarly to β particles. $β^+$ = positron.

vector is then introduced back into the wild-type strain. Induction of the transporter on the vector to high levels will allow relatively easy characterization of the cation selectivity and affinity because the majority of transport for the cation(s) in question will be through the induced system. Transport capacity

(V_{max}) measurements under such conditions have no physiological relevance of course. Second, one can, with perseverance, identify other, potentially all, transport systems for the cation in question. A strain can then be created carrying one and only one relevant transport system that can then be characterized without interference. This method was used to study the Mg^{2+} transporters of *S.* Typhimurium ***(8,9)***. Obviously, such an approach is useful for characterization of the basic transport properties of the system. It does not provide great insight into the physiology of the system or the cation. Nonetheless, such characterization is a prerequisite to the study of cation homeostasis.

1.5. Cation Selectivity and Uptake Kinetics

In examining the literature, several common problems emerge in evaluating reports of cation transport systems. By far, the most common problems are (1) inadequate characterization of the system's cation selectivity and (2) use of accumulation assays rather than initial rate assays to measure transport parameters.

1.5.1. Cation Selectivity

Just as consumers and physicians make the mistake of assuming that drugs have only a single action, scientists seem to commonly make the mistake of assuming that a transport system transports only a single cation. Although a few systems are undoubtedly highly selective for a single cation, Mother Nature's requirement for functionality is that a transporter be selective for a particular substrate under normal physiological conditions, not that it transport only a single substrate. For example, the three classes of prokaryotic Mg^{2+} transporters, CorA, MgtA/MgtB, and MgtE ***(10)***, transport only Mg^{2+} under physiological conditions. However, CorA can mediate influx of both Ni^{2+} and Co^{2+}, MgtA/B can mediate influx of Ni^{2+}, and MgtE can mediate influx of Co^{2+} in addition to Mg^{2+}. However, such transport would generally be considered nonphysiological because it occurs only at cation concentrations that do not normally occur physiologically and/or such transport is immediately toxic to the cells.

A recent example of the need for such characterization can be illustrated using prokaryotic Mn^{2+} transporters ***(11)***. One class of such prokaryotic systems is highly homologous to mammalian NRAMP transporters, which are known to mediate influx of ferrous iron. Likewise, the SitABCD ABC type transporters in prokaryotes are very similar to several iron-transporting ABC type transporters. Both types of systems were therefore initially assumed to be iron transporters. Actual transport data however shows that prokaryotic NRAMP class systems are usually high-affinity Mn^{2+} transporters ***(12)***, whereas some, though not all, SitABCD class systems are also highly selective for Mn^{2+} ***(13)***. Careful analysis of the transport properties of these systems in *S.* Typhimurium and *Escherichia coli* shows that they transport Mn^{2+} with an affinity of 0.1–0.5 μM, well within the concentration of Mn^{2+} encountered by biological systems. These transporters will transport Fe^{2+} as was initially assumed. However, their affinity for Fe^{2+} is 10–100 μM, orders of magnitude worse than their affinity for Mn^{2+} and far above the typical physiological concentration of Fe^{2+} seen by cells. Moreover, all such transporters tested to date transport Cd^{2+}, a highly toxic cation. Unfortunately, such information has not been obtained for the large majority of putative Mn^{2+} transporters published in the literature ***(14)***. The point is that without careful analysis of the range of cations that are either transported by a system or that inhibit a system's transport of other cations, relevant physiology may be missed or inferred incorrectly.

1.5.2. Kinetics of Transport

Many reports of transport processes use simple accumulation of isotope to characterize a presumed transporter. This leads to incorrect conclusions for several reasons. First, accumulation measurements are the sum of all influx and efflux processes that are operative during the measurement. Hence, by definition they are incapable of measuring transport by a single defined system unless extensive genetic manipulation has created strains or cell lines carrying only a single transporter for that cation. Second, activation or use of a transporter and/or the accumulation of a cation within the cell can trigger physiological responses which may in turn alter the properties of the transport system in question or regulate transport or other metabolic processes within the cell. Third, most prokaryotic organisms grow relatively rapidly. Because accumulation assays often are run for many minutes or even a few hours, not only is cation flux being mediated by multiple transport systems, these systems are in turn varying depending on the growth state of the cell. Finally, in eukaryotic cells, many cations, toxic or not, are often sequestered in intracellular vacuoles. This can lead to very high levels of uptake that reflect a combination of transport processes, not just those at the plasma membrane.

In many cases, the relative expression of transporters has been inferred from measurements of cation accumulation under different experimental conditions. This rarely has any direct correlation with the actual degree of expression of the system purportedly being measured. Not only, as noted, does an accumulation assay measure multiple transporters at once, the error is compounded when comparing two different experimental conditions because there is no guarantee that only the transporter in question changed its expression as a result of the experimental manipulation. In short, accumulation assays are rarely useful in defining a transport system and can lead to major problems in interpretation.

Just as with classic enzyme assays, it is initial rate measurements that are of most importance in defining properties of a transporter. Measurement of initial rates of uptake, the time period over which the rate of uptake is linear with time, obviates most of the issues noted above regarding accumulation assays. The first experiments with any system should determine the time period over which uptake is linear, and the time period for all subsequent assays should be chosen to be within that period of linearity. Moreover, experimental manipulation of cells may change this period of linearity. The careful investigator always determines whether the rate is still linear under all experimental conditions.

Because the characterization of transporters considered herein involves measurement of uptake in whole cells, it is rarely critical and sometimes not practical to measure detailed kinetic parameters under all conditions. The most critical parameter is determination of the apparent affinity ($K_{0.5}$) of the transporter for its (presumed) primary cation (*see* **Note 1**). This can be approximated by measuring the initial rates of uptake at multiple cation concentrations. A different and incorrect value will be obtained if total accumulation is measured as a function of cation concentration. Once a $K_{0.5}$ value is obtained, it is generally far easier to approximate the maximal rate of uptake by measuring the initial rate of uptake at a cation concentration equal to the $K_{0.5}$ value and doubling the rate to get the equivalent of V_{max} for an enzyme. The experimental manipulations required to actually calculate the maximal rate of uptake are extensive, and since, with whole cell experiments, one is generally interested in changes in the expression of the transporter rather than precise kinetic parameters, such an approximation introduces minimal error.

1.5.3. Measurement of Cation Selectivity

Once primary kinetic data are obtained for a transport system, it is then a simple matter to determine the selectivity of the system for various cations and their relative affinities. Using a total concentration of the radioisotopic tracer equal to the $K_{0.5}$ value, uptake is measured as a function of concentration of a second cation, i.e., an inhibition curve. Mass action considerations allow calculation of a K_i value for the second cation using the Cheng–Prusoff equation ***(15)***:

$$K_i = \mathrm{IC50}/(1 + D/K_d) \quad (1)$$

Under conditions where the total concentration of the primary cation (D) is equal to its $K_{0.5}$ value (K_d), the Cheng–Prusoff equation reduces to IC50 = 2^*K_i. Thus, if an IC50 value of 20 μM is measured, the apparent K_i for the competing cation would be 10 μM. It does not matter if there are other transport systems mediating uptake or efflux of the competing cation. The major caveats to this simplistic experimental approach are that competition be competitive and that initial rates are measured.

As already mentioned for Mn^{2+} transporters (*see* **subheading 1.5.1.**), it is critical that a complete series of cations be tested for each transporter to ensure that the physiological relevance of a transporter is properly understood. Likewise, it is critical that such assignments be made on the basis of direct transport assign rather than inference from phenotypic results. An example of the latter problem was the inference that the CorA Mg^{2+} transporter also transported Fe^{2+}, based on phenotypic growth assays ***(16,17)***. Direct assay of transport properties however clearly shows that CorA does not mediate uptake of Fe^{2+} ***(18)***.

1.6. Use of Metal Chelators

A constant problem in the study of metals in biology is the issue of cross-reactivity or cross-contamination. Biological systems require many different metal ions to function. Biological media, even when prepared with distilled, deionized water and pure reagents, always have some low concentrations of various metal ions. Thus, researchers often resort to the use of metal chelating agents to reduce metal concentrations. Unfortunately, metal chelators are almost always used incorrectly. Many workers are under the mistaken impression that some metal chelators are quite specific for a single cation or a very limited number of cations. An example is the use of EDTA and EGTA. The latter is commonly used to remove Ca^{2+} while leaving Mg^{2+} unchanged. This leads to the incorrect assumption that EGTA is a selective chelator of Ca^{2+}. In fact, EGTA has a poorer affinity for Ca^{2+} than does EDTA. What both of these chelators do is to bind *all* divalent (and trivalent) metal cations in the solution. The practical difference is that EGTA has a very poor affinity for Mg^{2+} compared with all other metal cations. But both chelators have far higher affinities for transition metal cations than for either Ca^{2+} or Mg^{2+}. Thus, for example, addition of 100 μM of either chelator to a solution would reduce the concentration of *all* transition metal cations to at least picomolar levels. Therefore, rather than selectively reducing Ca^{2+}, EGTA effectively removes *all* metal cations from solution with the sole exception of Mg^{2+}. Similar problems arise with virtually any other chelator. Chelating agents selective for a single metal ion and useful in a biological system essentially do not exist. This can be seen by examining the affinity constants for a variety of common chelating agents as summarized in **Table 2**. For example, although 2,2′-bipyridyl (dipyridyl) is commonly used as a selective agent to chelate ferrous iron (Fe^{2+}), its affinity for ferric iron (Fe^{3+}) is essentially equivalent. Under typical biological conditions, bipyridyl cannot distinguish between Cu^{2+}, Fe^{2+}, Fe^{3+}, and Zn^{2+} (also *see* **Note 2**).

1.6.1. Add-Back Experiments

Although it does not solve all problems with use of chelators, the better way to use chelators is by means of an add-back experiment. To determine the cation of interest, one could use EDTA (or EGTA) at a concentration slightly in excess of the total concentration of all divalent cations in solution. Then, the cation of interest can be added back on top of the chelator. This will ensure that the major divalent cation present free in the solution will be the cation of

Table 2
Metal-Ligand Stability Constants[a]

Ligand	H^+	Mg^{2+}	Ca^{2+}	Cu^{2+}	Fe^{2+}	Fe^{3+}	Mn^{2+}	Zn^{2+}
EDDA	9.6	4.0	?	16.2	10.7	22	7.0	11.1
MIDA	9.6	3.5	3.8	17.8	11.8	20.7	5.4	7.7
EDTA	9.5	8.8	10.7	18.8	14.3	25.1	13.9	16.5
EGTA	9.3	5.3	10.9	17.7	11.8	20.0	12.2	12.6
TAPEN	10.7	?	?	19.6	7.9	?	6.1	12.3
DPA	7.2	?	?	13.9	12.2	?	3.5	7.6
DTPA	10	9	10	21	16	28	15	18
2,2′-Bipyridyl	4.5-5	0.5	0.05	17.0	17.2	16.3	?	13.3
TPEN	7.1	?	?	20.2	14.4	?	10.1	15.4
1,10-Phenanthroline	5.0	1.6	1.1	21.0	21.0	14.1	17.1	10.1

EDDA, ethylenediiminodiacetic acid ($C_6H_{12}N_2O_4$); MIDA, *N*-methyliminodiacetic acid ($C_5H_9NO_4$); EDTA, ethylenedinitrilotetraacetic acid ($C_{10}H_{16}N_2O_8$); EGTA, ethylenebis (oxyethylenenitrilo) tetraacetic acid ($C_{14}H_{24}N_2O_{10}$); TAPEN, ethylenedinitrilotetrakis (trimethyleneamine) ($C_{14}H_{36}N_6$); DPA, iminobis (methylene-2-pyridine) (di-2-picolylamine) ($C_{12}H_{13}N_3$); DTPA, diethylenetrinitrilopentaacetic acid ($C_{14}H_{23}N_3O_{10}$);TPEN, ethylenedinitrilotetrakis (methylene-2-pyridine)(N,)N,N′,N′-tetra-2-picolylethylenediamine ($C_{26}H_{28}N_6$).

[a] From Smith and Martell, NIST Database #46 ***(19)***. The numbers are the negative log of the stability (dissociation) constant in molar. In many cases, the affinities are for a stoichiometric complex of 2:1 or 3:1 ligand:metal, e.g., 2,2′-bipyridyl or 1,10-phenanthroline form such complexes with iron.

interest. Mass action considerations of course dictate that the added cation will displace some amount of other cation from the chelator, but the large majority of free cation in solution will be the cation of interest. If, for example, one suspects that a process is Zn^{2+} sensitive, adding Zn^{2+} on top of the added chelator could restore functionality or the process. Conversely, adding back Mn^{2+} or Co^{2+} or another cation should not restore the process in question if it is actually selective for Zn^{2+}. Just as with characterization of transporter selectivity for cations, it is important to test multiple cations in such add-back experiments.

2. Materials

2.1. General Buffer Solutions

Transport assays in intact cells are best performed in a minimal medium containing only buffer, minimal salts to provide phosphate and nitrogen sources and a simple carbon source. High concentrations of phosphate should be avoided to prevent precipitation of cations. The buffer we use most commonly is N-minimal medium ***(20)***. Other similar media can be used, including HEPES and similar buffers, with or without minimal phosphate and monovalent cations. Minimal media are supplemented with glucose at 4 g/L (0.4% w/v). To avoid a stringent response and other unwanted physiological responses in bacteria, 0.1% (w/v) casamino acids are also added but can be omitted.

Wash buffer is used to rinse cells after initial filtration. Such buffer should always contain a relatively low concentration of EDTA (usually 0.5–1.0 mM) and a high concentration of Mg^{2+} (at least 10 mM), regardless of the cation being studied. The EDTA is to chelate Ca^{2+} and all transition metal divalent cations. We find that this minimizes background retention of radioisotope on the filters. The Mg^{2+} is useful to compete with other cations not only for binding sites on the filter itself but more importantly for cell surface binding of cations (*see* **Note 3**). The major reservoir of cellular Mg^{2+} and Ca^{2+} is the inner membrane and the cell wall or outer membrane in prokaryotes. These additions to the wash buffer should be tested during development of the assay using the particular species being studied. EDTA concentrations as low as 0.1 mM and as high as 5 mM may provide lower background depending on the filter used and the cell type being studied. Likewise, Mg^{2+} might optimally be 0.5 mM or as high as 50 mM. When studying the transport of transition metal cations, it usually does not help to add the metal of interest, e.g., Mn^{2+}, as this may lead to aggregation, leakiness, or even lysis of cells. Ca^{2+} should never be added to the wash buffer as even low concentrations can readily cause aggregation of cells and thus very high backgrounds.

1. N-minimal medium (5×): 100 mM Tris–HCl (65.7 gm/L Tris–HCl plus 0.98 gm/L Tris base), 1 mM KH_2PO_4, 5.0 mM KCl, and 7.5 mM $(NH_4)_2SO_4$. Adjust to pH 7.4 with KOH. Use reagent grade chemicals. For complete 1× medium, dilute fivefold and add glucose to 0.4% (w/v), casamino acids to 0.1% (w/v) if desired, and other supplements including cations as needed.
2. Wash buffer: 1× N-minimal medium plus 0.5 mM EDTA and 10 mM $MgSO_4$.
3. For radioisotopes emitting γ rays, filters can be placed in glass or plastic test tubes or scintillation vials as required by the γ-counter being used. For radioisotopes emitting β-particles, a scintillation fluid compatible with small amounts of water should be used or the filters must be completely dried before adding scintillation fluid and counting.
4. Filter apparatus (*see* **Note 4**): #1225 Sampling Manifold (Millipore Corporation, catalog #XX27 025 50). Filters: 25 mm #BA85 (catalog #10402578, Schleicher and Schuell, Keene, NH).
5. All cations are made up in water or N-minimal medium at 0.1 or 1.0 M stock concentrations and diluted as needed in water or N-minimal medium. The sole exception is ferrous iron (Fe^{2+}). It is made fresh immediately before use by making a stock solution of up to 10 mM in 10 mM ascorbic acid. All dilutions are made in 1 mM ascorbic acid regardless of the final iron concentration. See **Subheading 4** for notes on assay of iron and Ca^{2+} uptake.

3. Methods

3.1. Preparation of Cells

1. Cells are grown overnight in an appropriate medium. They are then washed by centrifugation at 4000 × *g* for 5 min and resuspended in the same medium at an OD_{600nm} of 0.05 and incubated with shaking at 37°C for 3–5 h until OD_{600nm} reaches at least 0.4 but not more than 1.0. This is mid-log phase for most common bacterial species. For study of basic transport properties, N-minimal medium with glucose and casamino acids is used. For studies of transport regulation, the medium could be a rich or specialized medium. Cells should not however be grown overnight in a rich medium and switched to a minimal medium in the morning as this results in a severe lag phase before the cells adapt to the minimal medium.
2. Once an appropriate OD_{600nm} is reached, cells are again harvested by centrifugation. From this point, sterility is unnecessary. The cells are rinsed twice with an equal volume of the same medium at 4°C. The rinse medium should contain glucose but no added divalent cation. The casamino acids may also be omitted. The cell pellets are then resuspended to an OD_{600nm} of 1.5–2.0 and placed on ice for no more than 1 h before assay. Alternatively, the cell pellets may be placed on ice for up to 3 h before resuspension and assay.

3.2. Transport Assay

1. Assay of ion uptake is initiated by adding 0.1 mL ice-cold cell suspension to 0.9 mL medium containing the cation and radioisotope of interest in a polypropylene 17 × 100 mm test tube (Fisher #14-956-7E or equivalent). Generally, we prepare the tubes by pipetting 0.8 mL medium plus 0.1 mL of the desired cation solution (or water as control) and pre-incubating the tubes at the desired assay temperature for at least 15 min in a shaking water bath (20–30 rpm) before addition of cells. The addition of the cold cells to the warm medium warms the cells almost instantly, and no lag period in uptake is generally observed (*see* **Note 3**).
2. Uptake is generally linear for 1–15 min depending on the transporter and must be empirically determined (*see* **Subheading 1.4.**). After the desired incubation period, 4 mL ice-cold wash buffer is added to a tube, and the mixture is immediately poured over a filter. Although filtration should normally be done immediately, for most transporters, the addition of ice-cold medium generally stops all transport processes sufficiently that filtration can be delayed for up to 1 min. Drainage through the filter takes 10–30 s. It is critical that the final amount of cells filtered does not exceed 0.2 OD_{600nm} (0.1 mL of 2.0 OD_{600nm} cells added to the assay tube originally). Otherwise the filter will clog and extremely slow drainage will occur leading to higher background values. Each filter is then washed twice with 4 mL ice-cold wash buffer. Once the initial filtration is performed, the washes can be delayed for up to several minutes while additional samples are filtered (*see* **Notes 5** and **6**).
3. After all filters are washed, the manifold head is removed but the vacuum is left on. This avoids backsplash of any liquid. Filters are removed with tweezers and placed in vials or other tubes for counting. While handling of the filters should be minimal, the cells appear quite firmly bound to the filter and touching the filter does not seem to elicit major problems.
4. Each experiment must include blank and/or zero-time controls. We find it most convenient to take duplicate or triplicate tubes, add the 4 mL ice-cold wash buffer, then add the 0.1 mL cell suspension, and immediately filter the mixture. This provides a composite zero-time value as well as a filter background. The value of this blank does not seem to vary significantly with the amount of cation present in the assay. The major exception to this is during detailed kinetic experiments where multiple concentrations of both radioisotope and nonradioactive cation are used. In this case, blanks are run for each different amount of radioisotope used in the assay (*see* **Note 3**).

4. Notes

1. Because the transport studies discussed herein are performed in whole cells rather than on isolated, purified transport proteins, we prefer the term $K_{0.5}$ rather than K_m or K_a, where $K_{0.5}$ is the cation concentration at which 50% of maximal uptake occurs. Unless the transporter in question is regulated in some manner at the

protein level, this value is constant regardless of the level of expression of the proteins comprising the transport system.

2. Measurement of iron and Ca^{2+} transport processes are more problematic than for most other cations. Under standard laboratory and cellular conditions, iron in aqueous solutions is readily oxidized to the ferric form (Fe^{3+}) in a matter of seconds to minutes. Thus, measurement of ferrous iron transport requires the continued presence of a reducing agent sufficiently strong to prevent oxidation of Fe^{2+} to Fe^{3+} but sufficiently weak to avoid disruption of cells and cellular processes. Bisulfite can be used to ensure that an iron solution remains in the ferrous state, but this compound can injure cells. The better reagent to use is ascorbate at about 1 mM. This concentration is sufficient to keep ferrous iron from oxidizing for 20–30 min in solution at a pH around 7, but not sufficiently strong to adversely affect cells. Nonetheless, control samples should always include ascorbate whether or not iron is present.

   Ca^{2+} presents a different problem. Although all cations are to some extent "sticky," Ca^{2+} is by far the worst in this regard. It is very difficult to keep Ca^{2+} (and $^{45}Ca^{2+}$) from sticking to even plastic tubes. Care must be taken to optimize the wash buffer and wash conditions in Ca^{2+} transport assays to minimize background. In some cases, use of silanized glass tubes and pipette tips may be necessary.

3. To discriminate between uptake of cation and the binding of cation to the cell, cell associated counts are compared after incubation of the radioisotope with cells at the assay temperature and at 4°C. Since uptake is an active process, it does not occur to a major extent in the cold. Therefore, counts retained by cells at 4°C are assumed to be adventitious binding. Mass action considerations predict that low-affinity binding to the cell surface at cation concentrations in the physiological range should come to equilibrium within a few seconds even in the cold. The validity of this assumption is supported by three observations: (1) cell-associated counts remain constant for at least 15 min at 4°C, (2) cell-associated counts measured at room temperature decrease to the level of those retained at 4°C in the presence of metabolic poisons, and (3) cell-associated counts show a minimal and nonsaturable dependence on cation concentration, consistent with a very large number of low-affinity binding sites.

4. The best filter head currently available is the sampling manifold made by Millipore Corporation, which uses 25 mm diameter filters of any type. Individual (single) filter heads can be purchased from various manufacturers that use filters of different sizes, but the Millipore manifold allows 12 samples to be analyzed at once. Generally, we use two such manifolds connected through a Y-connector to a single vacuum trap, which then connects to house vacuum or a vacuum pump. A vacuum trap is essential to avoid accidental spillover of (radioactive) filtrate into the vacuum pump or house vacuum line.The best filters we have found are #BA85 from Schleicher and Schuell, which are cellulose acetate based. Glass fiber filters tend to bind cations and result in high backgrounds. Millipore HAWP and Gelman

Metricel GN-6 filters are comparable in background retention to the BA85 filters but retain fewer cells and clog more readily.

5. The assay above can be readily adapted to measurement of efflux from cells as long as several parameters are controlled and some caveats understood. First, it must be understood that it is almost impossible, using whole cell assays, to assign the efflux of a cation to a single transport system. Many transporters can mediate cation:cation exchange under some conditions. In addition, to measure efflux using radioisotopes, it is necessary to have an influx system present in the first place to get the isotope inside the cell. However, once that isotope exits the cells, the influx system is still there and can take the cation back up into the cell, thus obscuring the efflux process. Finally, because the free concentration of cation within the cell is generally not known and cannot be controlled readily, kinetic parameters are impossible to determine. Despite these caveats, efflux can be measured and can give useful information about cation homeostasis.

   Efflux is generally measured in one of three ways. A relatively large aliquot of cells is initially grown in a medium (minimal or rich) containing the appropriate radioisotope with no or minimal added nonradioactive cation. High concentrations of nonradioactive cation should be avoided to increase specific activity within the cell. In the first approach, cells are washed with a minimal medium containing no divalent cation and placed on ice. At $t = 0$, the cells are rapidly and acutely resuspended in approximately their original volume of a minimal medium prewarmed to the desired temperature. Aliquots are then filtered at known time points after resuspension. The only limitation is that the filtered aliquot should not contain more than 0.2 OD_{600nm} of cells to avoid clogging the filter. It is relatively easy to take single aliquots at many different time points or replicate aliquots at a few discrete time points. Efflux is often rapid, and it is thus important to take quite a few time points during the first 5 min of efflux. Points should be taken until at least 50% of total isotope has been released from the cells. It is not necessary to add the test divalent cation to the medium, although this can be tested to determine if it increases the rate of efflux, thus potentially indicating a cation:cation exchange process.

   In the second approach, the cells grown in radioisotope are not centrifuged and washed but are diluted at least 100-fold using the growth medium. This dilutes the radioisotope, but not the actual cation concentration, assuming the major source of the cation being tested is the medium. This creates a gradient of isotope from inside to outside the cell. Thus, radioisotope will exit the cell allowing a measure of efflux through determination of cellular loss of radioisotope. This approach is somewhat more cumbersome because of the necessity to filter much larger volumes of medium. However, it disrupts growth by a different means than centrifugation and resuspension (in the first approach above). It also avoids the necessity to add a large amount of the test cation (see next paragraph).

   The third approach is similar to the second, but rather than dilution of the entire culture, a small aliquot of the cation being tested is added sufficient to raise the extracellular cation concentration at least 100-fold if not 1000-fold or greater. This dilutes the specific activity of the cation allowing radioisotope to exit the cell down its chemical gradient and thus efflux to be monitored by cellular loss of radioactivity. This approach of course suffers from the necessity to add a large concentration of a divalent cation that may disrupt cellular processes.

In general, efflux should be measured by all three methods. If the rate of efflux, determined by plotting the $\log_{10}$ of the radioactivity retained by the cells against time, is linear and roughly equivalent for the three approaches, one can be reasonably assured that the experimental manipulations have not altered the efflux transport process or altered the cell so that efflux is indirectly altered.

6. For cells that are relatively fragile, centrifugation through oil offers advantages. The assay is almost identical to that for bacteria described in **Subheading 3** above except that cell aliquots are layered over an oil mixture in a microcentrifuge tube and centrifuged to separate cells from the medium.

   The advantages of the approach are that more cell surface water is removed, lowering background, and that quite fragile cells can be analyzed. The disadvantages are that the method is more cumbersome, less volume of cells can be assayed per aliquot, and post-assay processing is lengthier. Nonetheless, the method is relatively rapid and accurate. In addition, total cation content can also be readily measured by this method. The oil mixture that is most commonly used is a 2:1 mixture of dibutyl- and dioctyl-phthalate. The density of the mixture is intermediate between that of water and most cells, bacterial or eukaryotic. Tests should be conducted to ensure that the density is appropriate for the cell type being examined. The ratio can be adjusted in either direction to provide an appropriate density. Up to 1.0 mL cells can be layered over 0.3 mL oil mixture in a 1.5-mL microcentrifuge tube. We generally use 0.8 mL cells layered over 0.5 mL oil mixture and centrifuge for 30 s. Care must be taken not to shake or tilt the microcentrifuge tubes to much. The aqueous layer is then removed by aspiration. The oil layer can be retained or it can be aspirated if the sample is to be counted for radioactivity. Regardless, the sides of the tube must be swabbed with a cotton-tipped applicator to remove any residual water. The cell pellet does not usually have to be resuspended for counting. The lid may be cut off and the entire tube placed in a scintillation vial. This approach is also optimal for determination of total cation content by atomic absorption. For this purpose, both the aqueous and oil layers are removed, the tubes swabbed and 0.1 mL 1.0 N HNO_3 added. The sealed tubes are the batch sonicated to solubilize the cation before assay using atomic absorption.

References

1. Rudolf, R., Mongillo, M., Rizzuto, R., and Pozzan, T. (2003) Looking forward to seeing calcium. *Nat. Rev. Mol. Cell Biol.* **4,** 579–586.
2. Tsien, R. Y. (2003) Breeding molecules to spy on cells. *Harvey Lect.* **99,** 77–93.

3. Aslamkhan, A. G., Aslamkhan, A., and Ahearn, G. A. (2002) Preparation of metal ion buffers for biological experimentation: a methods approach with emphasis on iron and zinc. *J. Exp. Zool.* **292,** 507–522.
4. Colvin, R. A., Fontaine, C. P., Laskowski, M., and Thomas, D. (2003) Zn^{2+} transporters and Zn^{2+} homeostasis in neurons. *Eur. J. Pharmacol.* **479,** 171–185.
5. Reynolds, I. J. (2004) Fluorescence detection of redox-sensitive metals in neuronal culture: focus on iron and zinc. *Ann. N.Y. Acad. Sci.* **1012,** 27–36.
6. Petrat, F., de, G. H., Sustmann, R., and Rauen, U. (2002) The chelatable iron pool in living cells: a methodically defined quantity. *Biol. Chem.* **383,** 489–502.
7. Grubbs, R. D., Snavely, M. D., Hmiel, S. P., and Maguire, M. E. (1989) Magnesium transport in eukaryotic and prokaryotic cells using magnesium-28 ion. *Methods Enzymol.* **173,** 546–563.
8. Hmiel, S. P., Snavely, M. D., Florer, J. B., Maguire, M. E., and Miller, C. G. (1989) Magnesium transport in *Salmonella typhimurium*: genetic characterization and cloning of three magnesium transport loci. *J. Bacteriol.* **171,** 4742–4751.
9. Snavely, M. D., Florer, J. B., Miller, C. G., and Maguire, M. E. (1989) Magnesium transport in *Salmonella typhimurium*: $^{28}Mg^{2+}$ transport by the CorA, MgtA, and MgtB systems. *J. Bacteriol.* **171,** 4761–4766.
10. Kehres, D. G. and Maguire, M. E. (2002) Structure, properties and regulation of magnesium transport proteins. *Biometals* **15,** 261–270.
11. Cech, S. Y., Broaddus, W. C., and Maguire, M. E. (1980) Adenylate cyclase: the role of magnesium and other divalent cations. *Mol. Cell. Biochem.* **33,** 67–92.
12. Kehres, D. G., Zaharik, M. L., Finlay, B. B., and Maguire, M. E. (2000) The NRAMP proteins of *Salmonella typhimurium* and *Escherichia coli* are selective manganese transporters involved in the response to reactive oxygen. *Mol. Microbiol.* **36,** 1085–1100.
13. Kehres, D. G., Janakiraman, A., Slauch, J. M., and Maguire, M. E. (2002) SitABCD is the alkaline Mn^{2+} transporter of *Salmonella enterica* serovar Typhimurium. *J. Bacteriol.* **184,** 3159–3166.
14. Papp-Wallace, K. M. and Maguire, M. E. (2006) Manganese transport and the role of manganese in virulence. *Annu. Rev. Microbiol.* **60,** 187–209.
15. Cheng, Y. C. and Prusoff, W. H. (1973) Relationship between the inhibition constant (Ki) and the concentration of inhibitor which causes 50 percent inhibition (I50) of an enzymatic reaction. *Biochem. Pharmacol.* **22,** 3099–3108.
16. Hantke, K. (1997) Ferrous iron uptake by a magnesium transport system is toxic for *Escherichia coli* and *Salmonella typhimurium. J. Bacteriol.* **179,** 6201–6204.
17. Chamnongpol, S. and Groisman, E. A. (2002) Mg^{2+} homeostasis and avoidance of metal toxicity. *Mol. Microbiol.* **44,** 561–571.
18. Papp, K. M. and Maguire, M. E. (2004) The CorA Mg^{2+} transporter does not transport Fe^{2+}. *J. Bacteriol.* **186,** 7653–7658.

19. Smith, R. M. and Martell, A. E. (2002) NIST critically selected stability constants of metal complexes database, NIST Standard Reference Database 46, Version 8.0. http://www.nist.gov/srd/nist46.htm.
20. Nelson, D. L. and Kennedy, E. P. (1971) Magnesium transport in *Escherichia coli*. Inhibition by cobaltous ion. *J. Biol. Chem.* **246,** 3042–3049.

15

Methods in Cell-to-Cell Signaling in *Salmonella*

Brian M. M. Ahmer, Jenee N. Smith, Jessica L. Dyszel, and Amber Lindsay

Summary

Many bacteria can sense their population density. This has been termed "quorum sensing." The bacteria use this information to coordinate their behavior, essentially behaving as multicellular organisms. The paradigm of Gram-negative quorum sensing is the LuxI/LuxR-type system employed by *Vibrio fischeri* to regulate luminescence. The LuxR transcription factor detects the presence of *N*-acylhomoserine lactones (AHLs) produced by LuxI. The AHL diffuses freely across the cell wall, and its accumulation signals a high population density within a confined space. Upon binding AHL, the LuxR transcription factor activates the luminescence genes. Homologous systems are used by numerous Gram-negative pathogens to regulate host interaction genes. The AHLs produced by different LuxI homologs can vary in the length and modification of their acyl side chain. In the first section of this chapter, we describe the use of bacterial biosensors to determine whether a particular bacterial species synthesizes AHLs. The second section describes how to identify AHL-responsive genes in *Salmonella typhimurium*, an organism that detects but does not synthesize AHLs. The approach described can be modified for use with any organism that responds to AHLs but does not synthesize them. The third section describes the use of recombination-based in vivo expression technology (RIVET) to study AHL detection in vitro and in vivo, in this case the mouse gut.

Key Words: *luxR*; *sdiA*; *N*-acylhomoserine lactones; biosensor.

1. Introduction

Many bacteria can sense their population density *(1)*. This has been termed "quorum sensing." The bacteria use this information to coordinate their behavior, essentially behaving as multicellular organisms. To date, the vast

From: *Methods in Molecular Biology, vol. 394: Salmonella: Methods and Protocols*
Edited by: H. Schatten and A. Eisenstark © Humana Press Inc., Totowa, NJ

majority of quorum-sensing systems are used to regulate genes involved with eukaryotic host interactions. These interactions can range from symbiotic to pathogenic depending on the bacterial species and host involved. Environmental bacteria are likely to use information concerning population density as well, but examples are currently scarce.

The paradigm of Gram-negative quorum sensing is the LuxI/LuxR-type system employed by *Vibrio fischeri* to regulate luminescence. The LuxR transcription factor detects the presence of *N*-acylhomoserine lactones (AHLs) produced by LuxI. The AHL diffuses freely across the cell wall, and its accumulation signals a high population density within a confined space. Upon binding AHL, the LuxR transcription factor activates the luminescence genes. This system allows *V. fischeri* to be luminescent at a high population density within its host, the Bobtail squid (*Euprymna scolopes*), but dark when drifting free in the ocean.

Homologous systems are used by numerous Gram-negative pathogens to regulate host interaction genes. The AHLs can vary in length and modification of their acyl side chain. Each LuxI/LuxR pair typically synthesizes and detects, respectively, one predominant species of AHL. LuxI of *V. fischeri* synthesizes an AHL of moderate acyl chain length, *N*-(3-oxo-hexanoyl)-L-homoserine lactone (oxoC6). *Pseudomonas aeruginosa* has two systems. The RhlI/RhlR pair synthesizes and detects, respectively, the small AHL, *N*-butanoyl-L-homoserine lactone (C4). The second pair, LasI/LasR, synthesizes and detects *N*-(3-oxo-dodecanoyl)-L-homoserine lactone (oxoC12).

Escherichia coli and *Salmonella enterica* serovar Typhimurium (*Salmonella typhimurium*) are unusual in that they encode a LuxR homolog, SdiA, but not a LuxI homolog *(2)*. It has been shown that *E. coli* and *S. typhimurium* use SdiA to detect exogenous sources of AHLs. How the detection of AHL benefits these organisms is not yet known; however, it is known that *S. typhimurium* regulates host interaction genes in response to the AHL production of other bacterial species. Oddly, *sdiA* expressed from its natural position in the chromosome can activate chromosomal fusions to its target genes only during growth in motility agar (at least this is currently the only laboratory condition known). SdiA will activate plasmid-based fusions under any condition. Therefore, experiments using SdiA and chromosomal fusions in this chapter will be performed in motility agar, but this growth condition can be changed for use with other systems.

In the first section, we describe the use of "biosensors" to determine whether a particular bacterial species synthesizes AHLs. These biosensors are simply *E. coli* carrying a plasmid that encodes both a LuxR homolog and a promoterless luciferase operon regulated by that homolog (*luxCDABE* from *Photorhabdus luminescens*). Because *E. coli* does not synthesize AHLs, the biosensor strain will luminesce only in the presence of AHLs from another source. Because luciferase activity can be affected by oxygen and metabolic factors, and because all promoters have some level of background activity, a second biosensor strain is used in parallel. This second strain lacks the *luxR* homolog so that the *luxR*-dependence of any particular response can be determined. Additionally, all the *E. coli* biosensor strains described here are mutated at the *sdiA* locus because *sdiA* can interfere with *luxR*-type biosensor systems *(3)*.

The second section describes how to identify AHL-responsive genes in *S. typhimurium*, but the approach can be modified for use with any organism that responds to AHLs but does not synthesize them. For example, in our laboratory we have identified AHL-responsive genes in *E. coli* K-12, *E. coli* O157:H7, *Klebsiella pneumoniae*, and *Enterobacter cloacae* (unpublished). Given that many AHL-responsive genes in these organisms appear to be horizontal acquisitions that are species specific, or even serovar specific, these genetic screens should be repeated by other laboratories working on related serovars. The approach described uses a transposon to generate random luciferase fusions, and these are screened for a response to AHL. This approach has the disadvantage that genes essential for growth cannot be screened, but it has the advantage that it can be used in organisms that do not yet have microarrays available. This approach can also be used with growth conditions that are not amenable to microarray analysis (we were unable to recover usable RNA from *S. typhimurium* grown in motility agar, for example).

The third section describes the use of recombination-based in vivo expression technology (RIVET) *(4)* to study AHL detection in vitro and in vivo, in this case the mouse gut. This same protocol can be modified to study AHL detection in essentially any environmental setting.

2. Materials

2.1. Screening Organisms for AHL Production

1. The six biosensor strains listed in **Table 1** and **Fig. 1** *(3)*.
2. Your bacterial strain(s) of interest.
3. Luria–Bertani (LB) broth Miller (EMD chemicals): Add 25 g LB broth powder to 1 L deionized water and autoclave. LB broth can be stored at room temperature (RT) (*see* **Note 1**).
4. LB agar Miller (EMD chemicals): Add 37 g LB agar to 1 L deionized water and autoclave. LB agar is cooled to 50°C before antibiotics are added and is poured into sterile Petri plates. Plates can be stored for 1 month at 4°C (*see* **Note 1**).

Table 1
Biosensors

Strain name	Quorum-sensing system	Fusion	Function
JLD271 + pAL101	Rhl system of *Pseudomonas aeruginosa*	*rhlRI'::luxCDABE*-encodes the RhlR protein and promoter of the *rhlI* gene	Detects C4
JLD271 + pAL102		*rhlI'::luxCDABE*-encodes only the *rhlI* promoter	Control for pAL101; does not detect AHL
JLD271 + pAL103	Lux system of *Vibrio fischeri*	*luxRI'::luxCDABE*-encodes the LuxR protein and promoter of the *luxI* gene	Detects oxoC6
JLD271 + pAL104		*luxI'::luxCDABE*-encodes only the *luxI* promoter	Control for pAL103; does not detect AHL
JLD271 + pAL105	Las system of *P. aeruginosa*	*lasRI'::luxCDABE*-encodes the LasR protein and promoter of the *lasI* gene	Detects oxoC12
JLD271 + pAL106		*lasI'::luxCDABE*-encodes only the *lasI* promoter	Control for pAL105; does not detect AHL

5. Syringes and syringe filters (0.22-μm cellulose acetate) (Life Science Products).
6. Tetracycline (tet) (Sigma-Aldrich), use at 20 μg/mL. The stock solution is 20 mg/mL in methanol. Cover container with aluminum foil, as tet is light sensitive, and store at 4°C.
7. Acidified ethyl acetate (EA). The EA stock solution is prepared by adding glacial acetic acid to EA (Sigma-Aldrich) at 0.1 mL/L. Store at RT.
8. Synthetic AHLs: *N*-butyryl-DL-homoserine lactone (a mixture of C4 isomers) (Sigma-Aldrich 09945), oxoC6 (Sigma-Aldrich K3007), and *N*-dodecanoyl-DL-homoserine lactone (a mixture of C12 isomers) (Sigma-Aldrich 17247). Make 1 mM stocks in EA and store at −20°C.
9. Intensified charge-coupled device (CCD) camera (*see* **Note 2**).

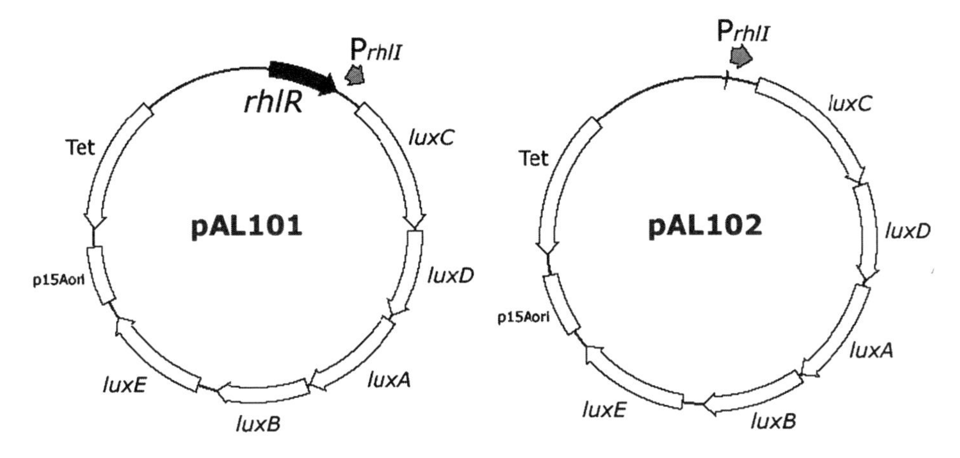

Fig. 1. Map of plasmids pAL101 and pAL102. The pAL101 plasmid encodes both RhlR and a promoter regulated by RhlR, P_{rhlI}. The *rhlI* promoter is fused to the promoterless *luxCDABE* operon. The pAL102 plasmid is identical except that *rhlR* is not present allowing the *rhlR*-dependence of any response to be determined. The *luxR*-based biosensors, pAL103 and pAL104, and the *lasR*-based biosensors, pAL105 and pAL106, are similar in organization.

10. Microtiter plate reader capable of recording luminescence and absorbance (*see* **Note 2**).
11. 96-well polystyrene microtiter plates with black walls and clear, flat bottoms (Corning #3651).

2.2. Screening for AHL-Regulated Genes

2.2.1. Generation of Random Luciferase Fusions Using a Transposon

1. *Escherichia coli* strain BW20767 + pUT-Tn5 *luxCDABE* Km2 (*see* **Table 2**).
2. Your bacterial strain of interest.
3. LB agar Miller (EMD Chemicals): Add 37 g LB agar to 1 L deionized water and autoclave. LB agar is cooled to 50°C before antibiotics are added and is poured into Petri plates. Plates can be stored for 1 month at 4°C.
4. LB broth Miller (EMD Chemicals): Add 25 g LB broth to 1 L deionized water and autoclave. LB broth can be stored at RT.
5. Kanamycin (kan) (Sigma-Aldrich), use at 50 μg/mL. The stock solution is 50 mg/mL in water. Filter-sterilize using a syringe and syringe filter. Store at 4°C.
6. Nalidixic Acid (nal) (Sigma-Aldrich), use at 50 μg/mL. The stock solution is 50 mg/mL in water. Bring to pH 11 to dissolve. Filter-sterilize using a syringe and syringe filter. Store at 4°C.

Table 2
Other Strains

Strain or plasmid	Genotype	Source or reference
BW20767	*Escherichia coli leu-63*::IS*10 recA1 creC510 hsdR17 endA1 zbf-5 uidA*(ΔMluI)::*pir*$^{+}$ *thi* RP4-2-tet::Mu-1kan::Tn7	*(5)*
GY4493	JB580v *yenI*::kan	Glenn Young, unpublished
JB580v	Wild-type *Yersinia enterocolitica* Serogroup O:8; nal^{r} *yenR* (R^{-}M^{+})	*(6)*
JNS3206	JS246 *srgE10::tnpR-lacZY* (kan^{r})	Unpublished
JNS3226	JS246 *srgE10::tnpR-lacZY sdiA1*::mTn*3* (kan^{r} amp^{r})	Unpublished
JS246	14028 zjg8103::*res1-tetRA-res1*	*(7)*
pUT-Tn5 *luxCDABE* Km2	pUT suicide vector (oriR6K, amp^{r}) carrying Tn*5 luxCDABE* Km2	*(8)*

2.2.2. Identifying AHL-Regulated Fusions

1. Agar agar (EMD chemicals).
2. LB motility agar (0.3 %) supplemented with kan and nal, and either 1 μM oxoC6 or the solvent control EA at a final concentration of 0.1 %. Motility agar is made by combining LB broth with 3 g Agar agar (EMD Chemicals) per liter and autoclaving. Cool to 50°C before adding antibiotics and either EA or oxoC6 (*see* **Note 3**).
3. Acidified EA. The EA stock solution is prepared by adding glacial acetic acid to EA (Sigma-Aldrich) at 0.1 mL/L. Store at RT.
4. Synthetic AHL of choice. In this protocol, we use *N*-(3-oxo-hexanoyl)-L-homoserine lactone (oxoC6) (Sigma-Aldrich K3007). Make 1 mM stock in acidified EA. Store at −20°C.
5. Kan (Sigma-Aldrich), use at 50 μg/mL. The stock solution is 50 mg/mL in water. Filter-sterilize using a syringe and syringe filter. Store at 4°C.

6. Nal (Sigma-Aldrich), use at 50 μg/mL. The stock solution is 50 mg/mL in water. Bring to pH 11 to dissolve. Filter-sterilize using a syringe and syringe filter. Store at 4°C.
7. 96-well polystyrene microtiter plates with black walls and clear, flat bottoms (Corning #3651).
8. Microtiter plate reader capable of recording luminescence and absorbance (*see* **Note 2**).

2.2.3. Determination of Transposon Insertion Point

1. GenElute Bacterial Genomic DNA Kit (Sigma, St. Louis, MO).
2. DNA sequencing primers that bind to the edge of the transposon oriented outward (*see* **Note 4**).
3. Applied Biosystems 3730 DNA Analyzer or appropriate fee-for-service facility (*see* **Note 5**).

2.3. Study of AHL Detection in the Environment Using RIVET

2.3.1. Measurement of Resolvase Activity In Vitro

1. *Salmonella enterica* serovar Typhimurium strains JNS3206 and JNS3226 (*see* **Table 2**). These are *sdiA*$^+$ and *sdiA*::mTn*3* [ampicillin-resistant (amp^r)] strains, respectively, that each carry a *srgE*::*tnpR-lacZY* fusion and a *res1-tetRA-res1* cassette.
2. LB agar Miller (EMD chemicals): Add 37 g LB agar to 1 L deionized water and autoclave. LB agar is cooled to 50°C before antibiotics are added and is poured into Petri plates. Plates can be stored for 1 month at 4°C.
3. LB broth Miller (EMD chemicals): Add 25 g LB broth to 1 L deionized water and autoclave. LB broth can be stored at RT.
4. Agar agar (EMD chemicals)
5. LB motility agar (0.3 %) supplemented with kan and either 100 nM oxoC6 or an equal volume 0.01 % EA. Motility agar is made by combining LB broth with 3 g Agar agar per liter and autoclaving. Cool to 50°C before adding antibiotics and either EA or oxoC6 (*see* **Note 3**).
6. Kan (Sigma-Aldrich), final concentration varies in the protocol. The stock solution is 50 mg/mL in water. Filter-sterilize using a syringe and syringe filter. Store at 4°C.
7. Ampicillin (amp) (Sigma-Aldrich), use at 150 μg/mL. The stock solution is 100 mg/mL in water. Filter-sterilize using a syringe and syringe filter. Store at 4°C.
8. Tetracycline (tet) (Sigma-Aldrich), use at 20 μg/mL. The stock solution is 20 mg/mL in methanol. Filter-sterilize using a syringe and syringe filter. Cover bottle with aluminum foil, as tet is light sensitive, and store at 4°C.
9. Acidified EA. The EA stock solution is prepared by adding glacial acetic acid to EA (Sigma-Aldrich) at 0.1 mL/L. Store at RT.

10. Synthetic AHL of choice. In this protocol, we use oxoC6 (Sigma-Aldrich K3007). Make a 1 mM stock in acidified EA. Store at −20°C.

2.3.2. Measurement of Resolvase Activity In Vivo

An animal-use protocol must be submitted and approved by your Institutional Animal Care and Use Committee or equivalent body before beginning any animal studies.

1. *Salmonella enterica* serovar Typhimurium strains JNS3206 and JNS3226 (*see* **Table 2**). These are *sdiA*$^+$ and *sdiA*::mTn*3* (amp^r) strains, respectively, that each carry a *srgE*::*tnpR-lacZY* fusion and a *res1-tetRA-res1* cassette.
2. Female CBA/J mice (8–10 weeks old) (Jackson Laboratories).
3. Animal Feeding Stainless Steel Biomedical Needles used for Oral Gavage of Mice (20 gauge and 1.5 inches) (Popper & Sons, Inc.).
4. Tuberculin 1 cc syringes with Slip Tip, BD injection systems (VWR).
5. Kan (Sigma-Aldrich), final concentration is 50 μg/mL in most plates, but in Xylose Lysine Desoxycholate (XLD) plates we use 150 μg/mL. The stock solution is 50 mg/mL in water. Filter-sterilize using a syringe and syringe filter. Store at 4°C.
6. Amp (Sigma-Aldrich), use at 100 μg/mL. The stock solution is 100 mg/mL in water. Filter-sterilize using a syringe and syringe filter. Store at 4°C.
7. Tet (Sigma-Aldrich), use at 20 μg/mL. The stock solution is 20 mg/mL in methanol. Cover container with aluminum foil, as tet is light sensitive, and store at 4°C.
8. AHL-producing organism *Yersina enterocolitica* and an isogenic *yenI* mutant (unable to synthesize AHLs) (*see* **Table 2**).
9. XLD agar (EMD Chemicals). Prepare by adding 55 g to 1 L water. Heat the agar to a boil to dissolve it completely. Cool agar in water bath rapidly to 50°C before pouring plates.
10. Tissue homogenizer (The least expensive solution is a microtube pestle from Life Science Products, Inc. A much more expensive solution is the Cyclone I.Q.[2] from Virtis.).

3. Methods

3.1. Screening Organisms for AHL Production Using the Cross-Streak Assay

1. Inoculate each biosensor strain in 5 mL LB tet and grow overnight at 37°C with shaking.
2. Inoculate bacteria to be tested in 5 mL LB and grow overnight at 37°C with shaking (*see* **Note 1**).

3. On three LB plates, add 20 μL overnight culture of the bacteria to be tested onto the side of the plate, just to the right of the center of the plate. Allow volume to "roll" entire way down plate, to form the horizontal cross-streak as seen in **Fig. 2**. If necessary, roll volume back and forth until distributed evenly. Allow this streak to dry completely. Mark the location of the cross-streak (*see* **Note 6**).
4. Use synthetic AHLs as positive controls for biosensor function. For each of the three AHLs, dilute at 1:100 in EA to make a 10-μM solution. Drip 20 μL of each 10 μM AHL solution across an LB plate, as done in **Subheading 3.1.3.** with the test bacteria (three plates total). The EA can cause the AHLs to roll much faster and dry very quickly, leaving an invisible streak. It is important to mark the location of the cross-streak before the AHL dries. However, using plates that have been poured and left to dry 2–3 days in advance makes for the best AHL and EA cross-streaks. A high moisture content in plates can cause these streaks to roll in random, multiple directions and to dry very slowly. If more recently poured plates are used, they can be dried quickly by placing them with their lids open in a laminar flow hood before the streaks have been applied.
5. On an additional three LB plates, use 20 μL EA to form a cross-streak. These will serve as solvent controls for the synthetic AHL plates. Mark the location of the cross-streak on each plate.
6. Once the cross-streaks are completely dry, slowly drip 10 μL of each biosensor pair perpendicular to the cross-streaks in duplicate. For example, JLD271 + pAL101 and JLD271 + pAL102 are dripped to have two streaks each on three different plates: one of the plates crossed with the test bacteria, one of the plates crossed with EA, and the only plate crossed with C4. This is repeated for the pAL103/104 and pAL105/106 sets of biosensor strains resulting in a total of nine plates (three plates each set). These perpendicular streaks cannot be rolled back and forth to make the streak uniform. It is important for the biosensors to pass through the cross-streak only once and that they dry quickly to avoid contaminating the entire streak.
7. Once the streaks are dry, incubate at 37°C. Remove the plates after 3, 6, and 9 h and image with the CCD camera (*see* **Fig. 2**).
8. Compare the EA and AHL plates. These plates demonstrate that the biosensors are working properly. There should be increased luminescence of the biosensor strain in the vicinity of the AHL streak but not the EA streak. The control biosensor that lacks the *luxR* homolog (pAL102, pAL104, and pAL106) should not show this differential light production. This demonstrates that the differential light production is dependent on the *luxR* homolog.
9. Examine the plates that were struck with the test bacteria. If the test strain is producing AHLs that are detectable by the biosensor, there will be an increase in light production in the vicinity of the test strain but not further away. The control biosensor that lacks a *luxR* homolog (pAL102, pAL104, and pAL106) should not show this differential light production (*see* **Fig. 2**).

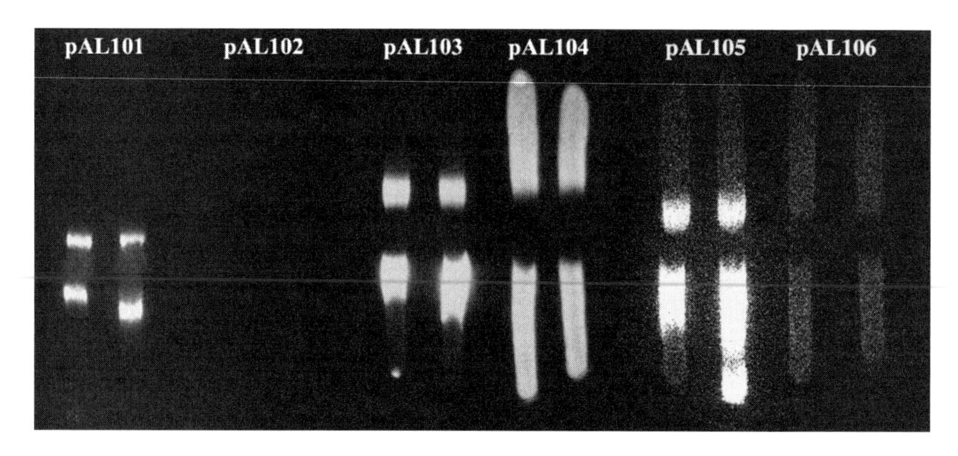

Fig. 2. The six biosensor strains in **Table 1** were cross-streaked in duplicate against *Pseudomonas aeruginosa* as described in the text. The *rhlR, luxR*, and *lasR* biosensors all respond to AHLs produced by *P. aeruginosa* as seen by the luminescence response observed near the cross-streak but not further away. The negative control plasmids show that this response requires the *luxR* homolog. The pAL104 plasmid has a higher level of basal luminescence than does pAL103. The luminescence from pAL104 is not localized to the region nearest the cross-streak and instead extends throughout the pAL104 streak showing that this is not a response to *P. aeruginosa*. The pAL104 strain is best used on a separate plate from pAL103. In this way, the sensitivity of the camera can be reduced until the basal luminescence emanating from the pAL104 strain is barely visible. A lack of increased luminescence nearest the *P. aeruginosa* is then easier to observe.

3.2. Screening Organisms for AHL Production Using a Liquid Assay

1. Inoculate the bacteria to be tested in 5 mL LB and grow overnight at 37°C with shaking (*see* **Note 1**).
2. The following day (day 2), subculture the test bacteria by inoculating 100 mL LB with 1 mL overnight culture and incubate at 37°C with shaking.
3. Harvest culture supernatant at three growth stages: log phase, late log, and stationary phase. To do this, remove 20 mL culture at each time point, centrifuge at 8000 × *g* for 10 min and collect the supernatant.
4. Filter the supernatant using a syringe and syringe filter. Store at 4°C.
5. On day 2, inoculate each biosensor strain in 5 mL LB tet and grow overnight at 37°C with shaking.
6. On day 3, dilute each of the six biosensor strains at 1:100 by adding 1 mL culture to 99 mL LB tet.

7. Using a separate 96-well microtiter plate for each of the biosensor strains, pipet 100 μL biosensor into every well of the plate.
8. Prepare a 10-fold dilution series of the three filtered supernatant samples and each of the synthetic AHLs (six samples). The easiest way to do this is to place the full strength samples in the first column of a 96-well plate. Then place 180 μL LB in all of the remaining wells. Pipet 20 μL of each full strength sample into the next column creating a 1:10 dilution. Repeat until all of the wells have been used except for the last column. The last column will serve as the "zero" concentration.
9. Pipet 25 μL of each sample well into the corresponding wells of each of the six biosensor plates.
10. Take a Time 0 reading using the Wallac Victor^2 1420 plate reader or a similar machine. Record both optical density at 550 nM and luminescence (relative light units or RLU) for each well of the six biosensor plates.
11. Incubate the plates at 37° C in a humid chamber (*see* **Note 7**). Record optical density and light production after 3, 6, and 9 h of incubation.
12. Use Excel or a similar graphing program to generate RLU/OD_{550} for each of the wells. Plot this versus AHL concentration for each biosensor strain. The synthetic AHL dilution series provides a standard curve of known AHL concentrations. These curves can be used to estimate the effective concentration of signal present in the filtered culture supernatant. A true signal will only activate one or more of the biosensor strains that contains a *luxR* homolog. The three strains that lack the *luxR* homolog provide information regarding how much luminescence can be expected in the absence of *luxR*-dependent transcription activation.

3.3. Generation of Random Luciferase Fusions Using a Transposon

1. Isolate a spontaneous nal-resistant (nal^r) mutant of your strain of interest. This drug resistance will be used as the counter selectable marker in the conjugation. Isolating the nal^r mutant can be done by plating bacteria on LB agar plates containing nal followed by overnight incubation at 37°C. Pick a colony and store it as a glycerol stock (1 mL liquid culture + 1 mL 50 % glycerol in a cryovial at −80°C). This strain will be referred to as the recipient strain. Alternatively, a suitable nutrient requirement can be identified by screening for growth on various carbon sources.
2. Grow the recipient strain in LB nal at 37°C overnight. Grow the donor strain, BW20767 + pUT Tn5 *luxCDABE* Km2, in LB kan amp at 37°C overnight.
3. The next day, centrifuge 1 mL of each culture at 8000 × *g* for 10 min, discard supernatant, and resuspend each pellet in 1 mL LB without antibiotics.
4. Mix together 100 μL of each strain and plate on an LB agar plate without antibiotics. Repeat this on 10 plates. Incubate the plates overnight at 37°C to allow the donor and recipient to conjugate. Also plate 100 μL of each strain on the LB kan nal plates that will be used the next day and incubate overnight at 37°C. This is done as a control to ensure that the antibiotics in the plates are working and neither the

donor nor recipient strain is able to grow on them. Any growth on these plates after incubation indicates that the following selection step will not be successful.

5. The following day, use 2 mL LB for each of the 10 plates to harvest the bacteria and remove to a microcentrifuge tube. Centrifuge at $8000 \times g$ for 10 min, discard supernatant, and resuspend each in 1 mL LB. From each tube, plate 10 μL on 10 LB kan nal plates and 100 μL on each of nine more LB kan nal plates. Repeat with all 10 conjugation plates, yielding 190 plates total, and incubate the plates overnight at 37°C. If you are not sure that *E. coli* will be able to conjugate with your strain of interest, one of the 10 conjugation plates can be tested first. Store the other nine plates at 4°C and plate the conjugation mix on the 19 LB kan nal plates as described above. This can save much time and many plates in the case that the procedure needs to be adjusted one or more times to get the desired results. The volumes plated may need to be adjusted so that roughly 300 colonies are obtained per plate. Each of the resulting colonies has the transposon inserted at a different random site in the chromosome and is thus referred to as a mutant. This process should be repeated until at least 10,000 mutants have been obtained (*see* **Note 8**).

3.4. Identifying AHL-Regulated Fusions

1. Prepare two solutions of LB motility agar, one containing 0.1 % EA and one containing 1 μM oxoC6. Pipet these solutions into alternating wells of a 96-well plate (200 μL per well) (*see* **Notes 3** and **9**). Let cool at RT for at least 20 min but no longer than a few hours. Label each 96-well plate with a number for identification during data analysis and mutant isolation.
2. With a sterile toothpick or pipet tip, pick a single Tn5 mutant colony from the LB kan nal plate and stab into the center of an EA well and the adjacent well of oxoC6. One person can easily patch 2000 colonies per day, but this step is best done by a group of four or more people so that the plates can all be placed in the incubator in a relatively short period of time (*see* **Note 10**).
3. Incubate plates at 37°C for 9 h in a moist environment (*see* **Note 7**).
4. For each plate, measure the luminescence generated from each well using the Wallac Victor^2 1420 plate reader, or a similar machine; include the plate number in the file name.
5. Use Microsoft Excel or a similar spreadsheet program to identify mutants carrying transposons in AHL-regulated genes by looking for differential light production in the EA and oxoC6 wells. It is important at this point to determine how the data has been organized for export by the plate reader so that the values from the appropriate wells can be compared.
6. To recover mutants of interest, obtain the appropriate 96-well plate and locate the wells in which it was stabbed by referring to the Excel sheet (e.g., plate 118, wells C3 and C4). Use a sterile toothpick or pipet tip to stab the EA well containing the mutant and streak to isolation on an LB kan nal plate. Incubate at 37°C overnight.
7. Repeat **steps 1–6** until at least 10,000 mutants have been screened (*see* **Note 8**).

8. To confirm that the mutants are truly responding to AHL, repeat the 96-well plate assay in triplicate for all recovered mutants (*see* **Note 11**).

3.5. Determination of Transposon Insertion Point

1. Prepare chromosomal DNA from an overnight culture of each AHL-responsive mutant using the GenElute Bacterial Genomic DNA Kit (Sigma) according to manufacturer's instructions.
2. Determine the insertion point of the transposon by DNA sequencing using one or more primers that bind to the end of the transposon oriented outward (*see* **Notes 4** and **5**).

3.6. Study of AHL Detection in the Environment Using RIVET

3.6.1. Measurement of AHL Detection In Vitro

1. Inoculate motility agar plates containing either 1 μM oxoC6 or 0.1 % EA (as the solvent control) with the RIVET strains (JNS3206 and JNS3226) by stabbing the strains individually into the center of the agar plates with a sterile toothpick or pipet tip. Inoculations should be done in triplicate. Incubate the plates overnight at 37°C (*see* **Note 3**).
2. Recover bacteria from the motility agar by stabbing the agar with a sterile toothpick or pipet tip and streak to isolation on LB kan plates.
3. The resolution of the *res1-tetRA-res1* cassette is measured by patching the colonies onto LB kan tet and recording the percentage that are tet sensitive. SdiA activity is indicated by a higher percentage of resolution in the *sdiA*$^+$ bacteria compared with the *sdiA* mutant bacteria.

3.6.2. Measurement of AHL Detection in Mice

1. Orally infect three groups of eight mice. Infect group 1 with an AHL-producing pathogen (*Y. enterocolitica*), group 2 with an isogenic mutant that does not produce AHL (a *yenI* mutant), and group 3 with a negative control (LB broth). Mice are given 10^7 cfu of the *Yersinia* strains (*see* **Note 12**).
2. After 24 h, orally infect all mice with 10^8 cfu of wild-type and *sdiA*-mutant RIVET strains.
3. On a daily basis, collect fecal samples from the mice (*see* **Note 13**). The fecal pellets (1–2) recovered are homogenized in 200 μL PBS and a dilution series is plated on XLD and XLD-kan plates (*see* **Note 14**). The XLD plates allow the growth of both *Salmonella* and *Yersinia* so that the numbers of organisms recovered can be recorded. The XLD-kan plates allow only the recovery of the *Salmonella* RIVET strains.
4. Patch isolated colonies from the XLD-kan plates onto LB kan amp and LB kan tet plates. The amp-sensitive colonies are *sdiA*$^+$ and the amp^r colonies are *sdiA* mutant. Changes in this ratio can be observed over time providing information on the fitness

phenotype of *sdiA*. The percentage of tet-sensitive colonies in each group provides a measure of *sdiA*-dependent activation of the *srgE::tnpR-lacZY* fusion.

4. Notes

1. It is assumed that LB is an appropriate medium to grow the organism to be tested for AHL production. All subsequent steps should be performed with a more suitable medium if LB is not appropriate for your specific organism(s) of interest. If this is the case, be sure to include the medium alone as a control in each experiment to test for activation of the biosensors. A media control is not included in this protocol because it is known that LB does not activate the biosensors. Additionally the temperature, time, and growth conditions may not be suitable for the organism being tested, so modify the protocol as necessary.
2. We use a Hamamatsu C2400-32 intensified CCD camera or a Wallac Victor^2 1420 microplate reader (Perkin Elmer).
3. Motility agar will become unstable after more than 48 h at 50°C. It is recommended that media be made the day before an experiment, poured the day of the experiment, and must be allowed to solidify *undisturbed* for 20 min to an hour before use. Antibiotics, EA, and AHLs should be added immediately before pouring.
4. We use at least two of the following four primers to provide verification of an insertion site:
   BA247 (GAGTCATTCAATATTGGCAGGTAAACAC),
   BA1090 (GAATGTATGTCCTGCGTCTTGAGTA),
   BA1496 (TCGGGAAAAATTTCAACCTGGCCGT),
   BA1497 (TTCCATCTTTGCCCTACCGTATAGA).
5. DNA sequencing using the bacterial chromosome as template can be performed by the Plant-Microbe Genomics Facility at Ohio State University (http://www.biosci.ohio-state.edu/~pmgf/).
6. This assay can be done using colonies rather than liquid cultures. Simply use sterile toothpicks or pipet tips to streak the colonies in the pattern described for liquid cultures. The liquid method tends to give clearer, more uniform results, but the colony method is much faster if screening hundreds or thousands of colonies.
7. The small volume of motility agar makes the wells very susceptible to drying. Plates should be stacked neatly in a plastic box or tray with wet paper towels and covered in plastic wrap during incubation.
8. Screening 10,000 mutants provides roughly 1× coverage of a typical genome containing 5000 genes (half of the insertions are in the wrong orientation). The Poisson distribution predicts that 1× coverage will sample 63 % of the genome. Given that not all 5000 genes can be hit by a transposon because some of them are essential for growth, 63 % is an underestimate. Obviously, increased coverage can be obtained with increased screening.
9. The wells can be filled with the medium of choice for the organism under study. LB motility agar was a special requirement for *sdiA* of *Salmonella*.

10. The luminescence of the bacteria can decrease dramatically after stationary phase so it is best to read the plates at about 9 h after inoculation. For practical reasons, this means that the plates need to be placed in the incubator early in the morning. This is why it is best for a group of people to rapidly patch 2000 colonies into the plates. A group of four people could patch all 10,000 colonies in a day, but somebody would need to be continuously reading batches of plates at the 9-h mark. This would go well into the night.
11. If your organism of interest is non-motile, the strain should be seeded into the motility agar for confirmation of response to AHL. Stabbing a non-motile strain can lead to differences in reading due to placement of the stab.
12. CBA/J mice are the preferred strain for long-term study of *Salmonella* in the gut as they are not susceptible to systemic *Salmonella* infection. BALB/c would be a better choice for studying *Salmonella* gene expression at systemic sites. Other mouse strains may be more appropriate for your organism of interest.
13. Fecal samples can be collected by placing an individual mouse on a clean paper towel then covering the mouse with a glass beaker with a side spout. The mice will defecate and fecal samples can be collected from paper towel. Clean both the bench top and beaker between collections with 70 % ethanol to avoid cross contamination.
14. One to two pellets can be resuspended in 200 μL PBS, serially diluted, and plated with good recovery of strains. Owing to the presence of particulate matter in the resuspended fecal samples, wide bore pipet tips should be used for preparing serial dilutions and plating when needed. The amount of sample collected should be consistent between mice at each time point. Fewer feces may be obtained over the course of the experiment as mice fall ill due to presence of the test pathogen (i.e., CBA/J mice are resistant to *Salmonella* but are sensitive to virulent *Y. enterocolitica* strains).

References

1. Waters, C. M. and Bassler, B. L. (2005) Quorum sensing: cell-to-cell communication in bacteria. *Annu. Rev. Cell Dev. Biol.* **21,** 319–346.
2. Ahmer, B. M. (2004) Cell-to-cell signalling in *Escherichia coli* and *Salmonella enterica. Mol. Microbiol.* **52,** 933–945.
3. Lindsay, A. and Ahmer, B. M. (2005) Effect of *sdiA* on biosensors of *N*-acylhomoserine lactones. *J. Bacteriol.* **187,** 5054–5058.
4. Slauch, J. M. and Camilli, A. (2000) IVET and RIVET: use of gene fusions to identify bacterial virulence factors specifically induced in host tissues. *Methods Enzymol.* **326,** 73–96.
5. Metcalf, W. W., Jiang, W., Daniels, L. L., Kim, S. K., Haldimann, A., and Wanner, B. L. (1996) Conditionally replicative and conjugative plasmids carrying *lacZ* alpha for cloning, mutagenesis, and allele replacement in bacteria. *Plasmid* **35,** 1–13.

6. Kinder, S. A., Badger, J. L., Bryant, G. O., Pepe, J. C., and Miller, V. L. (1993) Cloning of the YenI restriction endonuclease and methyltransferase from *Yersinia enterocolitica* serotype O8 and construction of a transformable R-M+ mutant. *Gene* **136,** 271–275.
7. Merighi, M., Ellermeier, C. D., Slauch, J. M., and Gunn, J. S. (2005) Resolvase-in vivo expression technology analysis of the *Salmonella enterica* serovar Typhimurium PhoP and PmrA regulons in BALB/c mice. *J. Bacteriol.* **187,** 7407–7416.
8. Winson, M. K., Swift, S., Hill, P. J., et al. (1998) Engineering the *luxCDABE* genes from *Photorhabdus luminescens* to provide a bioluminescent reporter for constitutive and promoter probe plasmids and mini-Tn*5* constructs. *FEMS Microbiol. Lett.* **163,** 193–202.

16

Development of *Salmonella* Strains as Cancer Therapy Agents and Testing in Tumor Cell Lines

Abraham Eisenstark, Robert A. Kazmierczak, Alison Dino, Rula Khreis, Dustin Newman, and Heide Schatten

Summary

Despite significant progress in the development of new drugs and radiation, deaths due to cancer remain high. Many novel therapies are in clinical trials and offer better solutions, but more innovative approaches are needed to eradicate the various subpopulations that exist in solid tumors. Since 1997, the use of bacteria for cancer therapy has gained increased attention. *Salmonella* Typhimurium strains have been shown to have a remarkably high affinity for tumor cells. The use of bacterial strains to target tumors is a relatively new research method that has not yet reached the point of clinical success. The first step in assessing the effectiveness of bacterial tumor therapy will require strain development and preclinical comparisons of candidate strains, which is the focus of this chapter. Several investigators have developed strains of *Salmonella* with reduced toxicity and capacity to deliver anti-tumor agents. Although methods for obtaining safe therapeutic strains have been relatively successful, there is still need for further genetic engineering before successful clinical use in human patients. As described by Forbes et al. in 2003, the main stumbling block is that, while bacteria preferentially embed within tumor cells, they fail to spread within the tumor and finish the eradication process. Further engineering might focus on creating *Salmonella* that remove motility limitations, including increased affinity toward tumor-generated chemotactic attractants and induction of matrix-degrading enzymes.

Key Words: *Salmonella* strain development; cancer therapy; *Salmonella* incorporation; cancer cells; *Salmonella*-containing vacuoles.

From: *Methods in Molecular Biology, vol. 394: Salmonella: Methods and Protocols*
Edited by: H. Schatten and A. Eisenstark © Humana Press Inc., Totowa, NJ

1. Introduction

Although several species of bacteria have been investigated as anti-cancer agents, most of the current approaches are focused on strains of *Salmonella*. Investigators are exploring avenues that include (1) direct attachment, invasion, and destruction of tumor cells; (2) bacteria as antigens for antibody production to specific tumor components; (3) bacteria as vectors to deliver foreign anti-tumor agents; (4) bacteria as agents that can metabolize accompanied drugs for greater effectiveness; and (5) bacteria as partners with other therapeutic agents, such as cisplatin or radiation. This chapter focuses on methods for development of genetically modified *Salmonella* strains that optimize tumor destruction without toxic effects. It is limited to the rationale and methods for the first step in strain development, i.e. testing of strains in tumor cell lines. After strain construction and observations of *Salmonella* anti-tumor phenotypes in tissue cultures, the obvious choice for an experimental animal is the mouse, a natural host for *Salmonella* serovars and therefore suited for anti-tumor efficacy and toxicity testing. Although the strains being studied are engineered to be non-toxic, these strains were originally developed from assumed ancestral murine (mouse) pathogens, some with a history of food poisoning in humans.

In any cancer therapy strategy, factors such as side effects, specificity, and a systemic delivery procedure to reach solid tumors must be considered. Advantages of developing special *Salmonella* strains as a delivery system include preferred migration to tumor cells versus non-tumor cells, non-toxicity, control with antibiotics, a large genomic reservoir for strain selection, engineering stimulation of the immune system at the tumor site, and the ability to deliver "foreign" genes by plasmid transfection and phage transduction. *Salmonella* characteristics such as motility ***(1,2)***, facultative anaerobiosis ***(3)***, and the ability to invade epithelial cells and engineered auxotrophies may contribute to successful interference of tumor growth (*see* **Table 1**).

There are over 2500 described *Salmonella* serovars. A large fraction of these have disease histories that eliminate them from consideration. The particular serovar that has appeal for these studies is *Salmonella enteritis* serovar Typhimurium. Typhimurium is a murine pathogen that also infects humans and many other organisms. Results obtained from studies of bacterial/host tumor cell interactions in experimental animal models could reasonably be expected to resemble therapeutic outcomes in humans.

Because the goal of *Salmonella* strain construction is to develop strain(s) for effective destruction of cancer tissue with minimal or no toxic effects, methods for achieving this goal depend on rational decisions, as described in this introduction. The specific methods will involve choice of strain, selection, and transfer of chosen gene insertions and deletions, and tagging methods for

Table 1
Examples of *Salmonella* Strains Considered Recently for Cancer Therapy Species

Strain	Genetic characteristics	References
Salmonella typhimurium VNP20009	*msbB* and *purI* deletion	***(4–12)***
S. typhimurium VNP20009, VNP20047, VNP20009	*msbB* and *purI* deletion, expressing CD	***(13)***
S. typhimurium VNP20009	Thymidine kinase (HSV1-tk) reporter gene	***(14)***
S. typhimurium VNP20009	TAPET-CD, 5-FC to 5-FU	***(15,16)***
S. typhimurium SL1344, SL3271	InvA-deficient, aroA	***(17)***
S. typhimurium SL3261	pcDNA3.1(+)-MG7/PADRE plasmid	***(4)***
S. typhimurium SL3261	AraA(−), CD gene	***(18)***
S. typhimurium SL7207	Inactivated *yej* operon	***(6)***
S. typhimurium SL7207 vs. SL3235	*aroA* strains	***(19)***
S. typhimurium SL7838	*YieF* NfsA gene delivery	***(20)***
S. typhimurium SL	α-ketal protein	***(21)***
S. typhimurium SL	*mIL-12*	***(22)***
S. typhimurium SL3261	To be filled in	***(23)***
S. typhimurium 4550	pIL2 pathogenicity island-2 gene Mutant and ASD deletion mutant	***(3,24)***
S. typhimurium X4550	Nco I site, MG7-AG mimotype, MG7/HbcAg fusion gene	***(25)***
S. typhimurium 4550	Nco I site, MG7-AG mimotype, MG7/HbcAg fusion gene	***(26)***
S. typhimurium X4550	Nco I site, MG7-AG mimotype, MG7/HbcAg fusion gene	***(25)***
S. typhimurium c4550 (pIL-2)	Contains interleukin-2 gene	***(27,28)***
S. typhimurium LB5000	C-Raf	***(29)***
S. typhimurium	Expressing *Escherichia coli* CD	***(30)***
S. choleraesuis ATCC15480	Transformed individually with pTCYLacZ, pTCYLuc, pRufCDES, and pRufCD. Vaccine strain	***(31,32)***
S. choleraesuis	Carrying thrombospondin-1 (TSP-1)	***(31)***
	aroA vaccine carrier strain	***(33)***

(Continued)

Table 1
(Continued)

Strain	Genetic characteristics	References
S. typhimurium 4550 (pIL2)	Carrying an asd+ plasmid for IL-2	***(3,34)***
S. typhimurium SLgp100	IL-2	***(35)***
S. typhimurium RE88	dam(−) and AroA(−)	***(36)***
S. typhimurium SF1591	Rough mutant of Ra chemotype	
S. minnesota	R595	***(37–39)***
S. typhimurium LT2(CRC1674)		Eisenstark et. al., unpublished
	his-2550 (plus suppressor mutation, DIIR49B), altered *rpoS* start signal (UUG), G to T mutation in position 168 in *rpoS* sequence, decreased HPI and HPII	
	180-kb duplication	
	aroA551::Tn10(TetR), ΔrfaD-rfaG::pKD4(KanS), ΔthyA::pKD4(KanR)	
	We also have a strain with ΔrfaH::pKD4(KanS) instead of the ΔrfaD-rfaG.	

CD, cytosine deaminase.
Abbreviations not explained in the above papers are listed in **Note 8.**

assaying effectiveness of attachment, invasion, destruction, and other specific alterations of cellular components by *Salmonella*.

1.1. How Do Salmonella *Attack Tumors Selectively?*

There are several ways in which *Salmonella* could attack tumors, each of which may require modified protocols. Among these are the following:

1. As a pathogen, *Salmonella* bacteria could stimulate local non-specific immune responses by preferentially accumulating at tumor sites ***(40–42)***.
2. *Salmonella* could interfere with the growth of tumor cells or compete with tumor cells for nutrients ***(43)***, thus impeding further tumor cell division and spread.

3. Lysis of bacteria attached to tumor cells might release locally acting anti-tumor cellular components.
4. Engineered *Salmonella* could be used as carriers for introducing genetic vectors that express anti-tumor components.
5. Tumor peptides released as a result of *Salmonella*-induced necrosis could also be used to stimulate production of anti-tumor antibodies.
6. Rapid growth of *Salmonella* in necrotic and hypoxic regions may allow *Salmonella* to multiply selectively in tumors.
7. Tumor-targeting *Salmonella* might be used to specifically express peptides at tumor sites. The peptides might have innate anti-tumor functions or be targets for other therapeutic chemicals.
8. *Salmonella* may be engineered to metabolize inactive pro-drugs into their active anti-tumor molecular structures at the tumor site, thus reducing toxic side effects to non-tumor cells.

1.2. Factors in the Design of Anti-Tumor Strain(s) of Salmonella

Saltzman *(3)* has emphasized that the first concern to be addressed is to choose *Salmonella* strains that would optimize success in each step of the journey of *Salmonella* cells to their tumor destinations and ultimate destruction or blockage of tumor growth, without subsequent adverse consequences of toxicity due to bacteria. As a result, strain selection for anti-tumor effects has focused mainly on the avoidance of subsequent adverse consequences through tedious removal of virulence genes. In this strategy, care must be taken to prevent loss of essential genetic elements required for therapeutic effectiveness. Thus, it is important to retain gene products that maximize tumor destruction.

1.3. Questions to be Addressed in Salmonella *Strain Development*

As pointed out in reviews of *Salmonella* anti-tumor effects (***3,44–50***), numerous new technical questions must be addressed in reaching successful use of *Salmonella* in cancer therapy. These include the following:

1. The chosen *Salmonella* strain(s) should have special clinical attributes. What are the genomic changes to be considered to develop superior strains?
2. Which experimental cancer cell model(s) should be selected in testing superior strains? What tumor cell lines should be used for preclinical tests?
3. What molecular biology techniques would be recruited to genetically engineer superior *Salmonella* strains?
4. What scoring method would be used to identify superior strains(s)?
5. How do candidate strains differ in attachment to surfaces, invasion, attraction to specific organelles, chromosomal destruction, etc.?

6. How can *Salmonella* cells be tracked in the course of experiments (e.g., fluorescent dyes, stains, etc.) (*see* **Table 2**) ***(50)***.
7. Can two different *Salmonella* cell lines be differentially marked in competition experiments?
8. Is *Salmonella* penetration necessary for anti-tumor effects, or does effectiveness merely result from surface contact?
9. If the anti-tumor effect mandates *Salmonella* penetration, does it also require bacterial metabolic activity? Multiplication?
10. Because the key to the ultimate goal is non-toxicity, what preclinical LD50 protocol should be used?
11. While mouse tests are standard for LD50 determinations, are there other model systems that could be used for toxicity testing? These model organisms could be useful at later stages when assaying mechanisms of tumor cell line targeting and destruction.
12. What method should be used to introduce *Salmonella* into the patient? Oral would be ideal for ease of administration, as *Salmonella typhimurium* finds its own route through organs to the tumor cells. Arguments against oral introduction would include poor control of inoculum concentration due to competition versus resident microflora and requirement to keep virulence genes required for invasion of the intestinal epithelium.
13. For effective therapy, should *Salmonella* infection be accompanied with other therapeutic measures? If so, which? What other anti-cancer treatments should be combined with *Salmonella* to optimize therapeutic effectiveness?
14. What assays should be used to determine success in clinical studies (***48,51–55***)?

1.4. Use of Salmonella *for Prostate Tumor Detection*

Although the objective of *Salmonella* targeting of tumor cells is to eradicate solid tumors, it might also have a quasi-preventive role. For example, the majority of older men test positive for prostate-specific antibody (PSA), an indication that prostate tumors may (but not necessarily) be present. However, conventional therapy upon observation of a high PSA titer is often postponed until significant increase in PSA levels is seen, because a high PSA does not always indicate an active tumor. This “watchful waiting” procedure could be enhanced by *Salmonella* administration and subsequent clinical testing for reduction in PSA levels.

1.5. Use of Single Strain Versus Multiple Strains

There may be an advantage to using a “cocktail” of two or more *Salmonella* strains. Genes for amplifying the tumor-killing phenotype of *Salmonella* could be put on a plasmid vector containing a stable cellular poison carried by a

Table 2
Methods of Tracking *Salmonella* in Tumor Attachment and Destruction

Technique	Product	Vendor/Reference	Target
Fluorescent protein expression	DsRed Monomer 586 nm	Clontech	In vivo expression
	DsRed2 582 nm	Clontech	
	DsRed-Express 579 nm	Clontech	
	AsRed2 592 nm	Clontech	
	HcRed1 618 nm	Clontech	
	AmCyan1 489 nm	Clontech	
	ZsYellow1 539 nm	Clontech	
	ZsGreen1 505 nm	Clontech	
	AcGFP1 505 nm	Clontech	
	EmGFP 509 nm	Invitrogen	
	YFP 527 nm	Invitrogen	
	CFP 505 nm	Invitrogen	
	BFP 440–447 nm	Invitrogen	
	Cycle 3 GFP 507 nm	Invitrogen	
	mPlum 649 nm	Tsien	
	mCherry 610 nm	Tsien	
	tdTomato 581 nm	Tsien	
	mStrawberry 596 nm	Tsien	
	J-Red 610 nm	Evrogen	
	mOrange 562 nm	Tsien	
	mKO 559 nm	MBL Intl.	
	Mcitrine 529 nm	Tsien	
	Venus 528 nm	Miyawaki	
	Ypet 530 nm	Daugherty	
	EYFP 527 nm	Invitrogen	
	Emerald 509 nm	Invitrogen	
	EGFP 507 nm	Clontech	
	CyPet 477 nm	Daugherty	
	mCFPm 475 nm	Tsien	
	Cerulean 475 nm	Piston	
	T-Sapphire 511 nm	Griesbeck	
Morphological staining	Live/Dead Stain CellTrace Stain	Molecular Probes	Cell membrane

(Continued)

Table 2
(Continued)

Technique	Product	Vendor/Reference	Target
Antibody labeling	Antibodies to O, H, or surface antigens with primary or secondary detection methods (immuno-gold labeling, NFPs)	Various suppliers	Surface antigens
Electron microscopy	Capture of high-detail snapshots during infection		Whole-cell imaging

"helper" *Salmonella* that expresses an unstable "antidote" protease from the chromosome, thereby reducing the chances of creating a more virulent strain in the host via transformation and/or recombination. Other variants of this strategy may be viable; however, care must be taken to ensure that both strains compete successfully for association and persistence at the target tumor site.

1.6. Choice of Salmonella *Strain[s] for Cancer Therapy*

Strategies for *Salmonella* anti-tumor effects were pioneered starting in 1978 (***39***,***56–66***). However, current experiments with this strategy received increased consideration with the use of strain VNP20009 (***46***,***48***,***67–71***). Several laboratories have used this strain either with or without genetic modifications, as may be seen in **Table 1**. VNP20009 was derived from ATCC14028. ATCC14028, listed as *Salmonella enterica* subspecies Enterica, was deposited as "*Salmonella typhimurium*, Castellani and Chambers. Isolation: tissue, animal (pools of heart and liver from 4-week-old chickens)." It has been circulated and used by many investigators in virulence and immunology studies and is often called just 14028.

The anti-tumor strategy with VNP20009 sparked our interest. We decided to compare the effectiveness of VNP20009 (ATCC14028 origin) with *S. enterica* serovar Typhimurium LT2, a strain of independent origin. We were curious whether the LT2 strain isolated about 60 years ago would also show affinity for tumor cells. LT2 was isolated by Lilleengen [Lilleengen Type (LT)], sent to Joshua Lederberg for P22 transduction studies in 1953, and shortly after sent to Miloslav Demerec for extensive genetic mapping and biochemical studies. It was the first *Salmonella* to be completely sequenced ***(72)*** and is deposited at ATCC as 70072. The LT2 sequence is the standard for microarray comparison of genomic changes in related strains ***(14)***. LT2 genetics have been extensively analyzed and thousands of mutant LT2 strains are officially curated by K. E. Sanderson at the Salmonella Genetic Stock Center (SGSC) (http://www.acs.ucalgary.ca/~kesander/).

Although the methods described in **Subheadings 1.7.–1.9.** center on our current usage of strain CRC1674 in the laboratory, the methods are applicable to investigations used in the past, as well as those that might be planned for the future. Our laboratory has several thousand replicas of these mutant strains from the original Demerec collection. Between the years 1997 and 2007, we have examined the diversity among LT2 strains that had been archived in agar stabs for over four decades ***(73–75)*** and are a half-century beyond its parental initial murine infection and isolation. Initially, we compared archival LT2 strains with VNP20009 with regard to interaction with breast cancer cells. One of our strains (CRC1674) destroyed breast cancer cells differently from VNP20009; thus, we have chosen this as our candidate strain. Its genotype is listed in **Table 1**. It stems from a pathogenic progenitor isolated in 1947. It has numerous genomic alterations as determined by microarray analysis ***(14)*** and other scoring methods. Among these are his-2550 (plus suppressor mutation, DIIR49B) ***(76)***. It has an altered *rpoS* start signal of UUG instead of AUG as well as a G to T mutation in position 168 in the *rpoS* sequence, which may account for decreased quantity of HPI and HPII hydroperoxidases regulated by *rpoS*. When analyzed by pulsed field gel electrophoresis (PFGE), it had the longest genome among 13 archived strains, with an extra band of 180 kb ***(77)***.

To further reduce the possibility of toxic shock, we have engineered additional genetic changes with the goal of developing an effective therapeutic strain. We have transduced strains now under investigation with *aroC* to reduce toxicity and deleted *thyA* (thymidylate synthetase) ***(78)*** to make it deficient in nucleotide synthesis. By selection of a "rough" phenotype that eliminates or reduces the amount of surface O-antigen, toxic shock events should be further reduced. We avoided the use of nitrosoguanidine to introduce auxotrophy, which could result in additional, but unknown secondary mutations. Because we are interested in determining the mechanisms of targeted attachment and invasion, we are also testing a strain with a shortened genome containing many random deletions. Construction of a therapeutic strain using this organism would be an additional strategy (i.e., adding desired genes) versus the subtraction strategy in current use (i.e., deleting undesired genes).

It is also recognized that there are many biological and immunological unknowns that take place in *Salmonella* invasion. It is exciting to greet the results of new experiments that lead to our understanding of the mechanisms behind *Salmonella* therapeutics. This method chapter is focused on genomic engineering of strains with various desired characteristics and their interaction with tumor cell lines.

1.7. Profiling S. typhimurium LT2 Expression When Co-Incubated with Cancer Cells; Isolating Tn5 Knockouts that No Longer Associate with Prostate Cancer Cell Lines

Most of the *Salmonella* strains developed for therapeutic study to date have originated with the ATCC14028 strain of *S. typhimurium*. We have genetically profiled both the LT2-derived CRC1674 and the ATCC14028-derived VNP20009 candidate strains by generating fluorescently labeled PCR products of >96% known open reading frames (ORFs) and measuring DNA–DNA hybridization on microarrays in collaboration with Michael McClelland at the Sidney Kimmel Cancer Center (San Diego, CA) to determine the genetic differences between the two strains [***(14)***, unpublished data]. We have determined that CRC1674 lacks nine genes found in the LT2 sequence and may have substantial changes (loss of highly similar sequences to promote DNA–DNA hybridization on microarrays) in 182 genes (4.4% of the genome) compared with the sequenced LT2 strain. In comparison with VNP20009, the CRC1674 strain possesses 61 genes that VNP20009 lacks, and 92 genes shared between CRC1674 and VNP20009 have substantial changes (unpublished data). The method used to determine the genomic profile of *Salmonella* is described in Chapter 6 (Porwollik and McClelland, this volume).

Although these mutations may be important factors for CRC1674's therapeutic activity on prostate cancer cells, these changes occurred during CRC1674's >50-year existence under starvation conditions in a stab vial and not co-existing with prostate cancer cell lines. Additionally, we do not know which genes in this list are expressed during co-incubation, invasion, and persistence within prostate cancer cells. In addition to making selected gene disruptions to reduce *Salmonella* toxicity, we have focused our efforts on two projects: (1) profiling of CRC1674 messenger RNA expression during initial stages of attachment and invasion of PC-3M prostate cancer cells and (2) identifying random Tn5 knockout mutants that can no longer associate with the PC-3M prostate cancer cell line but are normally expressed by CRC1674 during co-incubation with PC-3M. Genes that are specifically expressed during co-incubation with PC-3M but prevent attachment when disrupted with a Tn5 insertion are being studied to determine their mechanistic role in *Salmonella*–prostate cancer cell interactions; however, all genes shown to have a role during attachment and invasion are of use when designing therapeutic *Salmonella* strains.

1.8. mRNA Expression Profiles of CRC1674 to Identify Genes Specifically Expressed During Co-Incubation with Prostate Cancer Cell Line PC-3M

Important information can be gained by the initial profiling of mRNA expression levels in *Salmonella* that have been observed to invade and kill the PC-3M prostate cancer cell line. To do this, overnight cultures of CRC1674 are grown in Luria–Bertani (LB) broth and resuspended in Dulbecco's phosphate-buffered saline (PBS) (Gibco). A final concentration of approximately 1×10^6 bacteria/mL is added to cancer cells (in this case, PC-3M prostate cancer cells). Control experiments include incubation of CRC1674 in the same media without cancer cells or in the presence of a non-cancerous cell line.

Association of CRC1674 with the PC-3M line is typically seen within 1 h after inoculation (unpublished observation). At regular time points starting after 90 min of incubation in cell culture, cells and bacteria are removed from a cell culture vessel using Trypsin–EDTA and whole RNA isolated using the Qiagen RNeasy kit based on the guanidinium thiocyanate RNA extraction protocol reported by Chomczynski and Sacchi *(79)*. RNA isolated is treated with DNAse I to remove any residual genomic DNA. After inactivation of the DNAse I, the RNA is converted to a cDNA library by random priming with Ominscript and Sensiscript reverse transcriptase (Qiagen) using universal primers *(80)*.

To specifically isolate mRNA transcripts of *Salmonella* DNA and lower the *Salmonella* rRNA signal, we used the selective capture of transcribed sequences (SCOTS) technique described by Graham and Clark-Curtiss *(81)*. CRC1674 genomic DNA is prepared via sonication and photobiotinylated. PCR-amplified *Salmonella rrnA* sequence is mixed with the biotinylated genomic DNA and the cDNA transcripts isolated. After denaturation at 98 °C, the biotinylated *Salmonella* genomic DNA is specifically hybridized with *Salmonella* cDNA isolated from *Salmonella* mRNA transcripts. These genomic-cDNA hybrids are captured by the addition of streptavidin-coated magnetic beads and separated from the prostate cancer cell line cDNA transcripts. Excess *Salmonella* ribosomal RNA is captured by the excess of non-biotinylated *rrnA*

DNA. The cDNA can then be selectively eluted from the beads and amplified using fluorescently-tagged primers complimentary to the conserved regions of the universal primers and applied to a non-redundant microarray that currently covers 99% of all genes in the genomes of *Salmonella* serovars Typhimurium LT2, Typhimurium SL1344, Typhi CT18, Typhi Ty2, Paratyphi A SARB42, and Enteritidis PT4. In our laboratory, the microarray and subsequent mRNA analysis is done in collaboration with Michael McClelland at the Sidney Kimmel Cancer Center in San Diego, CA, or at the University of Missouri DNA Core Facility. Microarray analyses are now standard in large research core facilities.

These microarray data are analyzed for mRNA expression of CRC1674 *Salmonella* genes required during translocation to and attachment and invasion of PC-3M prostate cancer cells. Expression profiles of CRC1674 without prostate cancer cells allow screening out the expression of housekeeping genes and changes stimulated by translocation of the cells to new media. If necessary, cells can be grown in the PC-3M culture media to mid-log to remove expression profiles caused by the change in media types. This procedure can identify specific mRNA expression of genes involved in such phenotypes as chemotaxis, fimbrial biosynthesis, and formation of the "invasome" organelle structures with type III secretion systems and effectors (e.g., the *inv*, *spa*, or *hil* loci) shown to be involved in the attachment and invasion of the intestinal epithelium. Observation of some expected transcripts helps to assure that the mRNA extraction and amplification SCOTS procedure is working correctly. These data are of value when engineering superior therapeutic *Salmonella* strains; any characterized virulence genes not expressed during co-incubation with cancer cells are candidates for elimination to reduce the toxicity of therapeutic strain.

1.9. Creation and Isolation of Salmonella *CRC1674 Tn5 Knockout Pool(s) That No Longer Associates with Prostate Tumor Cells. Sequence Genes Flanking Tn5 Transposon Insertions that Prevent Association with PC-3M Cancer Cells*

To further compliment and focus on studies of genes specifically responsible for initial association and invasion of PC-3M prostate cancer cells by *Salmonella*, we have generated a random Tn5<KAN-2> knockout pool in our CRC1674 strain to isolate genes that can no longer associate with PC-3M prostate cancer cells. A Tn5<KAN-2> random knockout pool was made in Michael McClelland's laboratory using Epicentre's EZ-Tn5™ Transposome kit and generated 39,000 kanamycin-resistant mutants in the *Salmonella* strain

14028s background, approximately 10 independent knockout insertions in every gene of the 14028s *Salmonella* genome. We propagated bacteriophage P22 HT/*int* on 2% of the 14028s Tn5 pool and transduced the random Tn5 inserts into our CRC1674 strain, generating approximately 96,000 colonies to create a Tn5 knockout library in our therapeutic strain which we stored in 25% glycerol at −80 °C. If a Tn5 disruption pool created by transduction and assayed using the transposon footprinting method suggests that this Tn5 pool is limited, we suggest generating a random Tn5<KAN-2> pool directly in your candidate strain using the Epicentre EZ-Tn5™ Transposome kit (Bio-Rad).

Generation of a CRC1674 Tn5 disruption mutant sub-pool that cannot associate with PC-3M prostate cancer cells is performed by outgrowth of the CRC1674 Tn5 pool above to mid-log phase to allow recovery from −80 °C frozen stocks. Approximately, $1 \times 10^{6-7}$ cfu/mL of *Salmonella* washed in 1× PBS are mixed with PC-3M prostate cancer cells. Beginning at time points after 60 min of co-incubation, *Salmonella* in the culture supernatant are drawn off, spun down, and grown in LB broth containing 50 μg/mL kanamycin overnight for storage and analysis of the sub-pools using the "transposon footprinting" technique ***(82)***. Briefly, transposon footprinting is an assay that identifies specific Tn5 disruption mutants by creating a unique PCR amplification profile that depends on the site of Tn5 insertion and distance from *Nla*III restriction sites (5′-CATG^-3′). PCR products are produced by attaching linkers onto *Nla*III-digested genomic DNA and amplifying product using primers complimentary to the Tn5 sequence and a unique sequence in the attached linker.

To select genes that are absolutely required for association with PC-3M cells (and eliminate false positives), the sub-pool can be grown and the co-incubation procedure repeated until no reduction of the Tn5-mutant population is observed using the transposon footprinting assay. The genes disrupted by Tn5 transposon insertion are identified by sequencing of the transposon footprinting PCR products and evaluated for potential attenuation of downstream gene transcription.

In addition to finding genes with no assigned function, one can expect to find genes required for association of *Salmonella* with the intestinal epithelium and other cultured cells, including genes involved in chemotaxis, construction of fimbriae, secretion systems designed to release molecules that promote *Salmonella* phagocytosis, and transcriptional regulators that control these systems. Isolation of genes already characterized as important for association with other cell types will assist in further optimization of therapeutic strains; if genetic systems needed for association to other cell lines are not isolated

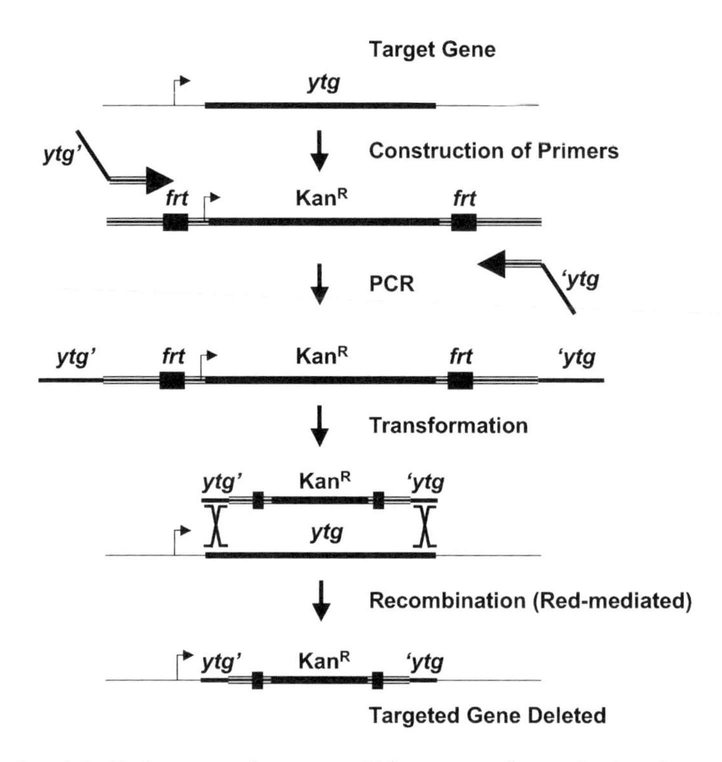

Fig. 1. Lambda Red gene replacement. This cartoon figure depicts the replacement of "your target gene" (*ytg*) with a Kanamycin resistance cassette. After identification and sequencing of the gene to be replaced, primers are designed that have 5′ ends with 35–50 bp homologous to the start (ytg') and end ('ytg) of the ytg coding sequence. The 3′ primer ends are homologous to sequence up and downstream of a selection cassette flanked by *frt* sites (in this figure, a Kanamycin resistance cassette). The selection cassette is amplified with the primers using PCR and the resulting linear product transformed into an organism that contains the targeted gene. Expression of the Red locus (Bet, Exo, and Gam) from a vector promotes recombination between the homologous ytg regions, resulting in deletion of the target gene. The resistance cassette can be removed by the addition of a vector that expresses Flp, a site-specific recombinase that will bind to and recombine DNA between the frt sites, looping out the resistance cassette.

in the pool, they become candidate genes that can be removed to increase the specificity and limit the toxicity of the therapeutic strain. Genes found in this strain that are specifically expressed during incubation with PC-3M cells are excellent candidates for further study.

1.10. Genetic and Cytological Characterization of Genes Required for Salmonella-Associated Prostate Cancer Cell Death

Once specific *Salmonella* genes are identified to be essential for targeting, attachment, invasion, inhibition, and/or death of tumor cells, there are many experimental options that can be explored. Characterization of the gene(s) selected for investigation should be used as a guide to further experimentation. Hypothetical functions assigned to uncharacterized genes in NCBI and other online databases suggest biochemical mechanisms used by these genes, and these assignments should be interpreted as guides for planning biochemical assays.

To confirm the importance of genes identified by mRNA expression and Tn5 selection, precise disruptions or deletions are made in the genes and the mutated *Salmonella* candidates tested for their ability to target, attach, invade, and kill tumor cells in your model. If you do not have access to pre-existing mutants, they can be constructed using the Lambda Red gene replacement system (often referred to as "Red-Swapping") pioneered by Datsenko and Wanner *(83)*. This protocol is graphically described in **Fig. 1**, and a brief description of the protocol follows below.

2. Materials

2.1. Lambda Red Gene Replacement Protocol for Targeted Gene Disruptions

1. Plasmids pKD46, pCP20 for gene expression. Plasmids pKD3, pKD4, and pKD13 for PCR templates.
2. LB, Ampicillin, Chloramphenicol, and Kanamycin media (liquid broth and solid plates).
3. A 30 °C and 42 °C incubator.
4. PCR machine with enzymes and disposables required to run PCR reactions.
5. Primers whose 5′ ends have perfect 35–50 bp homology to the DNA immediately upstream and downstream of the gene to be disrupted and 3′ ends have perfect 25 bp homology to the DNA flanking the desired selective gene used in the experiment (e.g., Kan^R, Cam^R).
6. Transformation equipment (for transformation by heat shock or electroporation).

2.2. Cell Lines and Culture Materials

2.2.1. Equipment and Consumables

1. Inverted microscope (American Optical Bio Star with ×20 objective or similar).
2. 75-cm^2 vented cap/canted neck cell culture flasks (Corning).
3. Centrifuge with 4 × 50ml swinging bucket rotor (IEC Centra MP4 or similar)

4. 10- and 25-mL disposable pipettes (Fisher Scientific).
5. 1-mL syringes (for gentamicin bottle) (B-D).
6. Costar cell scrapers (Corning INC.).

2.2.2. Cell Culture (see **Note 1**)

1. Keratinocyte Serum Free Medium (SFM) supplied with prequalified human recombinant Epidermal Growth Factor 1-53 (EGF 1-53) and Bovine Pituitary Extract (BPE) in separate packaging (Invitrogen).
2. RPMI 1640 Medium (1×) (Invitrogen).
3. F-12K Nutrient Mixture (Kaighn's Modification) (1×) (Invitrogen).
4. HyQ® Tase™ Cell Detachment Solution (1×) (trypsin-like enzyme without phenol red, regular trypsin also works, but HyQ® Tase™ is gentler.) (Hyclone).
5. Fetal bovine serum (FBS) (US Bio-Technologies, Inc.).
6. MEM Non-essential Amino Acid Solution 10 mM (100×) (Invitrogen).
7. L-Glutamine 200 mM (100×) (Invitrogen).
8. Gentamicin 40 mg/mL (American Pharmaceutical Partners, Inc.).

2.3. Fluorescence Microscopy of Prostate Cancer Cell/Salmonella Interactions

1. 60 mm × 15 mm Petri dishes (Fisher Scientific).
2. Lab-Tech II 2-well Chambered Slides (Nalge Nunc International).
3. 75 mm × 10 mm × 1 mm glass microscope slides (Fisher Scientific).
4. #1 Coverslips 24 mm × 40 mm (Fisher Scientific).
5. Prepared cell culture media without gentamicin.
6. PBS (1×) (Invitrogen).
7. Ethanol to clean slides (Fisher Scientific).
8. Clear nail polish.
9. BacLight™ Bacterial Viability Kit (Molecular Probes).
10. 3.7% paraformaldehyde (Supplied as 16%, Electron Microscopy Sciences).
11. Phalloidin-tetramethylrhodamine (Sigma, St. Louis, MO, cat#77418).
12. 15-mL centrifuge tubes (Biologix).
13. Spectraphotometer (Shimadzu UV160U or similar).
14. Vectashield mounting medium containing DAPI to stain DNA (Vector Laboratories, Burlingame, CA, cat# H-1200).
15. Access to Epifluorescence (Zeiss Axiophot or similar equipped with filters and Olympus DP70 digital camera) or Confocal Microscope System.

2.4. Electron Microscopy for Determination of Salmonella Incorporation into Vacuoles of Prostate or Other Cancer Cells (see Note 2)

1. 50% stock Glutaraldehyde, EM grade (Polysciences).
2. 0.1 M PBS.

3. Tannic Acid, EM grade (Polysciences).
4. Osmium tetroxide (OsO_4).
5. Ethanol (30, 50, 70, 90, 95, and 100%).
6. Embedding medium (Epoxy resin kit or similar) (Polysciences).

3. Methods

3.1. Lambda Red Gene Replacement Protocol for Targeted Gene Disruptions

1. Obtain plasmids pKD46 (42 °C temperature-sensitive, Amp^R plasmid that expresses the Lambda Red gene locus when induced by L-arabinose) and pCP20 (42 °C temperature-sensitive plasmid that constitutively expresses Flp).
2. Obtain plasmids pKD3 and pKD4 or other DNA that contains a selective gene flanked by *frt* sites for later removal of the selective gene.
3. Transform pKD46 into your transformation-competent candidate *Salmonella* strain and grow at 30 °C on LB + 50 μg/mL Amp to insure stability of the pKD46 plasmid. Archive a transformed colony for making later gene replacements in the strain (*see* **Note 3**).
4. Design your PCR primers as described in **Subheading 2.1**. **(item 5)**. **Fig. 1** shows this graphically. Be sure to BLAST the primer sequences against the complete genome sequence (if available) to determine if there is substantial homology to alternative sites. If this exists, engineer the 35–50 bp homologous region slightly upstream or downstream until the similarity disappears.
5. Perform PCR with your primers and the DNA template that has a selection or resistance gene flanked by FRT sites. We use pKD3 (Cam^R) or pKD4 (Kan^R) as templates.
6. Perform electrophoresis on the PCR reaction on agarose gels to confirm the presence of a PCR product of the predicted size. We typically run 5 μL of a 20–100 μL PCR reaction. If there are secondary products, purify the desired PCR product from the agarose gel. If there are no secondary products, we remove the salts from the finished PCR reaction using Qiagen Qiaquick PCR purification columns. Resuspend the DNA in deionized, sterilized H_2O.
7. Start an overnight culture of your candidate strain + pKD46 in 2 mL LB + 50 μg/mL Amp at 30 °C. Incubate at least 8 h.
8. Subculture the overnight culture into a fresh LB + 50 μg/mL Amp + 50 mM L-arabinose at a 1:100 dilution. Prepare heat shock or electrocompetent cells from this culture.
9. Transform the PCR product (typically 2 μL) into 20–50 μL of competent cells. We prepare electrocompetent cells and transform the DNA: cells in a 1-mm gap width electroporation cuvette on an Electroporator 2510 (Eppendorf) set to 1800 mV.
10. Immediately resuspend the transformed cells in 500 μL of LB broth and incubate at 30 °C for 1 h to allow for recovery and expression of the selective gene.

11. Spin down the cells and resuspend in 100 μL of LB broth. Plate the cells onto selective media (with the appropriate antibiotic). Incubate selection plates overnight at 37 °C. Use colony PCR or DNA sequencing to confirm the location of the putative gene disruption.
12. Resuspend a colony with a successful gene disruption in a 500-μL Eppendorf tube containing 100 μL LB. Incubate the tube at 42 °C for 8 h to dilute the pKD46 plasmid. Use a sterile stick or loop and streak the culture for isolation. Incubate at 37 °C. Patch 10–50 isolated colonies to LB + 50μg/mL Amp plates and LB plates to confirm the loss of pKD46.
13. If removal of the selective gene is desired, transform the Flp-expressing pCP20 plasmid into your strain, using Amp^R or Cam^R media for plasmid retention. Transformants can be patched onto LB plates and LB plates containing your original selective pressure to screen for loss of the selective gene. The pCP20 plasmid can then be lost using the procedure described in this step (**step 13**).

The above protocol is a modification of an excellently written Lambda Red Gene Replacement "Red-Swap" protocol available in pdf format at Professor Stanley Maloy's laboratory website (URL: http://www.sci.sdsu.edu/~smaloy/Research/protocols.htm). It is important to note that inducible promoters and other genetic elements can be swapped into the chromosome upstream of the gene(s) of interest to conduct experiments (e.g., looking at the effects of overexpression or for purification of gene products). You are not limited to gene disruption using this technique.

Construction of disruption and inducible overexpression mutations in your genes of interest can be co-incubated with your cancer model to confirm that the gene(s) are essential for therapeutic effect. These strains can be grown (and induced if examining the effect of overexpression) and co-incubated with prostate cancer cell lines on glass cover slips. Samples are fixed (and stained if desired) at set time points and examined using confocal, fluorescent, or electron microscopy. Analysis is aided by the use of morphological Live/Dead stains with confocal microscopy, fluorescent-tagged antibodies specific to *Salmonella*, or cancer cell receptors or sectioning of cancer cells to confirm invasion phenotypes of *Salmonella* using electron microscopy (*see* **Fig. 2b, c**). Microscopy can also be used in competition experiments between different candidate strains to determine which candidate has the superior therapeutic phenotype (or ensuring that both strains can colonize a tumor target if using a multiple-strain therapy). We use plasmids that constitutively express fluorescent proteins with different emission spectra to discern between different therapeutic strains in these competition assays (*see* **Table 2**). Many fluorescent proteins are commercially (e.g., BD Biosciences Clontech, Invitrogen) or non-commercially (Roger Y. Tsien's laboratory at UCSD has a wide selection of plasmids that express fluorescent

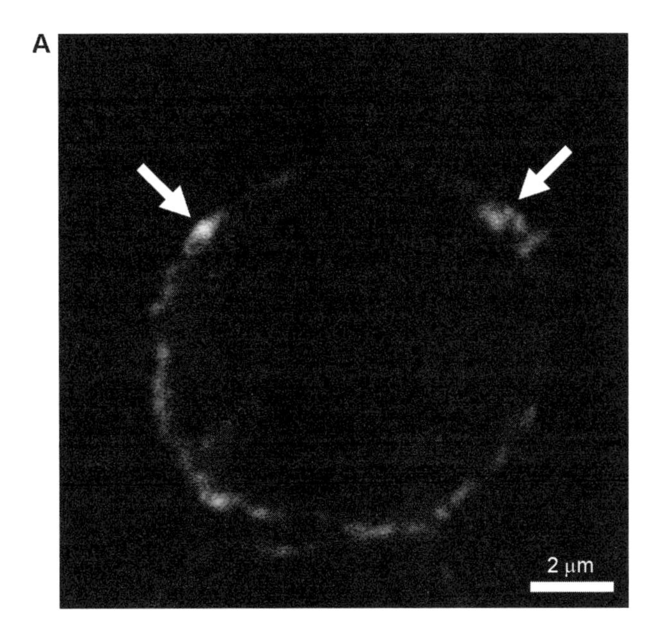

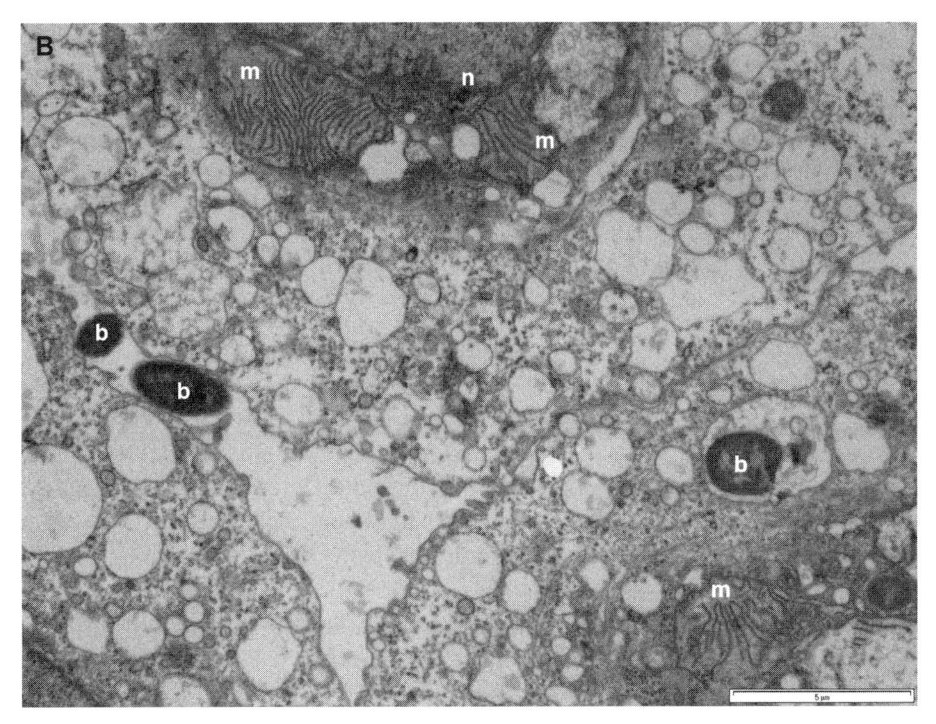

Fig. 2. (*see* legend on p. 343)

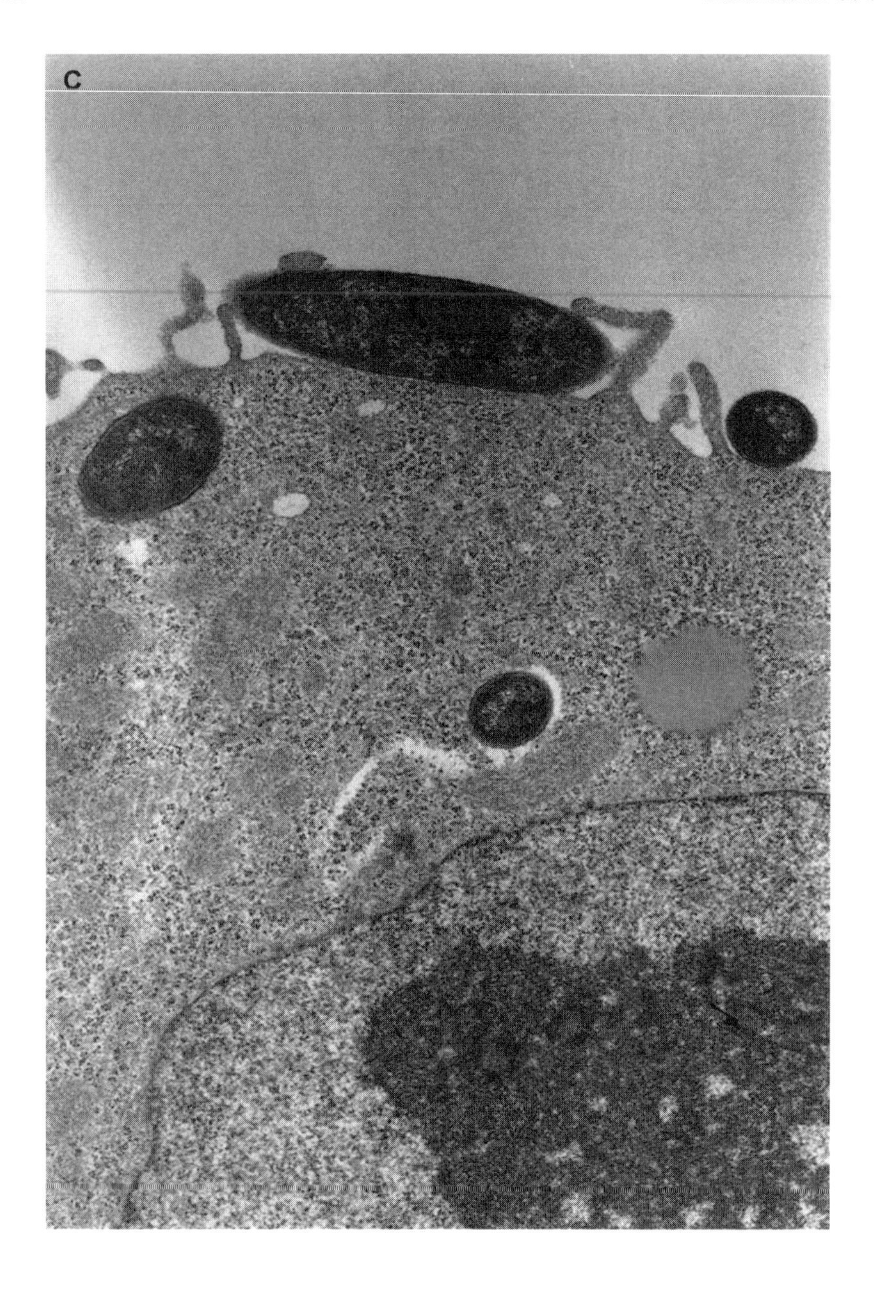
C

proteins) available ***(50)***. Strains that have clear competitive advantages can be quickly incorporated into efficacy and safety testing while continuing investigation of the mechanism(s) behind the genetic changes.

After confirming that *Salmonella* requires a particular gene or genes for anti-tumor therapy, different experiments can be designed to analyze the mechanism(s) for achieving the therapeutic effect. Besides analysis and annotation of the genetic sequence "*in silico*," biochemical purification and localization of the gene products using antibodies can be a good place to start analysis of what biochemical function(s) the gene products perform and where they are localized in the cell. These are general suggestions and further experiments can be designed depending on which function(s) the researcher thinks the gene products are performing during targeting, invasion, and/or killing of the tumor cells.

3.2. Culture of Cell Lines

The following three cell lines are maintained at 37 °C in a 5% CO_2/95% air atmosphere.

*3.2.1. RPWE-1 Normal Human Prostate Cell Line (*see **Note 4***)*

1. Maintain cells in Keratinocyte SFM with both supplied supplements (EGF 1-53 and BPE) and gentamicin added.
2. To culture, remove spent media from confluent culture and add 3 mL HyQ® Tase™ and 1 mL Keratinocyte SFM media.
3. Place in CO_2 incubator for 10 min.
4. Gently scrape cells making sure to keep the area being scraped lubricated with the liquid in the flask.

Fig. 2. **(A)** Rhodamine-phalloidin staining of breast cancer cell displays microfilaments at the cell surface. Arrows indicate increases in microfilament staining at the locations where *Salmonella* have invaded. Bar at bottom right represents 2 μm. **(B)** *Salmonella* invasion of prostate cancer cells. PC-3M cells incubated with *Salmonella* strain CRC1674 for 4 h. Note the presence of *Salmonella* containing vesicles indicative of successful invasion. The cancer cell is in a state of deterioration as indicated by the large vacuoles and the destruction of mitochondrial ultrastructure seen in the vicinity of the nucleus. b, bacteria; m, mitochondria; n, nucleus. Bar at bottom right represents 5 μm. **(C)** *Salmonella* attaching and invading prostate cancer cells. PC-3M cells incubated with modified *Salmonella* strain CRC1674. Note the *Salmonella* attaching, inducing membrane rippling (top) and successfully invading the prostate cancer cell line (magnification ×10,000)

5. Gently pipette culture into sterile 50ml centrifuge tube and centrifuge at 170 × g for 5min.
6. Carefully remove supernatant and resuspend pellet in 3ml fresh media.
7. Gently pull up medium containing cells and expel from pipette two to three times to mix and transfer half of volume to new 75-cm^2 flask containing 24 mL Keratinocyte SFM media (*see* **Note 5**).
8. Every 2 days, remove spent media and replace with new until cells appear confluent under an inverted microscope (7 days+).
9. Passage cells and repeat process. Passage day is when cells are transferred to dishes or other containers for experiments (*see* **Note 6**).
10. After about 15 passages, restart culture from frozen stocks.

3.2.2. PC-3 Human Prostate Carcinoma Cells (see **Note 7**)

1. Maintained cells in F-12K medium supplemented with 35 mL FBS, 5 mL L-glutamine, and 0.6 mL gentamycin.
2. Remove spent media from confluent culture and add 3 mL HyQ® Tase™ and place in CO_2 incubator for 1 min.
3. Spank the flask gently to loosen the cells and pull volume up and release from pipette to gently mix.
4. Gently pipette culture into sterile 50ml centrifuge tube and centrifuge at 170 × g for 5 min.
5. Carefully remove supernatant and resuspend pellet in 3ml of fresh media.
6. Transfer four drops of confluent culture into a new 75-cm^2 flask with 26 mL of prepared F-12K media. Every 3 days, remove spent media and replace with new until cells appear confluent under an inverted microscope (about 5–6 days).
7. Passage cells and repeat process. After about 17 passages, discard culture and restart from frozen stocks to insure quality. Passage day is when cells are transferred to dishes or other containers for experiments.

3.2.3. PC-3M Human Prostate Carcinoma Cells, Highly Metastatic Variant

1. Maintain cells in RPMI 1640 supplemented with 5 mL L-glutamine (2 mM), 5 mL Na pyruvate (1 mM), 50 mL FBS, 5 mL nonessential amino acids, and 0.6 mL gentamycin.
2. Remove spent media from confluent culture and add 3 mL HyQ® Tase™ and place in CO_2 incubator for 1 min.
3. Spank the flask gently to loosen the cells and pull volume up and release from pipette to gently mix.
4. Gently pipette culture into sterile 50ml centrifuge tube and centrifuge at 170 × g for 5 min.
5. Carefully remove supernatant and resuspend pellet in 3ml of fresh media.

6. Transfer two drops of confluent culture into a new 75-cm^2 flask with 26 mL of prepared RPMI-1640 media. Every 3 days, remove spent media and replace with new until cells appear confluent under an inverted microscope (about 5–6 days).
7. Passage cells and repeat process. After about 17 passages, discard culture and restart from frozen stocks to insure quality. Passage day is when cells are transferred to dishes or other containers for experiments.

3.3. Microscopy

3.3.1. Nomarski/DIC Imaging of Fixed Cells Grown on Coverslips

1. Start the *Salmonella* culture the day prior to experiments and place 1 mL overnight culture in a spectrophotometer. Take reading at 600 nm to estimate cell concentration.
2. Spin down the remainder of the culture at 2998 × *g* for 10 min, decant, and resuspend in an equal volume of PBS.
3. Add one drop of PC-3M or two drops of PC-3 confluent culture to 15-mm Petri dishes containing a 24 mm × 40 mm sterile coverslip in 5 mL of the appropriate cell culture media. Cells are allowed to grow into a monolayer for approximately 3 days and are fed on day 2 (24 h prior to experiment).
4. For the experiment, 1 mL culture is added to each Petri dish (except control). In the case of our usual time course experiments, dishes are returned to the CO_2 incubator and are pulled out and processed at their appropriate time (1–8 h).
5. Fix samples in 3.7% paraformaldehyde for 15 min and wash in PBS.
6. Mount on an ethanol-cleaned glass slide with a small drop of Vectashield.
7. Seal edges of the cover slip with clear nail polish.

3.3.2. Fixed Cell Imaging Microscope Settings

Images are obtained using Nomarski/DIC optics and a 60× UPlanapo (1.2 NA) objective together with a digital zoom of 3–4.

3.3.3. Live Cell Imaging in Tab-Tek II Chambered Slides

1. Start the *Salmonella* culture the day prior to experiments.
2. Place 1 mL overnight culture in a spectrophotometer and take a reading at 600 nm to estimate cell concentration.
3. Spin down the remainder of the culture at 2998 × g for 10 min, decant, and resuspend in an equal volume of PBS.
4. Add one drop of PC-3M confluent cell culture (2 for PC-3, 6 for RPWE-1) to 2 mL cell culture media in each slide chamber. Allow cells to grow in the chambered slides until a sparse monolayer forms (2 days PC-3 and PC-3M, 4 days RPWE-1).
5. Add 200 mL of this culture to each slide chamber (except controls) and allow to incubate at 37 °C or for a minimum of 60 min (time course as in confocal also

used). Stain samples with Molecular Probe's BacLight™ Live/Dead kit containing Syto 9 (stains live cells green) and propidium iodide (stains dead cells red).

3.3.4. Live Cell Imaging Microscope Settings

Images are obtained using the same objective described in **Subheading 3.3.2.** in an Olympic IX70 microscope coupled with a Bio-Rad, Radiance-2000 confocal system (Hercules, CA) under 488 nm excitation line of Kr/Ar laser (Syto 9) or 568 nm excitation line for propidium iodide. Transmitted light images are also obtained from the same region of interest.

3.3.5. Fluorescence Microscopy for Rhodamine-Phalloidin Staining

1. Wash cells in 0.1 M PBS.
2. Incubate cells with 25 μg/mL phalloidin-tetramethylrhodamine B (TRITC) in PBS for 1 h at room temperature.
3. Wash cell three times with PBS to remove unbound phalloidin conjugate.
4. Place 0.75 μg/mL DAPI solution on slide and place coverslip with cells facing the slide onto the slide.
5. View samples with Zeiss Axiophot equipped with epifluorescence microscopy attachment or similar instrumentation.

3.3.6. Transmission Electron Microscopy

1. Wash cells on coverslips with 0.1 M PBS.
2. Fix in 2% glutaraldehyde in 0.1 M PBS containing 0.2% filtered (0.1 μm filter) freshly prepared tannic acid for 30 min.
3. Rinse in 0.1 M PBS for 15 min.
4. Postfix in 0.5% OsO_4 in H_2O or in 0.1 M PBS for 15 min.
5. Wash in H_2O.
6. Dehydrate in ethanol series (30, 50, 70, 90, and 95%, 4× 100%) 10 min each.
7. Embed as indicated in Polysciences embedding kit instructions.
8. Remove coverslip from hardened sample with dry ice or liquid nitrogen.
9. Trim, section, and stain sections as available in microscopy laboratories.
10. Observe sections in transmission electron microscope.

4. Notes

1. A second supply of Keratinocyte SFM, RPMI-1640, and F-12K Nutrient Mixture (Kaighn's Modification) without the addition of gentamicin must also be available for culturing experimental samples.
2. Glutaraldehyde, Osmium tetroxide, and embedding media are toxic.
3. If your candidate strain grows poorly, you can transform pKD46 into a better-growing parental strain and introduce any gene replacements into your candidate strain using P22HT*int* transduction.

4. These cells must be passed heavily. They will not thrive in small numbers. These cells are very fragile. Plan ahead, as it may take some time to learn to culture these cells successfully.
5. It is advised to transfer the remaining volume into a second flask and maintain it as a backup.
6. Careful timing of culture maintenance ensures plenty of experimental material.
7. These cells grow at approximately half the rate of the PC-3M cells. Therefore, if PC-3 and PC-3M cells are required for an experiment on the same day, it is important to double the volume of PC-3 cells when setting up stock cultures or experimental samples.
8. Abbreviations not explained in key publications:
   msbB = myristoyl transferase in lipid A biosynthesis, suppressor of *htrB* (*lpxL*);
   HSV1 = positron emission tomography (PET) imaging of tumor-localized *Salmonella* expressing HSV1-TK;
   tk = thymidine kinase;
   SL = S (*Salmonella*) L (London Designation of strains);
   InvA = gene that codes for a possible secretory protein (associated with virulence);
   aroA = gene coding for a 3-enolpyruvylshikimate-5-phosphate synthetase;
   pIL2 = pathogenicity island-2 gene mutant;
   ASD = vectoring Campylobacter antigens, *Helicobacter pylori* region of adhesin antigen;
   Nco I site = sequence coding for MG7-Ag mimotope of gastric cancer;
   MG7-AG mimotype = a compound that mimics the structure of the MG7-AG epitope and provokes an additional antibody response;
   MG7/HBcAg fusion gene = gene encoding for hepatitis antigen and gastric cancer-specific tumor antigen;
   C-Raf = C-Raf-induced lung adenoma;
   pcDNA3.1(+)-MG7/PADRE plasmid = genes of the MG7-Ag mimotope plus Pan-DR epitope;
   yej operon = mutation induces superior CD8+ T-cell responses;
   CD and TAPET-CD = expressing *Escherichia coli* cytosine deaminase;
   cAMP = adenylate cyclase and cAMP receptor genes;
   pTCYLacZ = eukaryotic expression vector encoding *E. coli* β-galactosidase and the retroviral vector pRufCD;
   pTCYLuc = generated by excision of coding region of luciferase of pGL3 plasmid by digestion with HindIII and XbaI and cloned into the HindIII/XbaI sites of pTCY;
   pRufCDES = generated by cloning the coding region of murine endostatin in the BamHI/EcoRI sites of pRufCD;
   pRufCD = plasmid;
   TSP-1 = carrying thrombospondin-1; *araA* = expression of L-arabinose isomerase;
   pIL-2 = contains interleukin-2 gene (= pIL2) carrying an asd+ plasmid for IL-2;
   GFP-expressing = expression of the green fluorescence protein;

Ra = Regulation of cathelicidin gene expression: induction by lipopolysaccharide, interleukin-6; RA increases luciferase activity in PK-15 cells transfected with the PR-39 promoter-luciferase reporter.

References

1. Ciacci-Woolwine, F., Blomfield, I. C., Richardson, S. H., and Mizel, S. B. (1998) Salmonella flagellin induces tumor necrosis factor alpha in a human promonocytic cell line. *Infect. Immun.* **66,** 1127–1134.
2. Kasinskas, R. W. and Forbes, N. S. (2006) Salmonella typhimurium specifically chemotax and proliferate in heterogeneous tumor tissue in vitro. *Biotechnol. Bioeng.* **94,** 710–721.
3. Saltzman, D. A. (2005) Cancer immunotherapy based on the killing of Salmonella typhimurium-infected tumour cells. *Expert Opin. Biol. Ther.* **5,** 443–449.
4. Guo, C. C., Ding, J., Pan, B. R., et al. (2003) Development of an oral DNA vaccine against MG7-Ag of gastric cancer using attenuated salmonella typhimurium as carrier. *World J. Gastroenterol.* **9,** 1191–1195.
5. Hamazaki, Y. and Murao, T. (1979) Tumor induction in Swiss mice by filtrable agent and Salmonella typhimurium. *Acta Med. Okayama* **33,** 405–407.
6. Hummel, S., Apte, R. N., Qimron, U., Vitacolonna, M., Porgador, A., and Zoller, M. (2005) Tumor vaccination by Salmonella typhimurium after transformation with a eukaryotic expression vector in mice: impact of a Salmonella typhimurium gene interfering with MHC class I presentation. *J. Immunother.* **28,** 467–479.
7. Jia, L. J., Xu, H. M., Ma, D. Y., et al. (2005) Enhanced therapeutic effect by combination of tumor-targeting Salmonella and endostatin in murine melanoma model. *Cancer Biol. Ther.* **4**.
8. Murray, S. R., de Felipe, K. S., Obuchowski, P. L., Pike, J., Bermudes, D., and Low, K. B. (2004) Hot spot for a large deletion in the 18- to 19-centisome region confers a multiple phenotype in Salmonella enterica serovar Typhimurium strain ATCC 14028. *J. Bacteriol.* **186,** 8516–8523.
9. Pawelek, J. M., Sodi, S., Chakraborty, A. K., et al. (2002) Salmonella pathogenicity island-2 and anticancer activity in mice. *Cancer Gene Ther.* **9,** 813–818.
10. Platt, J., Sodi, S., Kelley, M., et al. (2000) Antitumour effects of genetically engineered Salmonella in combination with radiation. *Eur. J. Cancer* **36,** 2397–2402.
11. Thamm, D. H., Kurzman, I. D., King, I., et al. (2005) Systemic administration of an attenuated, tumor-targeting Salmonella typhimurium to dogs with spontaneous neoplasia: phase I evaluation. *Clin. Cancer Res.* **11,** 4827–4834.
12. Tjuvajev, J., Blasberg, R., Luo, X., Zheng, L. M., King, I., and Bermudes, D. (2001) Salmonella-based tumor-targeted cancer therapy: tumor amplified protein expression therapy (TAPET) for diagnostic imaging. *J. Control. Release* **74,** 313–315.

13. Mei, S., Theys, J., Landuyt, W., Anne, J., and Lambin, P. (2002) Optimization of tumor-targeted gene delivery by engineered attenuated Salmonella typhimurium. *Anticancer Res.* **22,** 3261–3266.
14. Porwollik, S., Wong, R. M., Helm, R. A., et al. (2004) DNA amplification and rearrangements in archival Salmonella enterica serovar Typhimurium LT2 cultures. *J. Bacteriol.* **186,** 1678–1682.
15. Fulton, A. M., Loveless, S. E., and Heppner, G. H. (1984) Mutagenic activity of tumor-associated macrophages in Salmonella typhimurium strains TA98 and TA 100. *Cancer Res.* **44,** 4308–4311.
16. Nemunaitis, J., Cunningham, C., Senzer, N., et al. (2003) Pilot trial of genetically modified, attenuated Salmonella expressing the E. coli cytosine deaminase gene in refractory cancer patients. *Cancer Gene. Ther.* **10,** 737–744.
17. Avogadri, F., Martinoli, C., Petrovska, L., et al. (2005) Cancer immunotherapy based on killing of Salmonella-infected tumor cells. *Cancer Res.* **65,** 3920–3927.
18. Li, Y. H., Xie, Y. M., Guo, K. Y., et al. (2001) Treatment of tumor in mice by oral administration of cytosine deaminase gene carried in live attenuated Salmonella. *Sheng Wu Hua Xue Yu Sheng Wu Wu Li Xue Bao (Shanghai)* **33,** 233–237.
19. Qimron, U., Madar, N., Mittrucker, H. W., et al. (2004) Identification of Salmonella typhimurium genes responsible for interference with peptide presentation on MHC class I molecules: Deltayej Salmonella mutants induce superior CD8+ T-cell responses. *Cell Microbiol.* **6,** 1057–1070.
20. Barak, Y., Thorne, S. H., Ackerley, D. F., Lynch, S. V., Contag, C. H., and Matin, A. (2006) New enzyme for reductive cancer chemotherapy, YieF, and its improvement by directed evolution. *Mol. Cancer Ther.* **5,** 97–103.
21. Chou, C. K., Hung, J. Y., Liu, J. C., Chen, C. T., and Hung, M. C. (2006) An attenuated Salmonella oral DNA vaccine prevents the growth of hepatocellular carcinoma and colon cancer that express alpha-fetoprotein. *Cancer Gene. Ther.*
22. Feng, K., Zhao, H., Chen, J., Yao, D., Jiang, X., and Zhou, W. (2004) Anti-angiogenesis effect on glioma of attenuated Salmonella typhimurium vaccine strain with flk-1 gene. *J. Huazhong Univ. Sci. Technol. Med. Sci.* **24,** 389–391.
23. Yuhua, L., Kunyuan, G., Hui, C., et al. (2001) Oral cytokine gene therapy against murine tumor using attenuated Salmonella typhimurium. *Int. J. Cancer* **94,** 438–443.
24. Barnett, S. J., Soto, L. J., 3rd, Sorenson, B. S., Nelson, B. W., Leonard, A. S., and Saltzman, D. A. (2005) Attenuated Salmonella typhimurium invades and decreases tumor burden in neuroblastoma. *J. Pediatr. Surg.* **40,** 993–997; discussion 997–998.
25. Meng, F. P., Ding, J., Yu, Z. C., et al. (2005) Oral attenuated Salmonella typhimurium vaccine against MG7-Ag mimotope of gastric cancer. *World J. Gastroenterol.* **11,** 1833–1836.
26. Feltis, B. A., Miller, J. S., Sahar, D. A., et al. (2002) Liver and circulating NK1.1(+)CD3(−) cells are increased in infection with attenuated Salmonella

typhimurium and are associated with reduced tumor in murine liver cancer. *J. Surg. Res.* **107,** 101–107.
27. Saltzman, D. A., Heise, C. P., Hasz, D. E., et al. (1996) Attenuated Salmonella typhimurium containing interleukin-2 decreases MC-38 hepatic metastases: a novel anti-tumor agent. *Cancer Biother. Radiopharm.* **11,** 145–153.
28. Saltzman, D. A., Katsanis, E., Heise, C. P., et al. (1997) Patterns of hepatic and splenic colonization by an attenuated strain of Salmonella typhimurium containing the gene for human interleukin-2: a novel anti-tumor agent. *Cancer Biother. Radiopharm.* **12,** 37–45.
29. Gentschev, I., Fensterle, J., Schmidt, A., et al. (2005) Use of a recombinant Salmonella enterica serovar Typhimurium strain expressing C-Raf for protection against C-Raf induced lung adenoma in mice. *BMC Cancer* **5,** 15.
30. King, I., Bermudes, D., Lin, S., et al. (2002) Tumor-targeted Salmonella expressing cytosine deaminase as an anticancer agent. *Hum. Gene Ther.* **13,** 1225–1233.
31. Lee, C. H., Wu, C. L., Tai, Y. S., and Shiau, A. L. (2005) Systemic administration of attenuated Salmonella choleraesuis in combination with cisplatin for cancer therapy. *Mol. Ther.* **11,** 707–716.
32. Lee, C. H., Wu, C. L., and Shiau, A. L. (2004) Endostatin gene therapy delivered by Salmonella choleraesuis in murine tumor models. *J. Gene Med.* **6,** 1382–1393.
33. Medina, E., Guzman, C. A., Staendner, L. H., Colombo, M. P., and Paglia, P. (1999) Salmonella vaccine carrier strains: effective delivery system to trigger anti-tumor immunity by oral route. *Eur. J. Immunol.* **29,** 693–699.
34. Soto, L. J., 3rd, Sorenson, B. S., Nelson, B. W., Solis, S. J., Leonard, A. S., and Saltzman, D. A. (2004) Preferential proliferation of attenuated Salmonella typhimurium within neuroblastoma. *J. Pediatr. Surg.* **39,** 937–940; discussion 937–940.
35. Weth, R., Christ, O., Stevanovic, S., and Zoller, M. (2001) Gene delivery by attenuated Salmonella typhimurium: comparing the efficacy of helper versus cytotoxic T cell priming in tumor vaccination. *Cancer Gene Ther.* **8,** 599–611.
36. Xiang, R., Mizutani, N., Luo, Y., et al. (2005) A DNA vaccine targeting survivin combines apoptosis with suppression of angiogenesis in lung tumor eradication. *Cancer Res.* **65,** 553–561.
37. Miragliotta, G., Di Vagno, G., Nappi, R., Fumarola, D., Jirillo, E., and Galanos, C. (1988) Evaluation of procoagulant activity production and other coagulative functions in cancer patients receiving acid treated Salmonella minnesota R 595 (Re). *Eur. J. Epidemiol.* **4,** 377–381.
38. Noriega, L. M., Van der Auwera, P., Daneau, D., Meunier, F., and Aoun, M. (1994) Salmonella infections in a cancer center. *Support. Care Cancer* **2,** 116–122.
39. Jirillo, A., Disperati, A., Balli, M., et al. (1987) Pilot study of intravenous administration of the acid-treated Salmonella minnesota R595 (Re) in cancer patients. *Tumori* **73,** 481–486.

40. Agrawal, N., Bettegowda, C., Cheong, I., et al. (2004) Bacteriolytic therapy can generate a potent immune response against experimental tumors. *Proc. Natl. Acad. Sci. USA* **101,** 15172–15177.
41. al-Ramadi, B. K., Al-Dhaheri, M. H., Mustafa, N., et al. (2001) Influence of vector-encoded cytokines on anti-Salmonella immunity: divergent effects of interleukin-2 and tumor necrosis factor alpha. *Infect. Immun.* **69,** 3980–3988.
42. Andreana, A., Gollapudi, S., Kim, C. H., and Gupta, S. (1994) Salmonella typhimurium activates human immunodeficiency virus type 1 in chronically infected promonocytic cells by inducing tumor necrosis factor-alpha production. *Biochem. Biophys. Res. Commun.* **201,** 16–23.
43. Zhao, M., Yang, M., Li, X. M., et al. (2005) Tumor-targeting bacterial therapy with amino acid auxotrophs of GFP-expressing Salmonella typhimurium. *Proc. Natl. Acad. Sci. USA* **102,** 755–760.
44. Bermudes, D., Low, B., and Pawelek, J. (2000) Tumor-targeted Salmonella. Highly selective delivery vectors. *Adv. Exp. Med. Biol.* **465,** 57–63.
45. Bermudes, D., Low, K. B., Pawelek, J., et al. (2001) Tumour-selective Salmonella-based cancer therapy. *Biotechnol. Genet. Eng. Rev.* **18,** 219–233.
46. Bermudes, D., Zheng, L. M., and King, I. C. (2002) Live bacteria as anticancer agents and tumor-selective protein delivery vectors. *Curr. Opin. Drug. Discov. Devel.* **5,** 194–199.
47. Chakrabarty, A. M. (2003) Microorganisms and cancer: quest for a therapy. *J. Bacteriol.* **185,** 2683–2686.
48. Pawelek, J. M., Low, K. B., and Bermudes, D. (2003) Bacteria as tumour-targeting vectors. *Lancet Oncol.* **4,** 548–556.
49. Rice, J. E., Makowski, G. S., Hosted, T. J., Jr., and Lavoie, E. J. (1985) Methylene-bridged bay region chrysene and phenanthrene derivatives and their keto-analogs: mutagenicity in Salmonella typhimurium and tumor-initiating activity on mouse skin. *Cancer Lett.* **27,** 199–206.
50. Shaner, N. C., Campbell, R. E., Steinbach, P. A., Giepmans, B. N., Palmer, A. E., and Tsien, R. Y. (2004) Improved monomeric red, orange and yellow fluorescent proteins derived from Discosoma sp. red fluorescent protein. *Nat. Biotechnol.* **22,** 1567–1572.
51. Heimann, D. M. and Rosenberg, S. A. (2003) Continuous intravenous administration of live genetically modified salmonella typhimurium in patients with metastatic melanoma. *J. Immunother.* **26,** 179–180.
52. Rosenberg, S. A., Spiess, P. J., and Kleiner, D. E. (2002) Antitumor effects in mice of the intravenous injection of attenuated Salmonella typhimurium. *J. Immunother.* **25,** 218–225.
53. Serganova, I. and Blasberg, R. (2005) Reporter gene imaging: potential impact on therapy. *Nucl. Med. Biol.* **32,** 763–780.

54. Toso, J. F., Gill, V. J., Hwu, P., et al. (2002) Phase I study of the intravenous administration of attenuated Salmonella typhimurium to patients with metastatic melanoma. *J. Clin. Oncol.* **20,** 142–152.
55. Yu, Y. A., Shabahang, S., Timiryasova, T. M., et al. (2004) Visualization of tumors and metastases in live animals with bacteria and vaccinia virus encoding light-emitting proteins. *Nat. Biotechnol.* **22,** 313–320.
56. Arnold, J. W., Niesel, D. W., Annable, C. R., et al. (1993) Tumor necrosis factor-alpha mediates the early pathology in Salmonella infection of the gastrointestinal tract. *Microb. Pathog.* **14,** 217–227.
57. Ashman, L. K., Cook, M. G., and Kotlarski, I. (1979) Protective effect of oral Salmonella enteritidis 11RX infection against colon tumor induction by 1,2-dimethylhydrazine in mice. *Cancer Res.* **39,** 2768–2771.
58. Ashman, L. K. and Kotlarski, I. (1978) Effect of Salmonella enteritidis 11RX infection on two-stage skin carcinogenesis in mice. *Aust. J. Exp. Biol. Med. Sci.* **56,** 695–701.
59. Eisenstein, T. K., Bushnell, B., Meissler, J. J., Jr., Dalal, N., Schafer, R., and Havas, H. F. (1995) Immunotherapy of a plasmacytoma with attenuated salmonella. *Med. Oncol.* **12,** 103–108.
60. Engelhardt, R., Otto, F., Mackensen, A., Mertelsmann, R., and Galanos, C. (1995) Endotoxin (Salmonella abortus equi) in cancer patients. Clinical and immunological findings. *Prog. Clin. Biol. Res.* **392,** 253–261.
61. Everest, P., Roberts, M., and Dougan, G. (1998) Susceptibility to Salmonella typhimurium infection and effectiveness of vaccination in mice deficient in the tumor necrosis factor alpha p55 receptor. *Infect. Immun.* **66,** 3355–3364.
62. Kumazawa, Y., Freudenberg, M. A., Hausmann, C., Meding-Slade, S., Langhorne, J., and Galanos, C. (1991) Formation of interferon-gamma and tumor necrosis factor in mice during Salmonella typhimurium infection. *Pathobiology* **59,** 194–196.
63. Kurashige, S., Kodama, K., and Yoshida, T. (1983) Synergistic anti-tumor effect of mini-cells prepared from Salmonella typhimurium with mitomycin C in EL4-bearing mice. *Cancer Immunol. Immunother.* **14,** 202–204.
64. Kurashige, S. and Mitsuhashi, S. (1982) Enhancing effects of mini-cells prepared from Salmonella typhimurium on anti-tumor immunity in sarcoma 180-bearing mice. *Cancer Immunol. Immunother.* **14,** 1–3.
65. Nigam, V. N. (1975) Effect of core lipopolysaccharides from Salmonella minnesota R mutants on the survival times of mice bearing Ehrlich tumor. *Cancer Res.* **35,** 628–633.
66. Vingelis, V., Ashman, L. K., and Kotlarski, I. (1980) Factors affecting the ability of various strains of Enterobacteriaceae to induce tumour resistance in mice. *Aust. J. Exp. Biol. Med. Sci.* **58,** 123–131.

67. Clairmont, C., Lee, K. C., Pike, J., et al. (2000) Biodistribution and genetic stability of the novel antitumor agent VNP20009, a genetically modified strain of Salmonella typhimurium. *J. Infect. Dis.* **181,** 1996–2002.
68. Low, K. B., Ittensohn, M., Le, T., et al. (1999) Lipid A mutant Salmonella with suppressed virulence and TNFalpha induction retain tumor-targeting in vivo. *Nat. Biotechnol.* **17,** 37–41.
69. Low, K. B., Ittensohn, M., Luo, X., et al. (2004) Construction of VNP20009: a novel, genetically stable antibiotic-sensitive strain of tumor-targeting Salmonella for parenteral administration in humans. *Methods Mol. Med.* **90,** 47–60.
70. Luo, X., Li, Z., Lin, S., et al. (2000) Antitumor effect of VNP20009, an attenuated Salmonella, in murine tumor models. *Oncol. Res.* **12,** 501–508.
71. Martinez-Lorenzo, M. J., Meresse, S., de Chastellier, C., and Gorvel, J. P. (2001) Unusual intracellular trafficking of Salmonella typhimurium in human melanoma cells. *Cell Microbiol.* **3,** 407–416.
72. McClelland, M., Sanderson, K. E., Spieth, J., et al. (2001) Complete genome sequence of Salmonella enterica serovar Typhimurium LT2. *Nature* **413,** 852–856.
73. Edwards, K., Linetsky, I., Hueser, C., and Eisenstark, A. (2001) Genetic variability among archival cultures of Salmonella typhimurium. *FEMS Microbiol. Lett.* **199,** 215–219.
74. Rabsch, W., Helm, R. A., and Eisenstark, A. (2004) Diversity of phage types among archived cultures of the Demerec collection of Salmonella enterica serovar Typhimurium strains. *Appl. Environ. Microbiol.* **70,** 664–669.
75. Tracy, B. S., Edwards, K. K., and Eisenstark, A. (2002) Carbon and nitrogen substrate utilization by archival Salmonella typhimurium LT2 cells. *BMC Evol. Biol.* **2,** 14.
76. Yourno, J., Barr, D., and Tanemura, S. (1969) Externally suppressible frameshift mutant of Salmonella typhimurium. *J. Bacteriol.* **100,** 453–459.
77. Sutton, A., Buencamino, R., and Eisenstark, A. (2000) rpoS mutants in archival cultures of Salmonella enterica serovar typhimurium. *J. Bacteriol.* **182,** 4375–4379.
78. Ahmad, S. I., Kirk, S. H., and Eisenstark, A. (1998) Thymine metabolism and thymineless death in prokaryotes and eukaryotes. *Annu. Rev. Microbiol.* **52,** 591–625.
79. Chomczynski, P. and Sacchi, N. (1987) Single-step method of RNA isolation by acid guanidinium thiocyanate-phenol-chloroform extraction. *Anal. Biochem.* **162,** 156–159.
80. Froussard, P. (1992) A random-PCR method (rPCR) to construct whole cDNA library from low amounts of RNA. *Nucleic Acids Res.* **20,** 2900.
81. Graham, J. E. and Clark-Curtiss, J. E. (1999) Identification of Mycobacterium tuberculosis RNAs synthesized in response to phagocytosis by human macrophages by selective capture of transcribed sequences (SCOTS). *Proc. Natl. Acad. Sci. USA* **96,** 11554–11559.

82. Kwon, Y. M., Kubena, L. F., Nisbet, D. J., and Ricke, S. C. (2002) Functional screening of bacterial genome for virulence genes by transposon footprinting. *Methods Enzymol.* **358,** 141–152.
83. Datsenko, K. A. and Wanner, B. L. (2000) One-step inactivation of chromosomal genes in Escherichia coli K-12 using PCR products. *Proc. Natl. Acad. Sci. USA* **97,** 6640–6645.

17

Further Resources for Molecular Protocols in *Salmonella* Research

Abraham Eisenstark and Kelly K. Edwards

The study of *Salmonella* spp. encompasses numerous disciplines and a wide range of specialties. *Salmonella* is exceptional among microorganisms in that it is capable of causing many common intestinal illnesses and yet has great potential as a molecular tool in the fight against disease.

Many classic techniques in molecular genetics were developed and tested in *Salmonella* including extensive studies in the effects of UV radiation and genetic selection. As a result, the *Salmonella* researchers of 2007 have built on this broad-based foundation of knowledge and expanded the applications for *Salmonella* as a tool for new discoveries in molecular genetics and as a powerful weapon in the area of targeted cell death. New modifications to the natural defenses of *Salmonella* organisms, such as lipopolysaccharide (LPS), are being circumvented to allow the safe introduction of large numbers of organisms targeted to specific malignant cell types.

An outstanding example of the contributions that the study of *Salmonella* has made to scientific endeavor in general is the Ames test. The Ames test is a biological assay used in genetic toxicology to test for the mutagenic potential of chemical compounds. The test was developed by Bruce Ames in the 1960s, but the principles behind the test stem from the history of microbiology. In this test, *Salmonella* histidine mutants are exposed to the potential carcinogen and then tested for their ability to recover from the histidine mutation. This type of testing reduces the need for animal-based testing in the early stages of pharmaceutical development and numerous other applications.

From: *Methods in Molecular Biology, vol. 394: Salmonella: Methods and Protocols*
Edited by: H. Schatten and A. Eisenstark © Humana Press Inc., Totowa, NJ

As noted by several authors in this volume, *Salmonella* illnesses are at top of the list among diseases. Thus, it is not surprising that many researchers and institutions have prepared procedures for detection, vaccine development, and basic molecular studies of molecular structure, activity, and interaction with eukaryotic organs and cells. A Google search (http://www.google.com) for "Salmonella Laboratory Manuals" yielded over 288,000 hits. A preponderant number of these were prepared by international, national, state, and local public health units, by organizations involved in food production and processing, by professors for appropriate laboratory courses, in microbiology textbooks, and by investigators in search of improved methods.

A narrowed Google search for "Salmonella Molecular Methods" surfaced 806 hits. Many of these are research papers for the development of improved methods. As might be expected, molecular methods have been introduced by public health organizations and medical diagnostic units. A recent publication ***(1)*** from the Centers for Disease Control and Prevention details several new microbiology tools and their application in the area of public health. The introduction of PCR in 1988 served as the genesis for new tools in molecular epidemiology and surveillance, whereas new applications for existing technologies, such as flow cytometry, has broadened the scope of disease diagnosis and treatment.

Of historical interest, the best of manuals for *Salmonella* and *Escherichia coli* genetic studies stemmed from courses taught at Cold Spring Harbor, NY, over the past half-century. These have been models for manuals used for advanced undergraduate and graduate level courses in universities throughout the world. This volume addresses many diverse molecular approaches to the study of *Salmonella*. There are, however, several sources for additional procedural information. Readers are reminded that some of these may be readily found on the Internet.

1. Some examples of Internet-available molecular procedures:
   http://www.sci.sdsu.edu/~smaloy/Research/protocols.htm
   This resource from the laboratory of Stanley Maloy includes Genetic Protocols, λ Red Swap, Basic Techniques for Bacterial Genetics, Challenge Phage, Antibiotics, P22 and Transduction, Electroporation, Enzymology protocols, In Vivo and In Vitro Proline Dehydrogenase Assay, In Vitro P5C Dehydrogenase Assay, Proline Transport Assay, Membrane Association Assay, Anti-PutA ELISA, PutA Purification, β-Galactosidase, PEG transformation, Macrophage Invasion Assay.
   Laboratory procedures from the John Roth lab (http://rothlab.ucdavis.edu).
2. Examples of molecular procedures published by the American Society for Microbiology (ASM).

An extensive menu of microbial procedures may be found in books published by the ASM (http://estore.asm.org). Titles include procedures for Applied Biotechnology, Environmental Bacteriology, Cell Biology, Clinical Microbiology, Immunology, Medical Microbiology, Molecular Biology, Genetics, DNA Repair and Mutagenesis.

3. Examples of molecular procedures published by Cold Spring Harbor Press (also indexed as CSHL):

   A Genetic Switch, Third Edition, Phage Lambda Revisited—Mark Ptashne, CSHL 2004.

   Discovering Genomics, Proteomics, and Bioinformatics—A. Malcolm Campbell and Laurie J. Heyer, CSHL 2003.

   Genes & Signals—Mark Ptashne and Alexander Gann, CSHL 2002.

   Lateral DNA Transfer: Mechanisms and Consequences—CSHL 2002.

   PCR Primer: A Laboratory Manual, Second Edition—Edited by Carl W. Dieffenbach and Gabriela S. Dveksler, CSHL 2003.

   Phage Display: A Laboratory Manual—Carlos F. Barbas III, Dennis R. Burton, Jamie K. Scott, and Gregg J. Silverman, CSHL 2001.

   Proteins and Proteomics: A Laboratory Manual—Richard Simpson, CSHL 2003.

   Purifying Proteins for Proteomics: A Laboratory Manual—Edited by Richard J. Simpson, CSHL 2004.

Reference

1. Robertson, B. H. and Nicholson, J. K. A. (2005) New microbiology tools for public health and their implications. *Annu. Rev. Public Health* **26,** 281–302.

Index

Printed in the United States of America